Random Fluctuations and Pattern Growth:
Experiments and Models

NATO ASI Series

Advanced Science Institutes Series

A Series presenting the results of activities sponsored by the NATO Science Committee, which aims at the dissemination of advanced scientific and technological knowledge, with a view to strengthening links between scientific communities.

The Series is published by an international board of publishers in conjunction with the NATO Scientific Affairs Division

A Life Sciences	Plenum Publishing Corporation
B Physics	London and New York
C Mathematical	Kluwer Academic Publishers
and Physical Sciences	Dordrecht, Boston and London
D Behavioural and Social Sciences	
E Applied Sciences	
F Computer and Systems Sciences	Springer-Verlag
G Ecological Sciences	Berlin, Heidelberg, New York, London,
H Cell Biology	Paris and Tokyo

Series E: Applied Sciences - Vol. 157

Random Fluctuations and Pattern Growth:
Experiments and Models

edited by

H. Eugene Stanley
Boston University,
Boston, Massachusetts, U.S.A.

and

Nicole Ostrowsky
University of Nice,
Nice, France

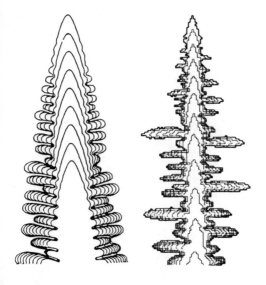

Kluwer Academic Publishers

Dordrecht / Boston / London

Published in cooperation with NATO Scientific Affairs Division

Proceedings of the NATO Advanced Study Institute on
Random Fluctuations and Pattern Growth: Experiments and Models
Cargèse, Corsica, France
18–31 July 1988

ISBN 0-7923-0073-4 (paperback)
ISBN 0-7923-0072-6 (hardback)

Published by Kluwer Academic Publishers,
P.O. Box 17, 3300 AA Dordrecht, The Netherlands.

Kluwer Academic Publishers incorporates the publishing programmes of
D. Reidel, Martinus Nijhoff, Dr W. Junk, and MTP Press.

Sold and distributed in the U.S.A. and Canada
by Kluwer Academic Publishers,
101 Philip Drive, Norwell, MA 02061, U.S.A.

In all other countries, sold and distributed
by Kluwer Academic Publishers Group,
P.O. Box 322, 3300 AH Dordrecht, The Netherlands.

Printed in The Netherlands

The subject of sandpiles and landslides was addressed in this School by Per Bak, as an example of self-organized critical behavior. Immediately after his lecture, several participants decided to experiment on their own. Shown here are two small sandpiles constructed on the Institute beach. Each was carefully prepared by slowly dropping grains of sand onto its peak. This process produces miniature landslides which occur when the sides of a sandpile become too steep.

Both sandpiles were built up to a "critical state" in which landslides of many different sizes were observed, just as discussed in Bak's lecture. Each landslide was triggered by the addition or displacement of just a few grains of sand. The sandpile on the right has just undergone such a landslide. The resulting shear profile, marked off by small sticks, surprisingly resembles an ellipse. This informal *hands-on* demonstration—performed by Guy Deutscher, Hans J. Herrmann and Greg Huber—produced an unexpected and interesting result!

• CONTENTS •

Course 8: Convection, Turbulence and Multifractals

Course 9: Random Fluctuations and Complex Systems

RANDOM FLUCTUATIONS AND PATTERN GROWTH:
Experiments and Models

• PREFACE •

The delicate leaf of a fern, the fetal spine—each a miracle to behold! As beautiful as sculpture, patterns such as these come not from the hands of artists, but rather from entirely natural processes as *random* as they are *ordered*.

Can we explain the way such systems 'choose their shapes'? Until recently, the answer to such questions has been a resounding NO!!!. Now, however, there is an increasingly widespread feeling that answers may be forthcoming. For example, we are learning to construct microscopic models that start from pure randomness yet produce recognizable patterns. These results—plus recent advances in both experimental work and abstract theory—are the source of a renewed interest in the field of random fluctuations and pattern growth.

The present book summarizes recent progress in this emerging field of inquiry; it is based on a NATO Advanced Study Institute which took place at the *Institut d'Etudes Scientifiques de Cargèse*, in Corsica, from 18-31 July 1988. We focussed on a vast range of topics: viscous fingering and dendritic growth, membranes and microemulsions, convection and turbulence, fracture and dielectric breakdown. Can one approach these subjects of classic experimental and theoretical difficulty with the same spirit that has been used in recent years to approach problems associated with phase transitions and critical phenomena? Can one discover parameters that can serve to quantitatively characterize the new forms? Can one build models based on assumptions of the form of the physics at the microscopic level, test these models with new experiments, and modify both as may be necessary? Answers characterized by optimism—guarded optimism at times—were the norm rather than the exception at this School.

The Organizing Committee consisted of P.-G. de Gennes (ESPCI, France), J. Kjems (Risø, Denmark), and L. Peliti (Napoli, Italy), in addition to the co-directors. In selecting lecturers, we placed a high premium on clarity of exposition—both oral and written. In direct response to serious concern from the ASI office of NATO that some ASI's were failing to function primarily as *schools*, we experimented with two novel ideas:

(i) Each of the nine courses that comprised our "School" was introduced with a *niveau zero* talk, designed to expose some fundamental concepts, identify some themes, and set a notation. These *niveau zero* presentations turned out extremely well, and some participants asked that every talk be in the style of the *niveau zero*!

(ii) We strove to directly involve all the participants in a series of *hands-on* demonstrations. Many lecturers and participants kindly arranged to develop—and transport to Corsica—suitable demonstrations that could be shown in lecture and tinkered with individually. Others brought software demonstrations that could be run on the Mac-II and two Mac-SE's that were available for all the participants. To our knowledge, this is the first systematic attempt to integrate the personal computer in teaching the fundamental concepts of a course in random phenomena and fluctuation theory. The personal computer can make a 'movie' of the underlying random process on one portion of the screen, while appropriate averages are continuously updated on the rest of the screen. Thus the student develops a hands-on appreciation for how physical laws (involving average quantities) develop from the microscopic events of the underlying random process. The overwhelming success of these "hands-on" efforts illustrated well the ancient adage

Tell me—I listen

Show me—I watch

Involve me—I understand!

Our sincere thanks are due to many. First and foremost, to those lecturers who put forth extra effort to make their presentations models of pedagogy and science, to those students whose frequent questions and perceptive observations opened new directions of pursuit, to Karine Ostrowsky, Serge Ostrowsky, and Chantal Ariano for 24-hour days of managerial help, to Joseph Antoine Ariano, Dan Ostrowsky, and Avraham Simievic for preparing a genuine *mechoui* for our Saturday night banquet, to the citizens of Cargèse for putting up with the "invading continentals" with good cheer and, most importantly, to all whose lightness of heart created that sort of relaxed atmosphere conducive to intellectual stimulation.

We are particularly grateful for the the the generous support of the Office of Naval Research, the National Science Foundation, E. I. duPont de Nemours, and especially the NATO Advanced Study Institute program under the inspiring leadership of Dr. Luis V. da Cunha. It is also a pleasure to record our immense debt to Darlene Carr, Harriet Matsushima, Renée J. Squier, John Stampfel, Jeri Urbanski, and especially to Jerry Morrow and Sonny Vu, who functioned as an admirable team in transforming a collection of microcomputer floppy discs and bitnet files into a polished manuscript. Laurent Delanoé and Idahlia Stanley kindly supplied most of the photographs that adorn the otherwise wasted space at the end of each lecture.

The multi-national offices of Kluwer Academic Publishers massaged the final product, and we wish to express our deep appreciation to Nel M. Pols, David J. Larner and Henny A.M.P. Hoogervorst for publishing simultaneously a hardbound and an inexpensive paperback edition. We also wish to thank all the other efficient and cheerful members of the Kluwer staff—most especially Tjaddie Ammerdorffer in Dordrecht and Patricia Simmons in the Boston office.

H. Eugene Stanley
Nicole Ostrowsky
Cargèse, 31 July 1988

SOME THEMES AND COMMON TOOLS

H. EUGENE STANLEY

Center for Polymer Studies and Department of Physics
Boston University, Boston, MA 02215 USA

This school concerns topics of classic difficulty: viscous fingering and dendritic growth, membranes and microemulsions, convection and turbulence, fracture and dielectric breakdown. In this opening talk, I wish to argue that we can approach these subjects of classic experimental and theoretical difficulty with the same spirit that has been used in recent years to approach problems associated with phase transitions and critical phenomena. This approach is to carefully choose a microscopic model system that captures the essential physics underlying the phenomena at hand, and then study this model until we understand 'how the model works.' Then we reconsider the phenomena at hand, to see if an understanding of the model leads to an understanding of the phenomena. Sometimes the original model is not enough, and a variant is needed. Fortunately the same underlying physics is often found to be common to both the model and its variants.

1. The Ising Model and Its 'Variants'

We begin, then, with the classic Ising model. Over 1000 papers have been published on this model, but only since 1977 have we known that if one understands the Ising model thoroughly, one understands the essential physics of many materials, since they are simply *variants* of the Ising model. For example, a large number of systems are related to special cases of the n-vector model, which in turn is a simple Ising model in which the spin variable s has not one component but rather n separate components s_j: $\mathbf{s} \equiv (s_1, s_2, \ldots, s_n)$.
   The Ising model solves the puzzle of how it is that nearest-neighbor interactions of *microscopic* length scale 1Å 'propagate' their effect cooperatively to give rise to a correlation length ξ_T of *macroscopic* length scale near the critical point (Fig. 1a). In fact, ξ_T increases without limit as the coupling $K \equiv J/kT$ increases to a critical value $K_c \equiv J/kT_c$,

$$\xi_T \sim A \left(\frac{K - K_c}{K_c} \right)^{-\nu_T}. \tag{1a}$$

The 'amplitude' A has a numerical value on the order of the lattice constant a_o. A snapshot of an Ising system shows that there are fluctuations on all length scales from a_o ($\cong 1$Å) to ξ_T (which can be from $10^2 - 10^4$Å in a typical experiment).

2. Random Site Percolation on a Lattice, and Its Variants

In percolation, one randomly occupies a fraction p of the sites of a d-dimensional lattice (the case $d = 1$ is shown schematically in Fig. 1b). Again, phenomena occurring on the local 1Å scale of a lattice constant are 'amplified' near the percolation threshold $p = p_c$ to a macroscopic length ξ_p.
   Here p plays the role of the coupling constant K of the Ising model. When p is small, the characteristic length scale is comparable to 1Å. However when p

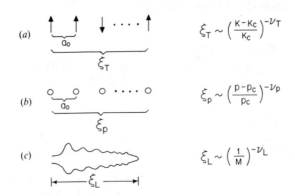

$$\xi_T \sim \left(\frac{K - K_c}{K_c}\right)^{-\nu_T}$$

$$\xi_p \sim \left(\frac{p - p_c}{p_c}\right)^{-\nu_p}$$

$$\xi_L \sim \left(\frac{1}{M}\right)^{-\nu_L}$$

Fig. 1: Schematic illustration of the analogy between (a) the Ising model, which has fluctuations in spin orientation *on all length scales* from the microscopic scale of the lattice constant a_o up to the macroscopic scale of the thermal correlation length ξ_T, (b) percolation, which has fluctuations in characteristic size of clusters *on all length scales* from a_o up to the diameter of the largest cluster—the pair connectedness length ξ_p, and (c) the DLA/DBM problem, whose clusters have fluctuations *on all length scales* from the microscopic length $d_o = \gamma/L$ (γ is the surface tension and L the latent heat) up to the diameter of the cluster ξ_L. Also shown, on the right side, is the analogy between the scaling behavior of the three length scales ξ_T, ξ_p, and ξ_L.

approaches p_c, there occur phenomena on all scales ranging from a_o to ξ_p, where ξ_p increases without limit as $p \to p_c$

$$\xi_p \sim A\left(\frac{p - p_c}{p_c}\right)^{-\nu_p}. \tag{1b}$$

Again, the amplitude A is roughly a lattice constant (~ 1Å).

Each phenomenon of thermal critical phenomena has a corresponding analog in percolation, so that the percolation problem is sometimes called a geometric or *connectivity* critical phenomenon. Any connectivity problem can be understood by starting with pure random percolation and then adding interactions, or whatever. Thus, e.g., we understand why the critical exponents describing the divergence to infinity of various geometrical quantities (such as ξ_p) are the same regardless of whether the elements interact or are non-interacting. Similarly, the same connectivity exponents are found regardless of whether the elements are constrained to the sites of a lattice or are free to be anywhere in a continuum.

3. The Laplace Equation and Its Variants

Just as variations in the Ising and percolation problems were found to be sufficient to describe a rich range of thermal and geometric critical phenomena, so we have

Table 1

A 'Rosetta stone' connecting the physics underlying (a) an electrical problem (dielectric breakdown), (b) a fluid mechanics problem (viscous fingering), and (c) a diffusion problem (dendritic solidification).

(a) electrical	(b) fluid mechanics	(c) dendritic solidification
electrostatic potential:	pressure:	concentration:
$\phi(r,t)$	$P(r,t)$	$c(r,t)$
electric field:	velocity:	growth rate:
$E \propto -\nabla\phi(r,t)$	$v \propto -\nabla P(r,t)$	$v \propto -\nabla c(r,t)$
conservation:		
$\nabla \cdot E = 0$	$\nabla \cdot v = 0$	$\nabla \cdot v = 0$
Laplace Equation:		
$\nabla^2 \phi = 0$	$\nabla^2 P = 0$	$\nabla^2 c = 0$

found that variants of the original Laplace equation are useful in describing puzzling patterns in fluid mechanics, dendritic growth, and various breakdown phenomena.

In the Ising model, we place a spin on each pixel (site) of a lattice. In percolation we allow each pixel to be occupied or empty. In fluid mechanics, we assign a number—call it ϕ—to each pixel. We might think of ϕ as being the pressure or chemical potential at this region of space.

The spins in an Ising model interact with their neighbors. Hence the state of one Ising pixel depends on the state of all the other pixels in the system—up to a length scale given by the thermal correlation length ξ_T. The 'global' correlation between distant pixels in an Ising simulation arises from the fact that neighboring pixels at i and j have a 'local' exchange interaction J_{ij}. Similarly, the correlation in connectivity between distant pixels in the percolation problem arises from the 'propagation' of local connectivity between neighboring pixels. In fluid mechanics, the pressure on each pixel is correlated with the pressure at every other pixel because the pressure obeys the Laplace equation.

One can calculate an equilibrium Ising configuration by 'passing through the system with a computer' and flipping each spin with a probability related to the Boltzmann factor. Similarly, one can calculate the pressure at each pixel by 'passing through the system' and re-adjusting the pressure on each pixel in accord with the Laplace equation. If we were to arbitrarily flip the configuration of a single pixel in the Ising problem (from $+1$ to -1), we would significantly influence the equilibrium configuration of the system out to a length scale on the order of ξ_T. Similarly, if we were to arbitrarily impose a given pressure on a single point of a system obeying the Laplace equation, we would drastically change the resulting pattern out to a length scale that we shall call ξ_L.

Does ξ_L obey a 'scaling form' analogous to Eqs. (1a) and (1b) obeyed by the functions ξ_T and ξ_p for the Ising model and percolation? We believe that the answer to this question is 'yes,' although our ideas on this subject remain somewhat tentative and subject to revision.

The best way to see the fluctuations inherent in structures grown according to the Laplace equation is to first introduce some specific models. There are two models that were at once thought to be fully equivalent, although it is now recognized that the actual patterns produced by each have a different 'susceptibility to lattice anisotropy.' The first of these models is diffusion limited aggregation (DLA)[1]. Here one releases a random walker from a large circle surrounding a seed particle placed at the origin. When the random walker touches a perimeter site of the seed, it 'sticks' (i.e., the perimeter site becomes a cluster site), and we have a cluster of mass = 2. A second random walker is then released. This process continues until a large cluster is formed.

In both thermal critical phenomena (or percolation) the length L introduced when we have a finite system size scales the same as the correlation lengths ξ_T (or ξ_p). Hence for DLA we expect that there will be fluctuations on length scales up to ξ_L, where ξ_L itself increases with the cluster mass according to

$$\xi_L \sim A\left(\frac{1}{M}\right)^{-\nu_L} \qquad \left[d_f \equiv \frac{1}{\nu_L} = \text{fractal dimension}\right]. \tag{1c}$$

Here the amplitude A is again on the order of 1Å. Note that (1c) is analogous to (1a) and (1b) if we think of $M \to \infty$ as being analogous to $K \to K_c$. Note also that $\nu_L = 1/d_f$ plays the role of the critical exponents ν_T and ν_p of (1a) and (1b). Suppose one tests this idea, qualitatively, by examining the largest DLA clusters in detail. One finds that indeed there are fluctuations in mass on length scales less than, say, the width W of the side branches. If one makes a log-log plot of W against mass M, one finds the same slope $1/d_f$ that one finds when one plots the diameter against M.[2]

Not only is d_f the same for the fluid mechanics problem and for the Laplace patterns, but *so also are the multifractal properties the same*.[3] Multifractals arise when one defines some quantity on all the pixel sites. Perhaps the simplest example is that of a charged needle: if we assign to every pixel a number equal to the electric field, then the set $\{E_i\}$ of field values for the perimeter sites of the needle form a multifractal set.

The same considerations apply to the fluid mechanics problem. Here the analog of the electric field $E \propto \nabla V$ is the growth probability $p_i \propto \nabla P$, where the index i runs over all perimeter sites i. Thus p_i is the probability that site i is the next to be added to the cluster. If we think of random walkers, then p_i is the hit probability (the probability that site i is the next to be hit by a random walker). Clearly the set p_i play a vital role in determining the dynamics of growth, since if we know all the p_i for every perimeter site i at a given time t, then we can predict (in a statistical sense) the state of the system at time $t+1$. Theoretical evidence has been advanced recently to suggest that the numbers p_i form a multifractal set: this set cannot be characterized by a single exponent (as in the case of the DLA aggregate itself) but rather an infinite hierarchy of exponents is required. The physical basis for this fact is that the hottest tips of a DLA aggregate grow much faster than the deep fjords; hence the *rate of change* of the p_i differs greatly when i is a tip perimeter site than when i is a fjord perimeter site.

4. "Dendritic Fluid Patterns"

By analogy with the Ising model and its variants, we can modify DLA to describe other fluid mechanical phenomena. One of the most intriguing of these concerns a variation of the viscous fingering phenomenon in which there is present anisotropy. Ben Jacob et al[4] imposed this anisotropy from by scratching a lattice of lines on their Hele-Shaw cell. They found patterns that strongly resemble snow crystals! If viscous fingers are described by DLA, then can the Ben Jacob patterns be described by DLA with imposed anisotropy?

Recently, we attempted to answer this question—specifically, we attempted to reproduce the Ben Jacob patterns with suitably modified DLA.[2] A scratch in a Hele-Shaw cell means that the plate spacing b is increased along certain directions, and the permeability coefficient k relating growth velocity to ∇P is proportional to b^2 ($k \propto b^2$). Hence we calculated DLA patterns for the case in which there was imposed a periodic variation in the k. It is significant that their simulations reproduce snow crystal type patterns, just like the experiments. These simulations relied for their efficacy on the presence of noise reduction.

'Noise reduction' means that we associate a counter with each perimeter site; each time that site is chosen, the counter increments by one. The perimeter site becomes a cluster site only after the counter reaches a pre-determined threshold value termed s.[2,6−7] When $s = 1$, we recover the original noisy DLA. Growth is dominated by the stochastic randomness in the arrival of random walkers. If s is very large, then growth is determined by the actual probability distribution. This fact is relevant to dendritic solid patterns, which we discuss next.

Of course, real dendritic growth patterns (such as snow crystals) do not occur in an environment with periodic fluctuations in $k(x, y)$. Rather, the *global* asymmetry of the pattern arises from the *local* asymmetry of the constituent water molecules. Can this local asymmetry give rise to global asymmetry? Buka et al[5] replaced the Ben Jacob experiment (isotropic fluid, anisotropic cell) by the reverse: isotropic cell but anisotropic fluid! To accomplish this, they used a nematic liquid crystal for the high viscosity fluid. Thus the analog of the water molecules in a snow crystal are the rod-shaped anisotropic molecules of a nematic.

5. "Dendritic Solid Patterns"

Dendritic crystal growth has been a field of immense recent progress, both experimentally and theoretically. Dougherty et al[8] have recently made a detailed analysis of stroboscopic photographs, taken at 20 second intervals, of dendritic crystals of NH_4Br (Fig. 2a). They have found three surprising results: (i) the sidebranches are non-periodic at any distance from the tip, with random variations in both phase and amplitude, (ii) sidebranches on opposite sides of the dendrite are essentially uncorrelated, and (iii) the rms sidebranch amplitude is an exponential function of distance from the tip, with no apparent onset threshold distance.

How can we understand these new experimental facts? Many existing models reflect the essential physical laws underlying the growth phenomena, but fail to find a tractable mechanism to incorporate the effects of noise on the growth. Growth of a dendrite from solution is controlled by the diffusion of solute towards

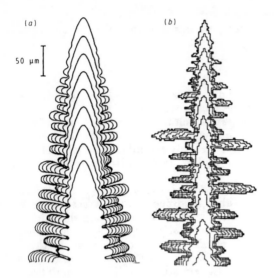

Fig. 2: (a) Experimental pattern of dendritic growth of NH_4Br (after Ref. 8). (b) DLA simulation with noise reduction parameter $s = 200$ (after Ref.9).

the growing dendrite. In the limit of small Peclet number, the diffusion equation reduces to the Laplace equation. The Laplace equation for a moving interface (the growing dendrite) brings to mind DLA. The simulations show that minute amounts of anisotropy become magnified as the mass of a cluster increases (Fig. 2b). Noise arises from the fact that there are concentration fluctuations in the vicinity of the growing dendrite, estimated to be roughly $\pm 10^5$ NH_4Br molecules per cubic micron.

The significance of the present findings is that the essential physics embodied in the DLA model seems sufficient to describe the highly uncorrelated dendritic growth patterns recently discovered from the experiments and quantitative analysis of Dougherty et al. More generally, we believe that it is worth exploring all the consequences of a straightforward physical model. Our optimism is based on the success of the Ising model and percolation in the past. We must be mindful that substantial variants of the original model may be called for. In our case, e.g., anisotropy must be introduced or else the pattern bears absolutely no resemblance to dendritic growth.

1. Witten, T. A., and Sander, L. M., *Phys. Rev. Lett.* **47**, 1400 (1981).

2. Nittmann, J., and Stanley, H. E., *Nature* **321**, 663 (1986).

3. Nittmann, J., Stanley, H. E., Touboul, E., and Daccord, G., *Phys. Rev. Lett.* **58**, 619 (1987).

4. Ben-Jacob, E., et al *Phys. Rev. Lett.* **55**, 1315 (1985).

5. Buka, A., Kertész, J., and Vicsek, T., *Nature* **323**, 424 (1986).

6. Tang, C., *Phys. Rev. A.* **31**, 1977 (1985).

7. Kertész, J., and Vicsek, T., *J. Phys. A* **19**, L257 (1986).

8. Dougherty, A., Kaplan, P. D., and Gollub, J. P., *Phys. Rev. Lett.* **58**, 1652 (1987).

9. Nittmann, J., and Stanley, H. E., *J. Phys. A* **20**, L981 (1987).

STRUCTURE, ELASTICITY AND THERMAL PROPERTIES OF SILICA NETWORKS

J. K. KJEMS and D. POSSELT

Risø National Laboratory
DK-4000 Roskilde
Denmark

ABSTRACT. The structural, mechanical and thermal properties of silica networks, like colloidal aggregates, aerogels and smoke-particle aggregates, are reviewed with emphasis on experimental studies. The importance of crossovers in lengths and frequency scales is stressed illustrated by the difference found between the Cauchy length and the density correlation length in aerogels and Cab-O-Sils.

1. Introduction

Silica networks have become important model systems for the study of fractal structures and their physical properties. This is to some extent due to the easy way in which they can be fabricated by skilled chemists but, equally important, because they allow for considerable variation in parameters like density, ρ, correlation length, ζ, and intrinsic particle size, a. These systems are therefore well suited for systematic studies of scaling relations between the structural parameters and properties like elasticity (Young's modulus), sound velocity, thermal conductivity, heat capacity and vibrational density of states. These properties are strongly affected by the morphology, i.e. the connectivity of the silica networks, and it is the purpose of this presentation to review the status of the experimental studies in this field hopefully with adequate reference to theoretical work.

Silica networks are made of aggregated clusters of spherical SiO_2 particles of radius, a, which can be made quite monodisperse. The networks result from processes like colloidal aggregation of SiO_2 particles in suspension, e.g. the commercially available[1] LUDOX^{TM}; flame hydrolysis of $SiCl_4$ to give products like Cab-O-Sil[2], polymerization of silanol molecules in non-aqueous solutions[3] leading to gel formation followed by hypercritical drying to give aerogels. The latter procedure produces homogeneous monolitic samples with densities in the range from ~ 0.05 g/cm^3 to ~ 2.0 g/cm^3, i.e. maximum densities near that of SiO_2-glass[3,4]. The morphology of the networks produced in the different manners seems to be similar to that of branched polymer networks in which the monomers correspond to the intrinsic silica particles of size, a.

2. Structures

It was Witten and Sander[6] who first perceived that the concept of fractal geometry introduced by Mandelbrot[7] could be used to characterize the structure of silica aggregates. They devised the famed DLA model that could generate similar structures with no natural length scale i.e. with power law density correlations over a wide range of distances. The pair correlation function g(r) is[8]

$$g(r) \propto \langle \rho_o \rho_r \rangle \propto r^{-A}, \qquad (1)$$

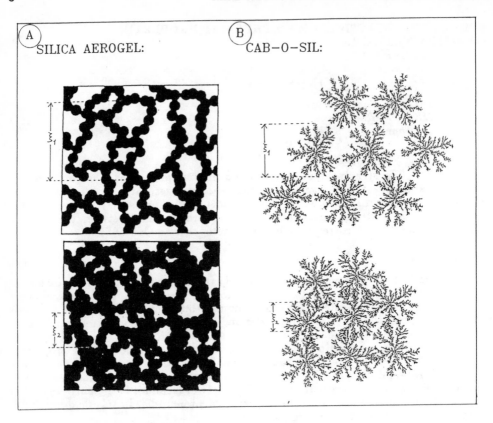

Fig. 1: Models of silica networks that illustrate how samples of different densities are obtained (A) by changing monomer concentration before aggregation (B) by compacting after aggregation.

where ρ is the mass density which for a mass fractal is a function of radius r. This leads to an expression for the scattered intensity, I, as function of momentum transfer

$$I_q \propto \int \langle \rho_o \rho_r \rangle e^{i\mathbf{q} \cdot \mathbf{r}} d\mathbf{r} \propto q^{-D}. \qquad (2)$$

Here $D = d - A$ is the Hausdorff or fractal dimension of the object and d is the dimension of the embedding space. This scattering law has been verified by x-ray, neutron and light scattering experiments[9,10,11] and the colloidal silica aggregates are found to be the systems with the largest range in length scales over which power-law correlations can be found. When analyzing scattering data it is very important to consider the crossover length scales. The most obvious is the finite particle size which normally can be included as a multiplicative form factor F_q for the particles so that

$$I_q \propto (1 + S_q)F_q, \qquad (3)$$

with S_q being the Fourier transform of the point pair correlation between particle centers.

A more subtle crossover length is the scale ξ_d where the density becomes uniform, i.e. where $\langle \rho_o \rho_r \rangle$ becomes constant. There is no universal description of how the constant level is approached but the following empirical form has proven to fit many experiments[10]:

$$\langle \rho_o \rho_r \rangle \propto r^{-A} e^{-r/\xi_d}, \tag{4}$$

which leads to

$$S_q = C(D-1)\xi_d^D \frac{\Gamma(D-1)\sin[(D-1)\arctan(q\xi_d)]}{[1+q^2\xi_d^2]^{(D-1)/2}(D-1)q\xi_d}. \tag{5}$$

A more detailed form that takes into account the cutoff in the correlation function at the near neighbour particle distance has also been devised by Freltoft et al.[10]

As pointed out by Courtens et al[11,13] it is important to know how the crossover length changes with sample preparation if one wants to make systematic studies with samples of different densities. It turns out that neutrally reacted samples of aerogels are mutually self-similar, i.e., $\xi_d \propto \rho^{1/D-d}$ whereas the compressed samples made from Cab-O-Sil powders[10,12] are not. This is illustrated in Fig. 2.

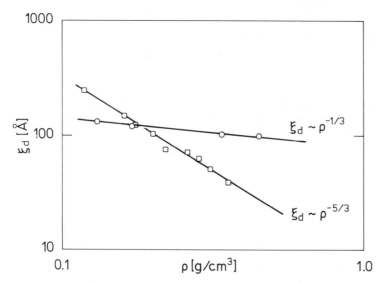

Fig. 2: Observed relation between macroscopic density and crossover length, for aerogels ($\square$)[13] and compressed Cab-O-Sil ($\circ$). Both sets are SANS (small angle neutron scattering) data. Note logarithmic scales.

3. Elasticity

The elastic properties of silica networks can be studied directly in samples of densities above ~ 0.1 g/cm³ by measurements of the bending or the torsional force

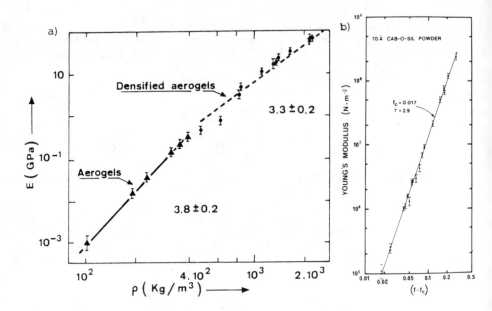

Fig. 3: Observed variation of the Young modulus E in (a) aerogels [5], (b) compressed Cab-O-Sil[14] determined by beam bending techniques.

constants. For both aerogels[5] and compressed powders of Cab-O-Sil[14] one finds power-law behaviour as function of $(\rho - \rho_c)$ or $(f - f_c)$, where ρ_c and f_c corresponds to the critical density or volume fraction for percolation, respectively.

The same information can be deduced from measurements of the sound velocity, v, in that

$$E \propto \rho v^2 \propto \rho^{1+2\alpha} \tag{6}$$

$$v \propto \rho^\alpha \tag{7}$$

$$E \propto (f - f_c)^\tau; \quad \rho \propto (f - f_c)^\beta; \quad 1 + 2\alpha = \tau/\beta. \tag{8}$$

For aerogels[11] one finds consistent values for τ/β and $1+2\alpha$. This will be discussed in more detail in the lecture by Courtens. The experimental values for the exponents have been used to make conclusions about the scalar or tensorial character of network elasticity,[15,16] but such deductions may depend on the mutual self-similarity, which is not found to hold for Cab-O-Sil samples with different densities.

4. Fractons

The elasticity and sound velocities represent the long wavelength limit of the vibrational excitation of the silica networks. When the wavelengths become comparable to the correlation length, ξ_d, the nature of the excitation changes from phonons to fractons[17] as illustrated in Fig. 4.

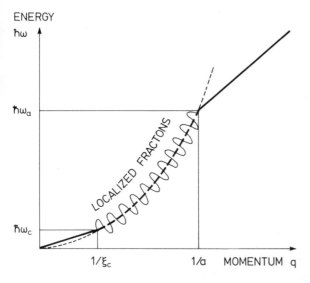

Fig. 4: Schematic representation of the dispersion of vibrational excitations in silica networks.

This shows the three regions that govern the elastic and thermal properties of such networks. In these regions the dominating excitation are phonons, fractons and phonons in the monomer particles respectively. The phonon regime and the phonon-fracton crossover have been studied extensively by Courtens et al. using Brillouin scattering techniques and will be discussed in the subsequent lecture.[18] We note one important point, namely that the crossover length scale ξ_c, also called the Cauchy-length, which is deduced from the analysis of the dynamical studies data, does *not* coincide with the length scale ξ_d that is deduced from the structural studies. In fact, ξ_c exceeds ξ_d by a factor of approximately five. This may be important for the analysis of amorphous materials like glasses and polymers where ξ_d is close to the atomic scale but where ξ_c may be large enough to warrant a description of the dynamics using the fracton concept.

In the fracton region the density-of-states is given by

$$Z(\omega) \propto \omega^{d_s - 1}, \qquad (9)$$

where d_s is the spectral dimension. The density-of-states can be measured directly by incoherent inelastic neutron scattering. This has been done for a sample of compressed Cab-O-Sil, $\rho = 0.2$ g/cm^3, by considering the difference between the scattering from a hydroxylated and a dried sample[19]. In this manner the incoherent scattering component can be selected under the assumption that the hydroxyl groups are tightly bound to the surface of the SiO_2-particles. The result is illustrated in Fig. 5. The fracton density of states corresponds to the low energy part of

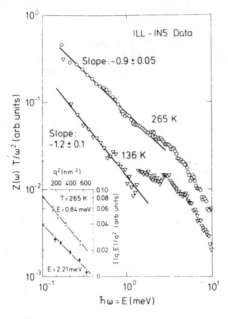

Fig. 5: Density-of-States $\times T/\omega^2$ versus energy for compressed Cab-O-Sil with $\rho = 0.2$ g/cm^3 at different temperatures. The slopes below $\hbar\omega = 1$ meV corresponds to $z(\omega)\alpha\omega^{d_s-1}$. The inset illustrates the extrapolation procedure.

the spectra and it is seen that there is an apparent change of exponent as function of temperature, probably due to anharmonic effects.

The crossover to the phonon region occurs at energies that are smaller than the instrumental resolution but the crossover to the particle region can be seen clearly at $\hbar\omega \sim 2$ meV in the 136 K data. The particle modes are the elastic modes of the SiO$_2$ particles.

5. Thermal Properties

The density-of-states is also directly reflected in the heat capacity, C_p. Low-temperature data for aerogels are consistent with the picture of the different regimes that are illustrated in Fig. 4. In particular, the values of C_p are found to be less than the asymptotic value C_D (debye limit at low energy) that is deduced from the elastic properties, even though the crossover temperature could not be reached in the experiment by Calemczuk et al.[20] The same authors also measured the thermal conductivity, K, of different aerogels and their data are shown in Fig. 6.

This data also shows evidence of three different regimes. At the lowest temperatures below 0.1 K one expects a phonon regime with $K \propto T^2$, this is followed by a fracton regime $0.1K < T < 5K$ where a linear temperature dependence is expected[21]. At still higher temperatures the internal modes in the SiO$_2$-particles may contribute and the thermal conductivity rises more rapidly. So far there has been no theoretical analysis of the high temperature regime so interested students may make original contributions here!

6. Summary

In this brief paper we have reviewed the structural and dynamical properties of silica networks. The scattering experiments have shown to give full information on the structural properties like fractal dimension, D, and the crossover length scales

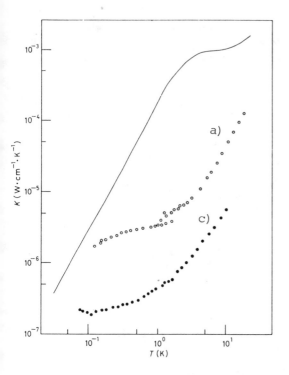

Fig. 6: Observed thermal conductivity of aerogels at $\rho = 0.87 \text{g/cm}^3$ (a) and $\rho = 0.27 \text{ g/cm}^3$ (c) from Ref. 20. The solid line represents bulk α-SiO$_2$.

ξ_d and a for a variety of systems. The dynamic properties reveal three frequency regimes separated by the crossover frequencies ω_c and ω_a. Below ω_c homogeneous phonon modes prevail whose life-times are dominated by disorder scattering. In the regime $\omega_c < \omega < \omega_a$ one finds the localized fracton modes with the characteristic spectral dimension d_s and the relatively poor thermal conductivity. Above ω_a the internal modes of the constituent particles make significant contributions to both the density-of-states and to the thermal conductivity. It is important to note that the crossover length scale, ξ_c, associated with ω_c does not necessarily coincide with the length scale, ξ_d, derived from structural studies of the same system.

In Table I the parameters describing the three different SiO$_2$-networks are compared.

	D	$\xi_d[\text{Å}]$	$a[\text{Å}]$	d_s
Cab-O-Sil (compressed)	2.5-2.6	100-132	20	1.7-2.1
Aerogel	2.4	40-320	10	1.3
Colloidal aggr.	1.5-2.2	100-1000	35-45	

Table I: Experimentally found values of structural and dynamic parameters describing different SiO$_2$ networks.

The experimental studies so far confirm the qualitative picture given here, but much more detailed and systematic work is needed before the analysis of scaling relations is on a firm footing in this field although the elegant work by Tsujimi et al.[22] is an important step in this direction.

Acknowledgements. This paper relies on the experience gained from fruitful collaboration with S.K.Sinha, T. Freltoft, D. Richter, and F.W.Poulsen.

1. LUDOX is a trademark of E. I. du Pont de Nemours and Co., Wilmington, Delaware, USA.
2. G. D. Ulrich and J. W. Riehl, J. Colloid. Interface Sci. **87**, 257 (1982).
3. S. Henning and L. Svensson, Physica Scripta **23**, 697 (1981).
4. C. J. Brinker, K. D. Keefer, D. W. Schaefer and C. S. Ashley, J. Non-Cryst. Solids **48**, 47 (1982).
5. T. Woignier, J. Phalippou, R. Sempere and J. Pelous, J. Phys., France **49**, 289 (1988).
6. T. Witten and L. M. Sander, Phys. Rev. Lett. **47**, 1400 (1981).
7. B. B. Mandelbrot, *The Fractal Geometry of Nature* (Freeman, San Francisco, 1983).
8. J. E. Martin and A. J. Hurd, J. Appl. Cryst. **20**, 61 (1987), gives the most recent review.
9. D. W. Schaefer, J. E. Martin, P. Wiltins and D. S. Cannell, Phys. Rev. Lett. **52**, 2371 (1984).
10. T. Freltoft, J. K. Kjems and S. K. Sinha, Phys. Rev. **B33**, 269 (1986).
11. E. Courtens and R. Vacher, Z. Phys. B Condensed Matter **68**, 355 (1987).
12. S. K. Sinha, T. Freltoft and J. K. Kjems, in *Kinetics of Aggregation and Gelation*, eds F. Family and D. P. Landau (North Holland, Amsterdam (1984), p. 87.
13. R. Vacher, T. Woignier, J. Pelous and E. Courtens, to be published.
14. J. Forsman, J. P. Harrison and A. Rutenberg, Can. J. Phys. **65**, 767 (1987).
15. S. Alexander, J. Phys. France **45**, 1939 (1984).
16. Y. Kantor and I. Webman, Phys. Rev. Lett. **52**, 1891 (1984).
17. S. Alexander and R. Orbach, J. Phys. (Paris) Lett. **43**, L625 (1982).
18. E. Courtens, these proceeding, and E. Courtens, J. Pelous, J. Phallipou, R. Vacher, T. Woigner, Phys. Rev. Lett. **58**, 128 (1987).
19. T. Freltoft. J. K. Kjems and D. Richter, Phys. Rev. Lett. **59**, 1212 (1987).
20. R. Calemcsuk, A. M. de Goer, B. Salce, R. Maynard and A. Zaremowitch, Europhys. Lett.   **3**, 1205 (1987).
21. O. Entin-Wohlman and R. Orbach, in *Time-Dependent Effects in Disordered Materials*, eds. R. Pynn and T. Riste (Plenum, New York, 1987), p. 243.
22. Y. Tsujimi, E. Courtens, J. Pelous and R. Vacher, Phys. Rev. Lett. **60**, 2757 (1988).

ROLAND
LENORMAND

ANOMALOUS DIFFUSION AND FRACTONS IN DISORDERED STRUCTURES

SHLOMO HAVLIN
Department of Physics
Bar-Ilan University, Ramat-Gan, Israel

In this lecture I review the developments of recent years in our understanding of transport properties in disordered *structures*. The breakthrough in this area came about largely by the discovery of *anomalous* diffusion[1-4] and by the concept of *fractons* introduced by Alexander and Orbach.[1] These are based on the fundamental concept of *fractals* introduced and developed by Mandelbrot[5] to characterize disordered structures in nature. Since then (1982), the interest of the scientific community in this area has increased dramatically, and as a consequence hundreds of papers have been published on this subject. The existence of anomalous diffusion for the case of singular transition rates on regular lattices was known earlier (see e.g., the review articles of Alexander et al[6] and Weiss and Rubin[7]). The fact that disorder in the *lattice structure* can lead to anomalies in the transport properties was the surprising result presented in Refs. 1-4. Although today there is a much better understanding of the dynamical properties of disordered structures, there still remain many open questions, and the static properties as well as the dynamical properties are not fully understood. In this lecture I will focus on several basic dynamical properties such as conductivity, diffusion and fractons. I will also discuss the concept of chemical distance and the probability density of random walks on fractals which contain much information about the dynamical properties. Relevant experiments will be mentioned only briefly since they will be discussed in detail in the talks of Jorgen Kjems and Eric Courtens. Effects of additional physical disorder which are different from the structural disorder will be discussed in the talk of Armin Bunde.

1. Percolation

The fractal features of percolation systems were found useful in characterizing disordered systems.[8] Consider a square lattice in which a fraction p of the sites are randomly occupied and $1 - p$ are empty. Nearest neighbors occupied sites form connected clusters. When p increases, the mean size of the clusters increases. There exists a critical concentration $p = p_c$ below which only finite clusters exist and an infinite cluster is formed above. This feature is represented by the correlation length ξ, which can be regarded as the mean distance between two sites on the same cluster. For $p \to p_c$, ξ diverges as

$$\xi \sim |p - p_c|^{-\nu}. \tag{1}$$

The percolation transition is characterized by the probability P_∞ that a site in lattice belongs to the infinite cluster. For $p < p_c$, $P_\infty = 0$ and for $p > p_c$, P_∞ increases with p as

$$P_\infty \sim |p - p_c|^\beta. \tag{2}$$

The exponents ν and β are universal and depend only on the space dimensionality d and not on the lattice structure. Percolation clusters are well described by fractals. Above the percolation threshold $(p > p_c)$ there exists an infinite cluster, but a finite correlation length $\xi \sim |p - p_c|^{-\nu}$. This correlation length $\xi(p)$ is interpreted as being a typical length up to which the infinite cluster is self-similar and can be regarded as a fractal. For example, for $r < \xi$ the mass M within a radius r scales as $M \sim r^{d_f}$ where d_f is the fractal dimension of the infinite cluster. For $r > \xi$ the self-similarity property is lost and the mean mass scales as for homogeneous object, $M \sim r^d$. Thus the infinite cluster is modelled by a regular lattice of fractal unit cells of size $\xi(p)$. At criticality $(p = p_c)$, $\xi \to \infty$ and the infinite cluster is self-similar in all length scales. The fractal dimension of percolation clusters has been shown to be $d_f = d - \beta/\nu$.

2. Anomalous Diffusion

Diffusion on percolation clusters is characterized by the mean square displacement $\langle R^2(t) \rangle$ of a random walker who can step only on occupied sites belonging to the cluster. Here, t denotes the number of steps. There exist three characteristic diffusion regimes. (i) At $p < p_c$, the largest clusters are of finite linear size $\xi(p)$ and $\langle R^2(t) \rangle \sim \xi^2(p)$ for $t \to \infty$. (ii) For $p > p_c$, the diffusion is anomalous and characterized by a crossover

$$\langle R^2(t) \rangle \sim \begin{cases} t^{2/d_w}, & R(t) < \xi(p); \\ t, & R(t) > \xi(p) \end{cases} \tag{3}$$

where d_w is the diffusion exponent. For $R > \xi(p)$, diffusion is regular $(d_w = 2)$ since the percolation structure is translationally invariant at these length scales and Fick's diffusion law is valid. (iii) At criticality, $p = p_c$, the incipient infinite cluster is self-similar in all length scales, and diffusion is anomalous for all $R(t)$. In this last regime, $p = p_c$, it is useful to distinguish between two cases. The first is diffusion only on the infinite cluster which is characterized by the exponent d_w of equation (3). The second case is diffusion on *all* clusters which is defined by averaging $\langle R^2(t) \rangle$ over all percolation clusters at criticality and characterized by d'_w. It can be shown that the two diffusion exponents are related by

$$\frac{d_w}{d'_w} = 1 - \beta/2\nu. \tag{4}$$

Thus $d'_w > d_w$ and diffusion is slower due to the finite clusters. For a recent review on diffusion on random structures see Ref. 9.

3. Conductivity and Diffusion

The conductivity of the percolation system $\sigma(p)$ behaves near p_c as $\sigma(p) \sim (p - p_c)^\mu$ where μ is the conductivity exponent. The conductivity is related to the diffusion constant $D \equiv \langle R^2(t) \rangle/t$, via the Einstein relation

$$\sigma = \frac{e^2 n}{kT} D \tag{5}$$

where $n \sim R^{d_f}/R^d$ is the carrier density. Using Eq. (1) and the above relation for σ it follows that $\sigma \sim R^{-\mu/\nu} \equiv R^{-\tilde{\mu}}$. Substituting these results into Eq. (5) yields $R^{-\tilde{\mu}} \sim R^{d_f-d}R^2/R^{d_w}$. Thus the relation between the diffusion exponent and the conductivity exponent is

$$d_w = 2 - d + d_f + \tilde{\mu} = d_f + \tilde{\zeta} = 2 + (\mu - \beta)/\nu \tag{6}$$

where $\tilde{\zeta} = 2 - d + \tilde{\mu}$ is the resistance exponent, characterizing the scaling of the resistance ρ with the size R of the system,

$$\rho \sim (R/R^{d-1})\sigma^{-1} \sim R^{-d+2+\tilde{\mu}} \equiv R^{\tilde{\zeta}}. \tag{7}$$

The most precise numerical values available for $d = 2$ are $d_w = 2.870 \pm 0.0015$ and $\tilde{\mu} = 0.9745 \pm 0.0015$.[10]

4. Fractons

The anomaly in the diffusion exponent d_w and the crossover in Eq. (3), is also reflected in a crossover in the density of states, $N(\epsilon)$, representing the coresponding vibrational problem.[1] For the infinite percolation cluster above p_c, $N(\epsilon)$ crosses over from a phonon behavior $N(\epsilon) \sim \epsilon^{d/2-1}$ for small ϵ (large t) to a fracton behavior $N(\epsilon) \sim \epsilon^{d_s/2-1}$ for large ϵ. This can be best understood from the Laplace transform relation between the density of energy states $N(\epsilon)$ and the probability $P_0(t)$ of a random walker to be at the origin after t steps.[1]

$$P_0(t) = \int N(\epsilon) \exp(-\epsilon t) d\epsilon. \tag{8}$$

Since $P_0(t) \sim R^{-d_f} \sim t^{-d_f/d_w}$ it follows that

$$N(\epsilon) \sim \epsilon^{d_f/d_w-1} \sim \epsilon^{d_s/2-1}, \tag{9a}$$

and d_s is called the fracton dimension[1] or the spectral dimension.[4] Since the energy eigenvalues ϵ are proportional to the square of the phonon frequency ω^2 it follows that the density of states is

$$N(\omega) = \omega^{d_s-1} F(\omega \tau) \tag{9b}$$

where $F(x) \sim x^{d-d_s}$ for $x \ll 1$, leading to the crossover discussed above. Around $\omega \sim 1/\tau$ an interesting step-like increase in $N(\omega)$ is expected due to the normalization condition.[11] The above disscussion was for modes on the infinite incipient cluster. When the density of states is averaged over all percolation clusters [similar to the case of Eq. (4)] the fracton dimension d'_s is $2d/d_w$.[11] It is interesting to note that the number of distinct sites averaged over all clusters scales as, $S_n \sim t^{d''_s}$ where

$$d''_s = \frac{d_s(d - 2\beta/\nu)}{d - \beta/\nu} = \frac{4}{d'_w}. \tag{10}$$

In recent years, there has been much interest in the Alexander-Orbach (AO) conjecture that $d_s = 4/3$ for percolation above $d = 1$.[1] Much theoretical and numerical effort has been directed to proving or disproving the AO conjecture and it is still regarded as an open question. A crossover from phonon behavior to fracton behavior has been observed experimentally in several systems. These include randomly diluted magnetic systems,[12] silica aerogels,[13] silica smoked-particle aggregates,[14] glassy ionic conductors,[15] and energy transfer experiments.[16]

5. Chemical Distance

The chemical distance ℓ between two sites is the length of the shortest path between the two sites. This concept[17] was found useful for transport properties. The mass of the cluster M scales with ℓ as $M \sim \ell^{d_\ell}$ where d_ℓ is the intrinsic fractal dimension. The exponent d_ℓ was found numerically very accurately (no theory yet!) by Herrmann and Stanley[18] for $d = 2$ percolation to be $d_\ell = 1.678 \pm 0.003$. (For the various exponents in percolation see Stanley.[19]) Since $M \sim \ell^{d_\ell} \sim R^{d_f}$ the Euclidean distance R can be related to ℓ,

$$R \sim \ell^{d_\ell/d_f} \equiv \ell^{\tilde{\nu}} \equiv \ell^{1/d_{\min}}. \tag{11}$$

When loops can be neglected the resistance is $\rho \sim \ell \sim R^{d_f/d_\ell}$ and $\tilde{\zeta} = d_f/d_\ell$. From Eq. (6), it follows[20] that $d_w = d_f(1 + 1/d_\ell)$ and the fracton dimension is $d_s = 2d_\ell/(d_\ell + 1)$ independent of d_f. Since the effect of loops is to enhance transport, it follows that these relations serve as rigorous upper bounds for $\tilde{\zeta}$, d_w and d_s. A dynamical quantity of interest related to $\tilde{\nu}$ is the velocity in which a fire front or an epidemic propagates (wetting velocity) by burning one chemical "shell" per unit time. The velocity of the fire in percolation system is[21]

$$v = \frac{dr}{dt} \equiv \frac{dr}{d\ell} \sim \ell^{\tilde{\nu}-1} \sim (p - p_c)^{-\nu(1-1/\tilde{\nu})}. \tag{12}$$

Since $\nu(1 - 1/\tilde{\nu}) \approx -0.16$ is very small (in $d = 2$), it follows that the increase of v upon crossing p_c is very steep. A fire that does not propagate at all at p_c will propagate *very* fast just above p_c.

6. Probability Densities

The probability density of random walks on fractals is of general interest since it contains much information on the dynamics. The probability density, $P(r, t)$, is the probability of finding a random walker at distance r at time t starting at the origin at $t = 0$. For diffusion in Euclidean space $P(r, t)$ is Gaussian. The question of the form of $P(r, t)$ on fractals is controversial and has recently attracted much interest. This is due due to the fact that $P(r, t)$ is relevant for localization of the wave function on fractals and it plays an important role in the Flory theory of self-avoiding walks on fractals. It is accepted that the average of $P(r, t)$ is of the form

$$P(r, t) = t^{-d_s/2} \Pi(r/t^{1/d_w}); \quad \Pi(y) = \exp(-y^u); \quad y > 1. \tag{13}$$

There is a controversy in literature about the value of u. It was shown recently[22] that the wave function of an impurity quantum state on a fractal decayes as $|\Psi(r)| \leq \exp -r^a$ with

$$a = \frac{u d_w}{u + d_w}. \tag{14}$$

If $u > d_w/(d_w - 1)$ as was suggested in the literature, then $a > 1$ and the function is "superlocalized." Analytical arguments on several fractals yield[9]

$$u = \frac{d_w}{d_w - 1}, \tag{15}$$

which leads to regular exponential decay for the wave function. Harris and Aharony[23] prove that $d_w/(\bar{d}_w - 1) \leq u \leq d_w/(\bar{\nu}d_w - 1)$. The lower bound is equal to u when $P(r, t)$ is averaged over all configurations, whereas u is equal to the upper bound when only typical configurations are taken. In a recent work Bunde et al[24] calculated the moments of $P(\ell, t)$ and $P(r, t)$ of random walks on the two dimensional infinite percolation cluster. They conclude that in the ℓ-space $u = d_w^\ell/(d_w^\ell - 1)$ for all moments and in the r-space $u = d_w/(d_w - 1)$ only for large moments. For small moments (below 2) the data indicate smaller values for u.

From Eqs. (13) and (15), it is possible to derive[9,25,26] a Flory theory for the end-to-end exponent ν of SAWs on percolation

$$\nu = \frac{2\bar{\nu}d_w - 1}{d_w(1 + \bar{\nu}d_f) - d_f},\tag{16}$$

where here d_f and d_w are the fractal dimension and the diffusion exponent of the *backbone*. This result, Eq. (16), is in good agreement with the existing data.

1. S. Alexander and R. Orbach, J. Phys. (Paris) Lett. 43, L625 (1982).
2. D. Ben-Avraham and S. Havlin, J. Phys. A15, L691 (1982); S. Havlin, D. Ben-Avraham and H. Sompolinsky, Phys. Rev. A27, 1730 (1983).
3. Y. Gefen, A. Aharony and S. Alexander, Phys. Rev. Lett. 50, 77 (1983).
4. R. Rammal and G. Toulouse, J. Phys. (Paris) Lett. 44, L13 (1983).
5. B. B. Mandelbrot, *Fractals: Form, Chance and Dimension* (Freeman, San-Francisco, 1977); *The Fractal Geometry of Nature* (Freeman, San-Francisco, 1982).
6. S. Alexander, J. Bernasconi, W. R. Schneider and R. Orbach Rev. Mod. Phys. 53, 175 (1981).
7. G. H. Weiss and R. J. Rubin, Adv. Chem. Phys. 52, 363 (1983).
8. D. Stauffer, *Introduction to Percolation Theory* (Taylor & Francis, London, 1985).
9. S. Havlin and D. Ben-Avraham, Adv. in Phys. 36, 695 (1987).
10. H. J. Herrmann , The EPS conference, April 1988 (Budapest).
11. A. Aharony, S. Alexander, O. Entin-Wohlman and R. Orbach, Phys. Rev. B31,2565 (1985).
12. M. B. Salomon and Y. Yeshurun, Phys. Rev. B36, 5643 (1987).
13. E. Courtens, J. Pelous, J. Phalippou, R. Wacher and T. Woignier, Phys. Rev. Lett. 58, 128 (1987).
14. D. Richter, T. Freltoft and J. K. Kjems, Phys. Rev. Lett. 59, 1212 (1987).
15. A. Avogadro, S. Aldrovandi and F. Borsa, Phys. Rev. B33, 5637 (1986).
16. P. Evesque, in *The Fractal Approach to the Chemistry of Disordered Systems, Polymers, Colloids, Surfaces*, ed. D. Avnir (John Wiley, 1988).
17. S. Havlin and R. Nossal, J. Phys. A17, L427 (1984).
18. H. J. Herrmann and H. E. Stanley, preprint (1988).
19. H. E. Stanley, in *On Growth and Form*, eds. H. E. Stanley and N. Ostrowsky (Martinus Nijhoff, Amsterdam, 1986).
20. S. Havlin, Z. V. Zjordjovic, I. Majid, H. E. Stanley and G. H. Weiss, Phys. Rev. Lett. 53, 178 (1984).
21. P. Grassberger, J. Phys. A18, L215 (1985); M. Barma, J. Phys. A18, L277 (1985).
22. Y. E. Levy and B. Souillard, Europhys. Lett. 4, 233 (1987).
23. A. B. Harris and A. Aharony, Europhys. Lett. 4, 1355 (1987).
24. A. Bunde, S. Havlin and E. Roman, preprint (1988).
25. J.-P. Bouchaud and A. Georges, preprint (1988).
26. A. Aharony and A. B. Harris, preprint (1988).

FRACTONS IN REAL FRACTALS

ERIC COURTENS[1] and RENÉ VACHER[2]

[1]*IBM Research Division, Zürich Research Laboratory*
8803 Rüschlikon, Switzerland

[2]*Laboratoire de Science des Matériaux Vitreux*
Université des Sciences et Techniques du Languedoc
F-34060 Montpellier, France

ABSTRACT. The vibrational excitations of well-characterized silica aerogels have been studied with Brillouin, Raman, and inelastic neutron scattering. Fractons are observed over a frequency range directly related to the fractal structure of these materials. The fractal dimension D, the spectral dimension d_s, a localization exponent d_ϕ, as well as the density of states $N(\omega)$, have been determined in these experiments. The overall behavior is in remarkable agreement with theoretical expectations.

1. Questions, Motivations, and Approaches

The vibrations of fractal networks have been theoretically described by new collective excitations, called fractons.[1] The corresponding scaling of the density of states, $N(\omega) \propto \omega^{d_s-1}$, is characterized by the spectral dimension d_s, which in scalar elasticity is related to the diffusion exponent θ, $d_s = 2D/(2+\theta)$.[1,2] In the real world, materials are fractal over a restricted length scale, $\xi > L > a$, between typical particle dimensions a and correlation length ξ. Beyond ξ, the density ρ is well-defined and the materials become homogeneous. To the above length limits, there are corresponding crossover frequencies between which fractons can be expected, $\omega_{co1} < \omega < \omega_{co2}$. Below ω_{co1}, one should observe usual long-wave phonons. Beyond ω_{co2}, particle modes should appear, starting at an ω_{min} which depends on the elasticity in the particles rather than on properties of the fractal network.

Until recently, many experimental attempts at the observation of fractons were in systems whose fractality is not well established,[3] or at frequencies possibly beyond ω_{co2}.[4,5] We were able to overcome these difficulties by performing various scattering measurements on silica aerogels. These are single phase, highly porous, silica based materials, which were known to be fractal over restricted length scales.[6] With somewhat different preparation conditions, materials of densities ranging from $\simeq 100$ to $\simeq 400$ kg/m^3, i.e., porosities from $\simeq 0.8$ to $\simeq 0.95$, were obtained.[7] Their mass fractality was demonstrated[7] by small-angle neutron scattering (SANS).[8] The lightest material is fractal over as much as two orders of magnitude in length scales, and all materials have the same D within experimental accuracy.[7] Moreover, we demonstrated that series of aerogels of different densities can be prepared whose structures are indistinguishable at scales where the materials are fractals. This property, which we called *mutual self-similarity*,[7] is a prerequisite for a meaningful scaling of results as a function of density. It implies that, for a series of samples, $\xi \propto \rho^{1/(D-3)}$.

With suitable choice of the materials and of the measurement techniques, we have now covered a large part of the frequency range of interest. Using Brillouin scattering, we investigated the region around the phonon-fracton crossover at ω_{co1}.[9,10] In view of the experimental limitations of this type of measurement, it is necessary to use a series of samples to cover the entire crossover region. In our samples, ω_{co1} falls around 0.5 to 5 GHz. The additional benefit of this approach is that it allows D and d_s to be determined from the scaling with ρ since the samples are mutually self-similar. At frequencies sufficiently higher than ω_{co1}, the coherent light scattering signal becomes weaker than the incoherent signal produced by local fluctuations. The latter can actually be singled out in depolarized backscattering, where the coherent intensity vanishes by selection rules, at least in first order. The full range between ω_{co1} and ω_{co2} was covered with Raman scattering spectroscopy at unusually low frequencies.[11] This provided the first spectroscopic observation of fractons over their full range of existence, as well as the measurement of a localization exponent d_ϕ. Above ω_{co2}, Raman scattering reveals particle modes, but the intensity of that signal remains difficult to interpret quantitatively.

To gain information on d_s from measurements on a single sample, it is necessary to obtain the density of states, $N(\omega)$. This can be done by incoherent inelastic neutron scattering.[12] So far, we used time-of-flight spectroscopy, the energy range of which covers the particle modes as well as the high-frequency end of the fracton region around ω_{co2}.[13] This also allowed an absolute determination of $N(\omega)$ which enabled the low-temperature thermal properties to be calculated in good agreement with existing data on similar materials.[14]

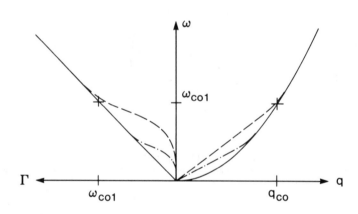

Fig. 1: Dispersion curves (right) and linewidth (left) for plane waves in the phonon-fracton crossover region, illustrated for two densities; q is the wave vector, ω the frequency, and Γ the broadening by elastic scattering as defined in Ref. 10. The values of ω_{co1} and q_{co} are indicated for one particular density. The solid lines are the universal curves for fractons.

2. The Phonon-Fracton Crossover

The situation at the phonon-fracton crossover is illustrated in Fig. 1. The universal dispersion relation for fractons is $\omega \propto q^\alpha$. For mutually self-similar samples, the same dispersion curve describes all samples in the fractal region. The exponent α can be derived from $N(\omega)d\omega \propto d^D q$, which yields $\alpha = D/d_s$. Phonons obey $\omega = c_0(\rho)q$, as illustrated for two particular densities in Fig. 1. The phonon dispersion intercepts the fracton dispersion at q_{co}, ω_{co1} for each particular sample. One also expects that $q_{co} \propto 1/\xi$. At frequencies below ω_{co1}, the phonon linewidth is determined by elastic Rayleigh scattering, $\Gamma \propto \omega^4$. At ω_{co1}, $\Gamma \simeq \omega_{co1}$, one reaches the Ioffe-Regel limit,[15] whose significance is that plane waves are not the fracton eigenmodes. In Brillouin scattering, the geometry of the experiment imposes plane waves of fixed q. Fractons of different frequencies ω do have Fourier components at this q, since they are highly localized excitations. This leads to the width observed in the spectra.

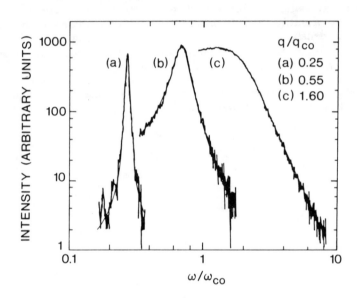

Fig. 2: Three typical Brillouin spectra with their fits to the EMA theory: (a) is in the phonon regime, (b) is near crossover, and (c), mostly above ω_{co1}, is in the fracton regime.

Brillouin spectroscopy was performed on a series of samples and at various q-values, as described elsewhere.[9,10] The spectra were fitted to a theoretical shape derived from an effective medium approximation (EMA).[16] The analytic expression for the shape is identical to that for usual phonon scattering, with the exception that both the linewidth Γ and the velocity c have unusual variations in ω, as illustrated in Fig. 1. We selected scaling expressions for $\Gamma(\omega)$ and $c(\omega)$, which have the proper asymptotic behavior and depend only on ω_{co1} and c_0. All spectra could be fitted with these two parameters plus the intensity. Examples are shown in Fig. 2. From the series of measurements, we extracted ω_{co1} and $q_{co} = \omega_{co1}/c_0$. This allowed the fracton dispersion curve to be traced for the first time, as shown in Fig. 3. From the scaling of $q_{co} \propto \rho^{1/(3-D)}$, one derives $D = 2.46$. This acoustical value is in satisfactory agreement with the mass value obtained from SANS, $D = 2.40$.[7] From

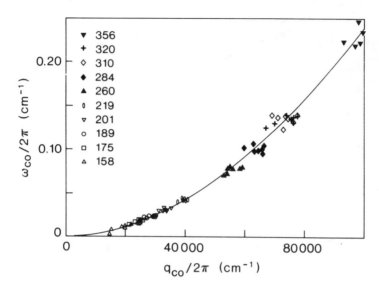

Fig. 3: The universal fracton dispersion curve and the points derived from the Brillouin measurements on a series of mutually self-similar samples. The different symbols label the sample densities, as indicated in kg/m^3. The points for each sample are derived at various scattering angles.

the dispersion curve, one then extracts $d_s = 1.3 \pm 0.1$. This happens to coincide with the value 4/3 predicted for scalar waves.[1,15] It does not necessarily mean that scalar elasticity[17] dominates the acoustic behavior of our samples.

3. Spectroscopy over the Full Fracton Regime

In addition to the coherent Brillouin signal, which decreases approximately like ω^{-4} at large ω,[16,10] one anticipates incoherent scattering originating in local polarization fluctuations. This leads to q-independent scattering, since most fractons have localization lengths much smaller than an optical wavelength. Fractons are expected up to a frequency of $\omega_{co2} \simeq \omega_{co1}(\xi/a)^{D/d_s}$, a relation derived from the dispersion law. This typically yields $\omega_{co2} \simeq 10$ cm^{-1}, independently of the sample density for a series of mutually self-similar samples for which a is approximately constant. Thus, Raman scattering from fractons in aerogels must be measured at unusually low frequencies compared with standard Raman spectrometry, a condition which was not met previously.[4]

To observe the Raman component alone, we performed depolarized scattering in the backward direction. The samples were oxidized to remove extraneous molecular groups that could otherwise contribute to the scattering. Tandem interferometry was used to achieve the high resolution required in the measurement.[11] A typical result is shown in Fig. 4. A power law is found in all samples in the fracton regime. Its slope does not depend significantly on ρ.[11] The end of the fracton region is marked by the onset of particle modes at $\omega_{min} \simeq 2\pi \times 0.83 v_t/2R$,[18] where v_t is the transverse sound velocity in the particles, and R their typical radius. It

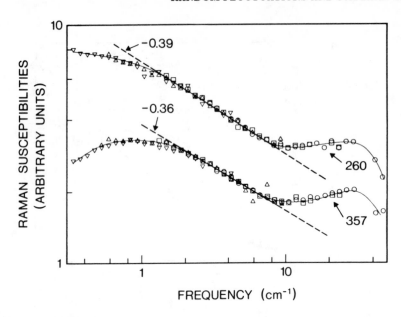

Fig. 4: Depolarized backscattering Raman intensities divided by the Bose-Einstein factor, *vs* frequency for two samples labelled by their densities. The straight lines are fits with the indicated slopes, while the thin curves are guides to the eye (after Ref. 11).

happens here that $\omega_{\min}$ approximately coincides with ω_{co2}. The curvature around ω_{co1} scales with ρ like ω_{co1} in Brillouin results.[10]

The scaling in ω of the Raman intensity I_i can be estimated as follows. First consider the scattering by a single fracton of typical Euclidean length $\ell \propto \omega^{-d_s/D}$. The Pockels piezo-optic coefficient in Euclidean space is proportional to the local density in the fracton volume, $\propto \ell^{D-3}$. The strain is the gradient of the displacement u, and it is conjectured[19] to be $\propto u\omega^{d_s d_\phi/D}$. Here, d_ϕ is a localization exponent, which should fall in the range $1 \le d_\phi \le \tilde{\zeta}_c$, where $\tilde{\zeta}_c$ is the "chemical" exponent characterizing the shortest connecting path between two particles.[19,20] The local polarization δP due to the fracton is the product of the piezo-optic coefficient, times the strain, integrated over the Euclidean volume ℓ^3. This yields $\delta P \propto u\ell^{D-d_\phi}$. The scattered intensity is the mean square amplitude, which contains a term $\langle uu \rangle$. Making use of the usual mode quantization, the integral of $\langle uu \rangle$ over space (here, fractal space), $\propto \langle uu \rangle \ell^D$, is proportional to the thermal population factor, $n(\omega)$, divided by ω. Grouping all terms, the scattering from a single fracton mode is $\propto \ell^{D-2d_\phi}n(\omega)/\omega$. Finally, this term must be multiplied by $N(\omega)$ to obtain the intensity I_i scattered by all fractons, since they scatter incoherently. Replacing ℓ by its scaling in ω, one obtains $I_i/n(\omega) \propto \omega^{-2+2d_\phi d_s/D}$. From the mean slope of $I_i/n(\omega)$, observed at several ρ's, and using D and d_s, one derives $d_\phi \simeq 1.6$.[11] Alternatively, specializing d_ϕ to the resistance exponent $\tilde{\zeta} = D(2 - d_s)/d_s$,[15] the

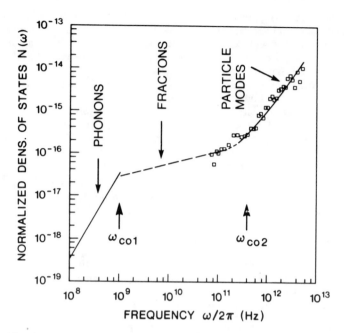

Fig. 5: Absolute density of vibrational states determined from Brillouin spectroscopy (phonon regime) and from neutron scattering (particle modes). The fracton line is traced with the slope $d_s - 1$ derived from measurements near ω_{co1}. The density of states is normalized to one upon integration over ω.

slope of $I_i/n(\omega)$ becomes $2(1 - d_s)$ which yields $d_s \simeq 1.2$. Since $\bar{\zeta}_c$ is always larger than $\bar{\zeta}$, we expect that $d_\phi = \bar{\zeta}_c$ could lead to a value of d_s very close to 1.3, as in Brillouin scattering.

4. The Vibrational Density of States

Incoherent neutron scattering from protons chemically bonded to the particle surfaces allows the density of states in porous media to be determined.[21] This method was applied previously to silica smoke.[5] Compared to silica smoke, the particles in our aerogels are more than two times smaller, which leads to correspondingly higher values of ω_{co2}. It is important in the measurement to obtain the pure incoherent contribution. We achieved this by taking the difference signal between two similar samples, one with protons and the other with deuterons attached to the particles.[13] This approach allows both samples to be thoroughly dried under identical conditions to avoid possible contributions owing to free moving water.

Measurements performed with the time-of-flight spectrometer Mibemol in Saclay revealed a nearly temperature-independent density of states, as explained in detail elsewhere.[13] Figure 5 shows the experimental points obtained at 60 K. The values ω_{co1} and ω_{co2} are derived from the Brillouin and SANS information. The experimental points above ω_{co2} have been fitted to a theoretical expression for the density of states of spherical grains.[22] The effective slope of the curve is between 1 and 2, the former corresponding to surface modes and the latter to bulk modes. This fit allows the ordinate scale to be fixed in absolute value,[13] since the main parameters of the grains are fairly well known. Using the Brillouin information, the density of states below ω_{co1}, i.e., in the phonon regime, can be calculated in the Debye model. The result is shown by the straight line of slope 2 in Fig. 5. Below ω_{co2}, one anticipates that $N(\omega)$ is dominated by fractons. Hence, we have drawn a line of slope $d_s - 1 = 0.3$, as given by the Brillouin results, through the lowest neutron data points. Remarkably, that line intercepts the phonon regime near ω_{co1}.

The combination of these low and high-frequency measurements leads thus to an overall quantitative picture of the density of states. To check the significance of the extended fracton range predicted in Fig. 5, we used the absolute $N(\omega)$ to calculate the thermal properties of silica aerogels at very low temperatures. The fractons produce a power-law region in the specific heat C, with $C/T^3 \propto T^{-d_s}$.[14] Our prediction agrees both in slope and absolute value with the measurements.[14,13] Similarly, we can calculate a plateau of the proper height in the thermal conductivity around 0.1 K.

The experimental work on which this review is based has been performed in collaboration with G. Coddens, J. Pelous, J. Phalippou, Y. Tsujimi, and T. Woignier. The neutron results were obtained at the Laboratoire Léon Brillouin, a joint laboratory Centre National de la Recherche Scientifique (CNRS) – Commissariat à l'Energie Atomique, Saclay, France. The authors thank Professors Aharony, Alexander, Entin-Wohlman, Orbach, and Teixeira for many discussions.

1. S. Alexander and R. Orbach, J. Phys. (Paris) Lett. 43, L625 (1982).

2. Y. Gefen, A. Aharony, and S. Alexander, Phys. Rev. Lett. 50, 77 (1983).

3. e.g., A. J. Dianoux, J. N. Page, and H. M. Rosenberg, Phys. Rev. Lett. 58, 886 (1987); A. Fontana, F. Rocca, and M. P. Fontana, Phys. Rev. Lett. 58, 503 (1987).

4. A. Boukenter, B. Champagnon, E. Duval, J. Dumas, J. F. Quinson, and J. Serughetti, Phys. Rev. Lett. 57, 2391 (1986).

5. T. Freltoft, J. Kjems, and D. Richter, Phys. Rev. Lett. 59, 1212 (1987).

6. D. W. Schaefer and K. D. Keefer, Phys. Rev. Lett. 56, 2199 (1986).

7. R. Vacher, T. Woignier, J. Pelous, and E. Courtens, Phys. Rev. B 37, 6500 (1988).

8. J. Teixeira, in On Growth and Form, H. E. Stanley and N. Ostrowsky, Eds. (Nijhoff, Dordrecht, 1986), p. 145.

9. E. Courtens, J. Pelous, J. Phalippou, R. Vacher, and T. Woignier, Phys. Rev. Lett. 58, 128 (1987).

10. E. Courtens, R. Vacher, J. Pelous, and T. Woignier, Europhys. Lett. 6, 245 (1988).

11. Y. Tsujimi, E. Courtens, J. Pelous, and R. Vacher, Phys. Rev. Lett. 60, 2757 (1988).

12. G. L. Squires, Thermal Neutron Scattering (Cambridge Univ. Press, Cambridge, 1978).

13. R. Vacher, T. Woignier, J. Pelous, G. Coddens, and E. Courtens, Europhys. Lett. (in press).

14. R. Calemczuk, A.M. de Goër, B. Salce, R. Maynard, and A. Zarembowitch, Europhys. Lett. 3, 1205 (1987).

15. A. Aharony, S. Alexander, O. Entin-Wohlman, and R. Orbach, Phys. Rev. Lett. 58, 132 (1987).

16. G. Polatsek and O. Entin-Wohlman, Phys. Rev. B 37,7726 (1988).

17. S. Alexander, J. Phys. (Paris) 45, 1939 (1984).

18. H. Lamb, Proc. Math. Soc. London 13, 187 (1882).

19. S. Alexander, O. Entin-Wohlman, and R. Orbach, Phys. Rev. B 32, 6447 (1985).

20. A. Brooks Harris and A. Aharony, Europhys. Lett. 4, 1355 (1987).

21. D. Richter and L. Passell, Phys. Rev. Lett. 44, 1593 (1980).

22. H. P. Baltes and E. R. Hilf, Solid State Commun. 12, 369 (1973).

ANOMALOUS TRANSPORT IN DISORDERED STRUCTURES: EFFECT OF ADDITIONAL DISORDER

ARMIN BUNDE

I. Institut für Theoretische Physik, Universität Hamburg
Jungiusstrasse 9, D-2000 Hamburg 36, W. Germany

ABSTRACT. We study how the mean-square displacement R^2 of a random walker is changed on fractal structures, in particular, if there is additional disorder in the system which can be caused either by non-uniform distributions of jump rates or by an external bias field. While the effect of a singular distribution of jump rates is to enhance the fractal dimension of the walk, an external bias field leads to a logarithmic time dependence $R^2 \sim \ln t$, which seems to hold for many fractal structures, including percolation and chain-like fractal structures in any dimension.

1. Static Properties of disordered structures

A simple model of a random two-component AB-mixture is a lattice, where each site has been occupied randomly with probability p (A-sites) or stays empty with probability $1 - p$ (B-sites). A- and B-sites stand for materials with very different physical properties. Here we want to assume that the A-component is conducting and the B-component is an insulator. Furthermore let us assume that electric current can flow only between nearest neighbor A-sites. Alternatively, using the language of diffusion, we can assume that Brownian particles can jump only between nearest neighbor A-sites, but they cannot enter the B-component.

At low conductor concentration p only *finite* clusters of nearest neighbor A-sites are formed and the mixture is an insulator. On the other hand, at large p, there exist many unblocked diffusion paths where particles can be transported from one end of the mixture to the other one, and we have a conductor. Between both concentrations there exists a threshold concentration p_c (*percolation threshold* or *critical concentration*), where for the first time a macroscopic diffusion path occurs (connecting opposite faces of the system), which allows the particles to *percolate*. The probability per site to belong to this "infinite" diffusion path (*infinite percolation cluster*) increases above p_c as $P_\infty \sim (p - p_c)^\beta$ and the conductivity grows as $\sigma \sim (p - p_c)^\mu$. Both below and above p_c, the linear size of the *finite* clusters is described by the correlation length ξ, which is defined as the mean distance between two sites on the same finite cluster. Close to p_c, we have $\xi \sim |p - p_c|^{-\nu}$. The exponents β and ν are universal, and depend only on the dimension d of the system, but not on minor structural details. In contrast, p_c depends strongly on the lattice structure (for recent reviews see, e.g., Refs. 1 and 2).

Close to p_c, large percolation clusters are *fractals* on length scales which are small compared with ξ and large compared with the smallest cluster length a: They are *self-similar* and their mass- length relation is described by a non-integer ("fractal") dimension d_f,

$$M(L) \sim L^{d_f}, \quad a \ll L \ll \xi, \tag{1}$$

where $d_f = d - \beta/\nu$ here. In two-dimensional systems, $\beta = 5/36$, $\nu = 4/3$ and thus $d_f = 91/48 \approx 1.896$, while in $d = 3$ one has $d_f \approx 2.5$. Eq. (1) holds also for the infinite cluster, which is fractal on intermediate length scales, but d-dimensional on large length scales, $L \gg \xi$. The selfsimilarity states that characteristic features of the structure (dangling ends, loops, etc.) occur on *all* length scales L with $a \ll L \ll \xi$, such that a magnification of a small part of the cluster cannot be distinguished from the original.

2. Anomalous Transport in Disordered Media

2.1. CONSTANT JUMP RATES

The question we want to address now, is, how the transport properties of the mixture are affected by the fractal structures in the system which occur close to p_c. To this end, we consider first the motion of a Brownian particle ("random walker") on the infinite percolation cluster at p_c, with constant jump rates between neighboring sites. Later we shall consider also the effect of varying jump rates. The infinite cluster represents a large labyrinth for the walker, with loops and dangling ends. Both, loops and dangling ends slow down the motion of the walker. Due to *self-similarity*, loops and dangling ends exist in all length scales $L \gg a$ and thus, compared with ordinary Euclidean systems, the walker is slowed down on *all* length scales. Hence Fick's law, $\langle r^2 \rangle \sim t$, is no longer valid. Instead, the mean-square displacement of the walker, averaged over many cluster configurations and starting points of the walker, becomes

$$\langle r^2 \rangle \sim t^{2/d_w}, \tag{2}$$

where d_w (the "fractal dimension of the walk") is roughly given by $d_w \approx 3d_f/2$.[3] The exponent d_w is universal and does not depend on the structure of the considered lattice nor on minor structural details. It depends only on the dimension of the lattice considered, just as d_f and the other static exponents do. But d_w can change considerably if we have additional disorder in the system, e.g. if the jump rates between neighbored cluster sites are not constant, but are randomly distributed. To study this effect, let us assume thermally activated processes, where the jump rate $W_{i,i+\delta}$ from site i to a nearest neighbor site $i+\delta$ is determined by the potential barrier $V_{i,i+\delta}$ between both sites,

$$W_{i,i+\delta} \propto e^{-V_{i,i+\delta}/k_B T}. \tag{3}$$

A random distribution of jump rates arises then from a random distribution of potential barriers. In the following we shall consider two simple models of disorder.

2.2. WAITING TIME DISTRIBUTION

In the first model, we assume that the maxima of the random potential stay constant, while the minima are chosen from a random distribution. Accordingly, the jump rates are solely determined by the depths of the potential minima. If the random walker falls into a deep minimum at site i, he has to wait on the average $\tau_i = 1/W_i$ time steps before he can jump to a nearest neighbor cluster site. Let us assume that deep potentials are exponentially rare, i.e. the probability $P_0(V)$ of finding a potential minimum of depth V satisfies

$$P_0(V) \propto e^{-V/V_0}, \tag{4}$$

where V_0 is the mean depth of the minima. Combining (3) and (4) we find that the probability P of finding a "waiting time" $\tau > 1$ scales as

$$P(\tau) \propto \tau^{\alpha-2}, \quad \alpha = 1 - k_B T/V_0. \tag{5}$$

For deriving d_w we follow Ref. 4.

Since $\langle r^2 \rangle$ *as a function of the N steps made by the walker* does not depend on how long the walker has to wait in each site, it scales as for constant waiting times,

$$\langle r^2 \rangle \sim N^{2/d_w}. \tag{6}$$

On the other hand, the elapsed time t is $t = N\bar{\tau}$ where $\bar{\tau}$ is the mean time the walker stays in one site. We can evaluate $\bar{\tau}$ from the number S of distinct sites visited so far by the walker,

$$\bar{\tau} = \frac{1}{S} \sum_{i=1}^{S} \tau_i = \int_1^{\tau_{max}} d\tau\, \tau P(\tau). \tag{7}$$

For $\alpha < 0$, the integral converges to a finite value when N and τ_{max} tend to infinity. Consequently, $\bar{\tau}$ stays constant for large N and the asymptotic behavior of $\langle r^2(t) \rangle$ is the same as for constant waiting times.

For $0 < \alpha < 1$, $\bar{\tau}$ diverges as $\bar{\tau} \sim (\tau_{max})^{\alpha}$ for $\tau_{max} \to \infty$, and the asymptotic behavior of $\langle r^2(t) \rangle$ is changed. To see how τ_{max} depends on S and N, let us consider first how the waiting times τ_i are chosen from random numbers x_i, $0 < x_i < 1$. Since the x_i are homogeneously distributed, we have $P(\tau)d\tau = dx$ and thus $\tau_i \sim x_i^{1/(1-\alpha)}$ satisfies (5). Now consider S random numbers x_i. Again, since the random numbers are homogeneously distributed, their minimum value scales as $x_{min} \sim 1/S$. Accordingly, τ_{max} scales as $\tau_{max} \sim X_{min}^{1/(\alpha-1)} \sim S^{1/(1-\alpha)}$ and we obtain from (7)

$$\bar{\tau} \sim S^{\alpha/(1-\alpha)}, 0 < \alpha < 1. \tag{8}$$

On the infinite cluster, the number of visited sites S scales with the number of steps N as[9]

$$S \sim N^{d_f/d_w}. \tag{9}$$

From (8) and (9) we obtain easily the elapsed time $t = \bar{\tau}N$ as a function of N. Then using (6), we find $\langle r^2 \rangle \sim t^{2/\bar{d}_w}$ where now $\bar{d}_w$ is given by

$$\bar{d}_w = \begin{cases} d_w, & -\infty < \alpha < 0, \\ d_w + d_f \alpha/(1-\alpha), & 0 \le \alpha < 1. \end{cases} \tag{10}$$

Note that this relation is general and holds for *all* fractals where (9) is valid, i.e., for the spectral dimension $d_s < 2$. It can be easily generalized to include $d_s \ge 2$ and Euclidean systems.[2,4]

2.3. JUMP RATE DISTRIBUTION

In the second model, the potential minima are constant, but the potential barriers are chosen from an exponential distribution [see Eq. (4)], where now V_0 represents the mean potential barrier in the system. This leads to a power law distribution of jump rates,

$$P_0(W) \sim W^{-\alpha}, \qquad W \leq 1, \tag{11}$$

where $\alpha = 1 - k_B T / V_0$, as before. In one dimension, there is no difference between both models. In higher dimensions, both are very different. In the first model, a random walker cannot circumvent deep potential minima, while here he can circumvent large barriers. Hence the question how d_w depends on α cannot be answered generally in this case, but the answer depends drastically on structural details. In 2 and 3 dimensions, large barriers can be circumvented easily, and the mean-square displacement follows Fick's law asymptotically, $\bar{d}_w = d_w = 2$.

On large percolation clusters, there exist sites which are only singly connected and which a random walker has to pass in order to come from one part of the cluster to a distant one. Along these one-dimensional pieces (which can be assigned to a fractal dimension $d_f = 1/\nu$—see Ref. 5), the distribution of jump rates has large effects and hence these singly-connected sites dominate the transport for singular distributions, $\alpha > 0$. There exists no rigorous solution for $\bar{d}_w$ as in the simpler model before, but rigorous bounds have been derived,[6-8]

$$d_f + \frac{1}{\nu(1-\alpha)} \ \leq \ \bar{d}_w \ \leq \ d_w + \frac{\alpha}{\nu(1-\alpha)}, \qquad 0 < \alpha < 1. \tag{12}$$

The upper bound can be obtained from (10) with $d_f = 1/\nu$ the fractal dimension of the relevant singly-connected sites. It has been found numerically[7] on d=2 percolation clusters that $\bar{d}_w$ approaches the lower bound for $\alpha > 0.5$, and is close to the upper bound for $\alpha > 0.5$. For $\alpha \leq 0$, we have $\bar{d}_w = d_w$.

So far, we have considered only random walkers on the infinite percolation cluster at criticality. In reality, mobile particles move on all clusters, also on the finite ones, and experiments cannot be done exactly at p_c which in most cases is not known exactly. But these facts can be taken into account easily. If the walker can start with equal probability on all A-sites in the lattice, then d_w is modified into an "effective" $\bar{d}_w$,[1,2]

$$\bar{d}_{w,eff} = \frac{\bar{d}_w}{1 - \beta/2\nu}, \tag{13}$$

where $1 - \beta/2\nu = 91/96$ in $d = 2$ and ≈ 0.8 in $d = 3$.

If we are above p_c, then fractal structures occur only on length scales $L \ll \xi$. On large length scales $L \gg \xi$ there exists only the infinite cluster which has the dimension d of the random mixture and diffusion obeys Fick's law. Consequently, the anomalous time dependence of $\langle r^2(t) \rangle$ can be found only for $\langle r^2(t) \rangle \ll \xi^2$, where the particles explore the fractal labyrinths of the clusters.

The exponent $k = 2/\bar{d}_{w,eff}$ should show up in the frequency dependent conductivity $\sigma(\omega)$. To a good approximation, $\sigma(\omega)$ is related to $\langle r^2 \rangle$ by[9]

$$\sigma(\omega) \sim \omega^2 \int_0^\infty \langle r^2(t) \rangle \exp(i\omega t) dt. \tag{14}$$

Hence we expect an anomalous frequency dependence

$$\sigma(\omega) \sim \omega^{1-k}, \qquad \omega(\xi) \ll \omega \ll \omega(a), \qquad (15)$$

where $\omega(\xi) \equiv \xi^{-d_w}$,""" tends to zero for $p \to p_c$.

In physics, nonuniform distributions of potential barriers can have different origins. *Continuous percolation systems* like the random void model can be mapped onto lattice systems with a power law distribution of jump rates. If diffusion is continous, then $\alpha = -1$ in $d = 2$ and $\alpha = 1/3$ in $d = 3$.[6] If diffusion is noncontinous (finite jump distance), then in a large time regime $\alpha = 1/3$ in $d = 2$ and $\alpha = 5/7$ in $d = 3$.[10] *Long range interactions* between the diffusing particles also can cause, on certain length scales, random distributions of jump rates ("cage-effect," see Ref. 11). Thus we expect in ionic mixtures such as mixed alkali conductors or dispersed ionic conductors[12] conductivity exponents $1 - k$ larger than in normal percolation, in agreement with experiments.[13]

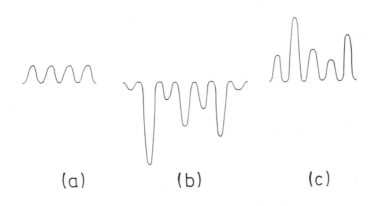

(a)   (b)   (c)

Fig. 1: Illustration of the three models considered here. (a) Constant potential maxima and minima, (b) constant maxima and variable minima, and (c) constant minima and variable maxima.

2.4. BIASED DIFFUSION

Additional disorder can be created in a disordered structure also by an external bias field. A bias field (e.g., in x-direction) gives the random walker a higher probability $P_+ \propto 1 + E$ of moving along the field and a lower probability $P_- \propto 1 - E$ of moving against the field, where $0 \leq E \leq 1$ represents the strength of the bias field. In the infinite percolation cluster at criticality, loops and dangling ends exist on all length scales. A random walker is not forced to follow the shortest path between two distant sites, but is driven into loops and dangling ends where he can get stuck. Both loops and dangling ends act as random delays on the motion of the walker, similar to the effect of a bias field in a random comb which pushes a walker into the teeth of the comb. The probability of finding large random delays determines the transport behavior. At criticality, the distribution of the effective "trapping"-lengths of loops and dangling ends should follow a power law $P_0(L) \sim L^{-(1+\alpha)}, \alpha > 0$ (see Ref. 14) and this yields

$$\langle r^2(t) \rangle^{1/2} \sim \left(\frac{\log t}{\log[(1 + E)/(1 - E)]} \right)^{\alpha}. \qquad (16)$$

It has been found numerically that $\alpha \approx 1.$[14] Under the influence of the bias field, the laws of diffusion are changed drastically, from a power law Eq. (2) to a *logarithmic* time dependence. We have the "paradoxical" situation that the motion is slowed down when the bias is enhanced, just the opposite of what happens in homogeneous Euclidean systems.

The simple logarithmic time dependence, $R \equiv \langle r^2 \rangle^{1/2} \sim \log t$, is even more general. If *topologically linear* fractal structures are considered, e.g., self-avoiding walks in any dimension or the external perimeter of percolation clusters in $d = 2$, then from rigorous treatments and numerical investigations[15] exactly the same simple logarithmic time dependence is obtained. Thus we may question whether this law describes generally biased diffusion on random fractals, i.e., if the "universal" law $R \sim t$ valid in uniform systems is changed into the "universal" law $R \sim \log t$ in random fractals.

If we are above the critical concentration, then the situation becomes more complicated: large "trapping" lengths are exponentially rare $[P_0(L) \sim \exp(-L/L_0)]$ and transport is again characterized by power law relations and, in addition, by dynamical phase transitions. Below a critical bias field $E_c = \tanh(1/2L_0)$ diffusion is classical and $R \sim t$. Above E_c, diffusion is anomalous, $R \sim t^{2/d_w(E,L_\sim)}$ where $d_w(E, L_0) = L_0 \log[(1 - E)/(1 + E)]$ (see Ref. 16); L_0 is a mean "trapping" length, which tends to infinity for $p \to p_c$ and decreases monotonically for p approaching unity. We have again the paradoxical situation that the motion of a walker is slowed down (above E_c) when the bias field is enhanced.

1. D. Stauffer, *Introduction to Percolation Theory* (Taylor and Francis, London, 1985); S. Havlin, this book.
2. S. Havlin and D. Ben-Avraham, Adv. in Phys. 36, 695 (1987), and references cited therein.
3. S. Alexander and R. Orbach, J. Phys. , Paris, 43, L625 (1982).
4. H. Harder, S. Havlin, and A. Bunde, Phys. Rev. B 36, 3874 (1987).
5. H. E. Stanley, J. Phys. A 10, L211 (1977); A. Coniglio, Phys. Rev. Lett. 46, 250 (1981).
6. B. I. Halperin, S. Feng, and P. N. Sen, Phys. Rev. Lett. 54, 2391 (1985); S. Feng, B. I. Halperin, and P. N. Sen, Phys. Rev. B 35, 197 (1987).
7. A. Bunde, H. Harder, S. Havlin, Phys. Rev. B 34, 3540 (1986).
8. P. M. Kogut and J. P. Straley, J. Phys. C 12, 2151 (1979); J. P. Straley, J. Phys. C 15, 2333 and 2343 (1982).
9. H. Scher and M. Lax, Phys. Rev. 10, 4491 (1974).
10. J. Petersen, H. E. Roman, A. Bunde, and W. Dieterich, preprint.
11. K. Funke, Solid State Ionics 18/19, 183 (1986) and 27, 197 (1988) and references cited therein; A. Bunde and W. Dieterich, Phys. Rev. B 31, 6012 (1985).
12. A. Bunde, W. Dieterich, and E. Roman, Phys. Rev. Lett. 55, 5 (1985); H. Harder, A. Bunde, and W. Dieterich, J. Chem. Phys. 85, 4123 (1986).
13. K. Funke, Ref. 11; J. Bates, priv. commun.
14. S. Havlin, A. Bunde, A. Glaser, and H. E. Stanley, Phys. Rev. A 34, 3492 (1986); A. Bunde, H. Harder, S. Havlin, and H. E. Roman, J. Phys. A 20, L865 (1987).
15. H. E. Roman, M. Schwartz, A. Bunde, and S. Havlin, Europhys. Lett. (in press)
16. A. Bunde, S. Havlin, H. E. Stanley, B. Trus and G. H. Weiss, Phys. Rev. B 34, 8129 (1986).

INFORMATION EXPONENTS FOR TRANSPORT IN REGULAR LATTICES AND FRACTALS

G. L. BLERIS and P. ARGYRAKIS

Department of Physics 313-1
University of Thessaloniki
GR-54006 Thessaloniki, Greece

ABSTRACT. The information exponent D_I is presented for random walks in regular lattices and on fractal structures. It is shown that this exponent gives an alternate way of describing the transport properties in lattices providing more information than is customary. It is analogous to the thermodynamic entropy, as it bears a $P \ln P$ character. We show that D_I is equivalent to the other well-known critical exponents with the only difference being that for fractals it takes unusually long times to reach such behavior. We discuss the implications of this behavior, and show the need for knowing this function in terms of experimental results that one needs to explain.

1. Introduction

It is well established that random walks have been very helpful in the last few years in studies of transport properties in homogeneous and random systems such as fractals. This models based on random walks and their properties have actually opened the way for specific solutions to problems in these systems that were thought of as being too complicated to solve just a few years ago. For example, the range of transport is matched very well by the number-of-sites-visited parameter (S_n), both theoretically and in numerical computer experiments. All this work is described in detail in several reviews or conference proceedings.[1-3] Here we introduce still another exponent, the information exponent D_I, which in addition to customary exponents also gives information about the "entropy state" of the system as a function of the inhomogeneity and randomness present. It gives more information about the areas of frequent revisitation, and we show that it is related to the general problem of multiple visits on lattices, a problem that was recently solved. Let us consider the events:

$$E : \quad E_1, E_2, E_3, \ldots, E_n \tag{1}$$

of a complete system together with their probabilities of occurrence:

$$P : \quad P_1, P_2, P_3, \ldots, P_n. \tag{2}$$

Then we say that we have a finite scheme:

$$\begin{pmatrix} E \\ P \end{pmatrix} = \begin{pmatrix} E_1 E_2 E_3 \ldots E_n \\ P_1 P_2 P_3 \ldots P_n \end{pmatrix}. \tag{3}$$

Every finite scheme describes a state of uncertainty. We define the quantity $I(P)$ as the measure of the amount of uncertainty associated with a given finite scheme:

$$I(P_1, P_2 \ldots P_n) = -\sum_{k=1}^{n} P_k \ln P_k. \tag{4}$$

The function $I(P)$, also known as the "ignorance function," is a basic thermodynamic extrinsic property that gives information concerning the number of states available to the system as a function of some thermodynamic variables, such as the temperature, the energy, etc. We may call $I(P)$ the entropy function. It is directly related to the microcanonical ensemble. If $W(P_k)$ is the configurational entropy obtained by taking nP_1 configurations of type 1, nP_2 configurations of type 2, etc.[1] on a lattice of n sites in all possible ways, then, disregarding overlaps:

$$W(P_k) = \frac{n!}{\prod_k (nP_k)}$$
$$\ln W(P_k) = \ln n! - \ln \prod_k (nP_k)! \tag{5}$$

and by using the Stirling approximation we have:

$$\ln W(P_k) = -n \sum_k P_k \ln P_k, \tag{6}$$

which is the same as (4), apart from the factor n. This approach has already been used in the past[4] to derive the Cluster Variational Method in studies of order-disorder phase transformation. The analogy with random walks is simple: P_k is the probability that the particle visits the kth site on the lattice structure. This is equal to the number of times a site has been visited per time unit. The summation index k runs over all sites vistied at least once, i.e., it goes from $k = 1$ to $k = S_n$. Then, in analogy to all other fractal exponents, we define as the information dimension:

$$D_I = \frac{I(P)}{\ln N}. \tag{7}$$

This form is very general and it is not restricted to random walks only. It is independent of the underlying lattice, i.e., it is valid both for perfect and imperfect lattices. However, since we expect a considerably larger number of visited sites in the case of a perfect lattice than in the fractal lattice, we also expect a correspondingly larger D_I. The exact solution for the P_k series is simply the problem of multiple visits on lattices, i.e., the analytical formalism that will give answers to how many sites have been visited exactly (and at least) a certain number of times as a function of time. We recently solved this problem,[5] the solutions being:

$$S(r, N) = N(1 - w)w^{r-1}$$
$$V(r, N) = N(1 - w)^2 w^{r-1}, \tag{8}$$

where $S(r, N)$ $[V(r, N)]$ is the number of sites that have been visited at least [exactly] r times in N steps, and $w = 1 - S(1, n)/N$. These functions now can be directly used to calculate the D_I exponent. The above work is limited, however, in that it deals only with perfect lattices, since it is only in this limiting case where we know the generating function for the visitation probability and derive (8). Our preliminary work[6] showed that the D_I exponent is in the same class of exponents as the spectral (fracton) dimension exponents for perfect and fractal lattices, with the only difference being that for fractal lattices it takes longer than the usual amount of time to reach such behavior. The method of calculationns is presented in Sec. 2, and our results are given in Sec. 3 together with the pertinent discussion.

2. Numerical Methods

The numerical simulation techniques are described in detail in our earlier work.[6] Lattices are generated in 2-dim and 3-dim geometries with lattice sizes 600×600 and $100 \times 100 \times 100$, respectively. The fractal lattices are generated by using percolation clusters exactly at the critical point ($p_c = 0.5935$ for the 2-dim case and $p_c = 0.315$ for the 3-dim case, for site percolation). This is done using the CMLT method.[6] All small clusters are erased and only the largest percolating cluster is maintained. We monitor a random walk process by using an index on each lattice site, signifying the number of visits for this site at any given instance. This index is initially zero for all sites. A particle is placed at random on the lattice and starts performing a random walk. Every time it steps on a site the occupancy index is incremented by one for this particular site, and the process is repeated. The value of this index at the end of each realization is used to derive the P value. Similarly, the Sierpinski gasket is generated to 9th order (29526 sites). The same visitation index is utilized as for the lattices. All other details, such as direction of motion, cyclic boundary conditions, random number generators, etc. are the same as in our past studies of random walks.[6] The average usually is taken after 1000 realizations.

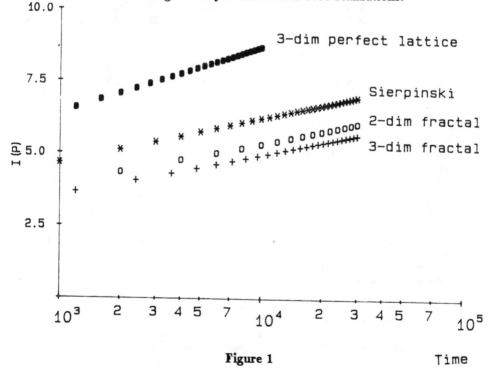

Figure 1

3. Results and Discussion

Our results are given in Fig. 1, which shows $I(P)$ as a function of time. As expected all lines are perfectly linear, since the structures shown are either perfect lattices or exhibit fractal behavior. The numerical values of the D_I dimension are derived from the slopes of the several straight lines. Since we deal with critical exponents it is rather important to have exact numerical values for these. This is because they appear in exponential form and considerably influence the overall function's behavior. It is also true that strictly speaking only at time = infinity do these exponents approach their asymptotic values and have their usual meaning. However, from past work we know that in numerical calculations it does not take that long of a time to reach this limit. Typically the range 1000 to 2000 times steps suffices. Sometimes much smaller is enough. But as we see in this work this does not happen here. One, of course, would wonder what is the meaning of the non-asymptotic values, and if they have some information to offer us. The answer is yes! In comparing with experimental results in certain cases one measures lifetimes of states which are certainly finite, i.e., they correspond to transport over a small, finite number of steps, and this necessitates knowing the numerical values of all relevant parameters in this time range. It is precisely for this reason that we calculate not only the asymptotic time limit of D_I but also the way it reaches this limit.

n	3-dim perfect latt (calc)	(exp)	Sierpinski $D = 0.685$	2-dim (fract) $D = 0.65$	3-dim (fract) $D = 0.666$
1000	0.970	0.983	0.638	0.60	0.57
10000	0.985	0.994	0.661	0.62	0.607
30000	0.993	0.996	0.667	0.632	0.665
100000			0.685	0.648	

The exact values of D_I are given in the Table as a function of time. Here we see that for the perfect lattice the expected value is reached fairly soon, while for the fractals it takes a somewhat longer time period to reach the expected value. It is thus obvious that D_I attains the spectral dimension value only in the asymptotic time limit.

1. H. E. Stanley and N. Ostrowsky, eds., *On Growth and Form* (Martinus Nijhoff Publishers, Dordrecht, 1986).

2. S. Havlin and D. Ben-Avraham, Adv. Phys. 36, 695 (1987).

3. A. Blumen, J. Klafter, and G. Zumofen, in *Optical Spectroscopy of Glasses*, I. Zschokke, ed. (Reidel, 1986).

4. R. Kikuchi, Phys. Rev. 81, 988 (1951); J. Chem. Phys. 19, 1230 (1951).

5. G. L. Bleris and P. Argyrakis, Z. Phys. B 72, xxx (1988).

6. P. Argyrakis, Phys. Rev. Lett. 59, 1729 (1987).

7. A. I. Khinchin, *Mathematical Foundations of Information Theory* (Dover, New York, 1957).

ANOMALOUS TRANSPORT IN RANDOM LINEAR STRUCTURES

ARMIN BUNDE,[*†] SHLOMO HAVLIN,[‡] H. EUGENE STANLEY[†]

[*]*I. Institut für Theoretische Physik, Universität Hamburg
Jungiusstrasse 9, D-2000 Hamburg 36, W. Germany*

[†]*Center for Polymer Studies and Department of Physics
Boston University, Boston, MA 02215 USA*

[‡]*Department of Physics, Bar-Ilan University
Ramat Gan, Israel*

ABSTRACT. We consider transport properties of disordered one-dimensional systems and discuss the effect of hard-core interactions on the diffusion properties of Brownian particles as well as effects of spatial correlations along the chain.

1. Introduction

In recent years, the problem of transport in random linear structures has received much interest (for review articles, see e.g., Refs 1-3). The study of one-dimensional type systems is motivated by two facts: First, there exist many realizations of systems where quasi-one-dimensional transport is important. These include ionic and electronic conductors, conductive polymers, semiconductor glasses, and composite materials. Second, in one-dimensional systems many problems can be solved analytically, and their solution sheds some light on the behavior in higher dimensions.

Here, we consider two types of disorder (random transition rates and random local fields) on the transport properties in linear chains. One leads to anomalous diffusion following a *power law* in time, the other one leads to "ultra" slow diffusion, yielding a *logarithmic* time dependence of the mean-square displacement of a Brownian particle. We discuss in detail the effects of hard-core interactions between the particles and effects of spatial correlations between the random fields.

2. Random Transition Rates

Consider a single particle (random walker) jumping between nearest neighbor sites on the chain. The sites are at the potential minima and are separated by random barriers which determine the transition rates $\{W\}$ between neighbored sites. We assume that their distribution follows a power law

$$P(W) \sim \frac{1}{W^\alpha}, \qquad \alpha < 1. \tag{1}$$

This type of distribution can be obtained, for example, if the heights of the barriers are exponentially distributed (large barriers are exponentially rare) and the jump process is thermally activated, leading to Boltzmann factors in the transition rates. In this case, $\alpha = 1 - k_B T/V_0$ where V_0 is the mean height of the barriers in the chain.

For constant transition rates ($\alpha = -\infty$, corresponding to the limit of high temperatures), the mean square displacement of the particle scales as $R^2 \equiv \langle x^2(t) \rangle \sim t$.

In case of many particles with hard core interaction (only one particle can be on one site),[4] the r.m.s. displacement (range) of a tagged particle scales with the number of time steps t as $R \sim t^{1/4}$. For finding how R is changed by the distribution of rates it is important to distinguish between the number of steps N made by the walker and the total elapsed time t. Since R as a function of N does not depend on the time spent on each site, it scales as for constant barriers, i.e.,

$$R \sim N^{1/d_w^0} \tag{2}$$

where $d_w^0 = 2$ for noninteracting particles and $d_w^0 = 4$ for hard-core tagged particles.

The effect of the distribution (1) is to provide additional random delays. After N steps, the elapsed time t is

$$t = N\bar{t} \tag{3}$$

where $\bar{t}$ is the average time the particle spends in one site,

$$\bar{t} = \frac{1}{R} \sum_{i=1}^{R} \frac{1}{W_i} = \int_{W_{\min}}^{1} dW \frac{1}{W} P(W). \tag{4}$$

Here $W_{\min}$ is the minimum transition rate encountered by the particle when travelling the distance R. From (1) we obtain $W_{\min} \sim R^{-1/(1-\alpha)}$ (for a derivation, see, e.g., Refs. 3 or 6). Hence, from (4), $\bar{t} \sim R^{\alpha/(1-\alpha)}$ for $\alpha > 0$ and $\bar{t} = \text{const}$ for $\alpha \leq 0$. Using (3) and (4) we obtain asymptotically $R \sim t^{1/d_w}$, where[5]

$$d_w = \begin{cases} d_w^0, & -\infty < \alpha < 0, \\ d_w^0 + \alpha/(1-\alpha), & 0 \leq \alpha < 1. \end{cases} \tag{5}$$

The result for noninteracting particles was obtained by an exact Green's function technique by Alexander et al.,[1] and applied to explain the anomalous temperature dependence of the dynamical conductivity of hollandite. For interacting particles, Eq. (5) is strongly supported by numerical simulations.[5]

Next consider the distribution function $P(x,t)$, the probability density to find the particle at time t at distance x from its starting point ($x = 0$). The Fourier transforms of $P(x,t)$ with respect to x and t represent the coherent scattering function (if noninteracting particles are considered) and the incoherent scattering function (for a tagged hard-core particle) which are accessible in scattering experiments. A simple scaling form for $P(x,t)$ for both independent and hard-core particles is

$$P(x,t) = P(0,t)G\left(\frac{x}{R(t)}\right), \tag{6}$$

where $G(u) = 1$ at $u = 0$. The probability of return to the origin $P(0,t)$ is proportional to $1/R(t)$ asymptotically. We expect $G(u)$ to have the form of a stretched exponential,

$$G(u) \sim \exp(-u^\delta), \qquad u \gg 1. \tag{7}$$

For constant transition rates ($\alpha = -\infty$), $G(u)$ is Gaussian in u,[4] i.e., $\delta = 2$, for both interacting and noninteracting particles. For the distribution (1) of transition rates, it has been found by Koscielny-Bunde et al.[7] that δ varies strongly with α. For α negative or zero, $G(u)$ is Gaussian and *independent* of α. For $\alpha = 1/2$, $\delta \approx 1.4$ and for $\alpha = 2/3$, $\delta \approx 1.2$. In all cases, the scaling function $G(u)$ does *not* depend explicitly on the concentration of the hard-core particles. Hence, the concentration dependence of $P(x,t)$ is determined solely by the concentration dependence of the range $R(t)$.

3. Random Local Fields

3.1. UNCORRELATED FIELDS

First we consider the case of uncorrelated random bias fields. A particle at site i has the probability $p_+ = (1 + E_i)/2$ to step to the right and the probability $p_- = (1 - E_i)/2$ to step to the left. The local bias field E_i can accept the values $+E$ or $-E$ with equal probability and $0 \leq E \leq 1$. As found by Sinai,[8] diffusion now is logarithmically slow and the range of a single random walker is proportional to $(\log t)^2$. To study the effect of hard core interaction on R we make use of the fact that the random field problem can be mapped[3] onto a diffusion problem with random transition rates W, which are chosen from the distribution

$$P(W) \sim \frac{1}{W \log^3 W}. \tag{8}$$

In order to find the asymptotic behavior of $R^2(t)$ we follow the scaling arguments outlined in the foregoing section, Eqs. (2)-(5). The minimum transition rate encountered by the particle when travelling the distance R is[2,3] $W_{\min} \sim \exp(-R^{1/2})$. From (4) and (8) we find that asymptotically $\bar{t} \sim \exp(R^{1/2})$ and hence

$$t = N\bar{t} \sim R^{2/d_w^0} \exp(R^{1/2}).$$

This yields asymptotically $R \sim \log^2 t$, independent of d_w^0, and the logarithmic time dependence is the same for independent and hard-core particles. This prediction is supported by numerical simulations: The hard-core interaction only modifies the prefactor of $\log^2 t$, which tends to zero if the concentration of hard-core particles approaches unity.[5]

   In contrast, the asymptotic behavior of the distribution function [characterized by the exponent δ in (7)] seems to depend on the hard-core interaction. It has been found numerically that for noninteracting particles[9] $\delta \approx 1.25$, while for hard-core particles $\delta \approx 1.5$.[5] In both cases, the exponent is independent of the value of E. The numerical result $\delta = 1.25$ for independent particles may be contrasted with the result $\delta = 1$ for the case of homogeneously distributed fields.[10]

3.2. CORRELATED BIAS FIELDS

Next we discuss the effect of correlations between the random bias fields. In many real disordered materials, such as polymers, porous materials, and amorphous systems, the spatial disorder is *correlated*. For example, if we model the permeability of a porous rock by an array of resistors whose resistances are chosen *randomly*, then it is possible to find huge resistances neighboring tiny resistances. Such configurations cannot occur in nature since the permeability of a "crack" cannot fluctuate arbitrarily. The spatial disorder is *correlated*.

In general, the problem of diffusion in the presence of random fields is related to the following correlated resistor model.[11] Consider a set of N resistors in series, where the resistance R_j of resistor j changes in a *correlated* fashion,

$$R_{j+1} \equiv (1 + \epsilon)^{\tau_j} R_j. \tag{9}$$

Here $\epsilon > 0$ is arbitrary, and τ_j is chosen randomly to be $+1$ or -1. We consider the case in which $\{\tau_j\}$ have long-range spatial order. Because neighboring resistors may only differ by a factor of $(1 + \epsilon)$, this model insures a smooth spatial variation of the resistance. From Eq. (9), the resistance of bond ℓ is

$$R_\ell = R_1 (1 + \epsilon)^{\sum_{j=2}^{\ell} \tau_j}. \tag{10}$$

This model represents a *random multiplicative process*, in contrast to the familiar *random walk model*, which is a random additive process.

In the following we will be interested in the result of a typical measurement of $R_{\text{tot}} \equiv R_{\text{tot}}(N)$ the total resistance of the N-resistor chain. The *typical* value is represented by the logarithmic average. We may estimate the resistance of a typical resistor

$$R_{\text{typ}} \sim \exp[\langle \log R_\ell \rangle] \sim (1 + \epsilon)^{X(N)}, \tag{11}$$

where $X(N)$ represents the r.m.s. displacement in the N-step walk defined by the $\{\tau_j\}$, $X(N) \equiv \langle (\sum_{j=1}^{N} \tau_j)^2 \rangle^{1/2}$. The number of such typical resistances (i.e., typical walks) is of the order of N, so from (11) the logarithmic average $\langle \log R_{\text{tot}} \rangle$ scales as

$$\langle \log R_{\text{tot}} \rangle \sim \log(N \times R_{\text{typ}}) \sim X(N) \log(1 + \epsilon) + \log N, \tag{12}$$

If $\{\tau_j\}$ are uncorrelated, $X(N) = \sqrt{N}$. If $\{\tau_j\}$ have long range spatial order, $X(N)$ depends on the details of the correlation. In general, the $\{\tau_j\}$ may be correlated through a power-law relation in Fourier space

$$\langle \tau_q \tau_{-q} \rangle \sim \frac{1}{q^\lambda} \quad \text{for small } q. \tag{13}$$

For the special case of uncorrelated $\{\tau_j\}$, $\lambda = 0$. If two neighboring τ_j tend to be of the same sign (*ferro* case), then $\lambda > 0$ while if two neighboring τ_j tend to be of opposite sign (*antiferro* case), then $\lambda < 0$.

It can be shown[11] that

$$[X(N)]^2 \sim N^{1+\lambda} \quad [\lambda > -1] \tag{14}$$

Combining (14) and (9), we find that the typical resistance scales as

$$R_{\text{typ}} \sim N \exp(N^{(1+\lambda)/2}) \quad [\lambda > -1]. \tag{15}$$

This result is supported by numerical simulations.[11]

Study of the correlated resistor chain also provides insight into diffusion in the presence of quenched *correlated* disorder. This generalizes the classic Sinai model for diffusion in the presence of uncorrelated random fields. Consider a random walker on a topologically one-dimensional system. The probabilities $W_{j,j\pm1}$ of hopping from site j to its two neighbors are proportional to the inverse of the corresponding resistances between the sites. Hence

$$\frac{W_{j,j-1}}{W_{j,j+1}} = (1+\epsilon)^{\tau_j}, \tag{16}$$

where now ϵ plays the role of a *local bias field*. From the normalization condition $W_{j,j+1} + W_{j,j-1} = 1$ we obtain $W_{j,j\pm1} = (1 \pm E)/2$ where $E \equiv \epsilon/(2+\epsilon)$.

The mean logarithm of the time the random walker takes to travel a distance L along the chain is proportional[11,12] to the fluctuations of the field biased against the walker,

$$\langle \log t \rangle \sim X(L). \tag{17}$$

Accordingly, the first passage time in the diffusion problem plays a similar role as the resistance in the resistor problem. Correspondingly, when taking into account correlations between the local bias fields (which are determined by the correlations between the τ, see Eq. (13)) we obtain on substituting (14) into (17)

$$\langle \log t \rangle \sim L^{(1+\lambda)/2}. \tag{18}$$

The Sinai result is obtained for the average of the displacement for fixed t. If we assume that our result (18) will hold for fixing t and averaging L, we recover the Sinai result in the particular case $\lambda = 0$ (uncorrelated fields).

This work has been supported by MINERVA and NATO.

1. S. Alexander, J. Bernasconi, W. R. Schneider, and R. Orbach, Rev. Mod. Phys. **53**, 179 (1981).
2. J. W. Haus and K. W. Kehr, Phys. Rep. **150**, 263 (1987).
3. S. Havlin and D. Ben-Avraham, Adv. Phys. bf **36**, 695 (1987).
4. T. E. Harris, J. Appl. Prob. **2**, 323 (1965); P. M. Richards, Phys. Rev. B**16**, 1363 (1977); P. A. Fedders, Phys. Rev. B**17**, 40 (1978); S. Alexander and P. Pincus, Phys. Rev. B**18**, 2011 (1978).
5. E. Koscielny-Bunde, A. Bunde, S. Havlin, and H. E. Stanley, Phys. Rev. A**37**, 1821 (1988).
6. A. Bunde, this book.
7. E. Koscielny-Bunde, A. Bunde, S. Havlin, and H. E. Stanley, to be published.
8. Y. Sinai, Theor. Prob. Appl. **27**, 256 (1982).
9. A. Bunde, S. Havlin, H. E. Roman, G. Schildt, and H. E. Stanley, J. Stat. Phys. **50**, 1271 (1988).
10. M. Nauenberg, J. Stat. Phys. **41**, 8103 (1985); H. Kesten, Physica 138A, 299 (1986).
11. S. Havlin, R. Blumberg-Selinger, M. Schwarts, H. E. Stanley, and A. Bunde, Phys. Rev. Lett. **61**, 1438 (1988).

MORPHOLOGICAL TRANSITIONS IN PATTERN GROWTH PHENOMENA

JANOS KERTÉSZ

Institute for Technical Physics, HAS
Budapest P. O. Box 76, H-1325, Hungary

ABSTRACT. Pattern formation in Laplacian processes leads to diverse morphologies depending on the physical parameters during the growth. Analogies with equilibrium phase changes are emphasized and characteristic features of specific morphologies as well as transitions between them are discussed.

1. Introduction

The basic concept in explaining the fascinating multifariousness manifested in nature is broken symmetry. The equations which describe phenomena and processes are usually highly symmetric and the solutions are responsible for our very non-symmetric world.

In physics the mechanism of breaking symmetry was first understood in the theory of phase transitions like the transition from the paramagnetic to the ferromagnetic phase (see Ref. 1). For example the solutions of a rotationally invariant Hamiltonian reflect the transition from the symmetric (paramagnetic) state to the state where the total magnetization is nonvanishing and points into a specific direction (ferromagnet). The solution with the symmetry of the Hamiltonian becomes unstable at a critical value of the control parameter (temperature) and new branches of stable solutions appear. Due to some small initial perturbations or to the fluctuations always present in the system one of the stable solutions is chosen.

Patterns or spatial structures are by definition states with broken symmetry. They are often produced via far-from-equilibrium processes which, however, bear an apparent resemblance to thermodynamic phase transitions.[2] Again, in analogy with thermodynamic phase transitions, it may happen that a phase shows stability in a certain range of the control parameter only and a sequence of instabilities characterizes the system. Laplacian growth to be discussed here in detail provides with good examples of this kind.

We start this lecture with a brief summary of the description of equilibrium phase transitions. In the third part an overview is presented on morphological phases and transitions in Laplacian growth. In part four experimental results are collected. Finally, in the conclusion, particular emphasis is given to open questions requiring further study.

2. Thermodynamic transitions[1,3]

2.1. BOLTZMANN'S ORDERING PRINCIPLE AND LANDAU THEORY OF PHASE TRANSITIONS

Why do different equilibrium phases exist? At fixed temperature, number of particles and volume the equilibrium state of matter is determined by the minimum of the free energy F, where

$$F = E - TS, \qquad (1)$$

with E, T, S being the energy, temperature and entropy, respectively. In a gas the energy is large; this is compensated by the large entropy term multiplied by a large

T. At low temperatures the TS term has minor contribution and therefore it is "worth" creating the crystalline state with broken symmetry—i.e., low entropy—in order to achieve very low energy. Both entropy and energy are intermediate in the liquid state of matter. This is a well-known manifestation of Boltzmann's ordering principle.

The so-called Landau theory provides the starting point for the well established theory of thermodynamic phase transitions. The basic concept is the *order parameter* which is a quantity measuring the grade of order. In the Landau theory it is supposed that the free energy F can be expanded in a power series of the order parameter q:

$$F(q,T) = a_0(T) + a_1(T)q + \left(\frac{a_2}{2}\right)(T - T_c)q^2 + \left(\frac{a_3(T)}{3!}\right)q^3 + \left(\frac{a_4(T)}{4!}\right)q^4 \dots \quad (2)$$

The standard mean field result for transitions with continuous change of the order parameter (second order transition) is recovered if the odd terms in (2) can be neglected. The mechanism is that crossing T_c from above the minimum in F at $q = 0$ becomes unstable while new minima corresponding to the phase with broken symmetry ($q \neq 0$) appear and one of them is realized due to infinitesimal perturbations. Using symmetry arguments, Landau theory is able to estimate the order of the transition. If, e.g., the third order term cannot be neglected, the transition is expected to be of first order (discontinuous change of the order parameter).

It is interesting to investigate how the order parameter develops in time. The equation of motion is generally assumed in the following form:

$$\dot{q} = -\frac{\partial F}{\partial q}. \quad (3)$$

Using this relation the way of approaching equilibrium can be studied or statements about the stability of given states can be formulated. Let us consider the case $F = a(T - T)q^2/2 + bq^3/3! + cq^4/4!$ as an example! For $T > T_c$ the terms linear in q (*linear stability analysis*) suggest local stability of the $q = 0$ solution although this is not the absolute minimum of F already for $b^2/6ac > T - T_c (> 0)$: Hysteresis characteristic for first order (discontinuous) transitions occurs. If $b = 0$, the new minimum appears *via* a bifurcation at T_c and $q = 0$ (continuous transition).

2.2. THE ROLE OF FLUCTUATIONS AND THE CRITICAL STATE

In a first order transition the stability criterion provides a theoretical limit for the existence of the "old" mode.

What happens in second order transitions? Since there is no nucleation barrier, the new phase gradually develops so that at the critical point a special situation emerges: The fluctuations become enormous here. The spatial extent of the fluctuations are characterized by the so-called correlation length ξ which diverges at T_c as

$$\xi \propto |T - T_c|^{-\nu}, \quad (4)$$

with $\nu > 0$. That means that just at T_c there is no finite physical characteristic length in the system (except of the microscopic cutoff). The fluctuations fundamentally influence the phase transition; Landau or mean field theory turns out to be an approximation only since it is not able to handle the fluctuations in a proper way. It is the renormalization group (RG) theory[1] which gives the well founded description about the transitions.

The basic idea in renormalization group (RG)[1] is the elimination of degrees of freedom using the scale invariance of critical systems. Due to the absence of a characteristic length the scale transformation carried out at T_c does not alter the physics and the important information about the transition can be obtained if the transformation is carried out in the vicinity of T_c (or more precisely near to the fixed point of the transformation).

Perhaps the most important result from RG theory is that it put a clear picture behind the experimental observation of universality. Universality means that some characteristic like the so-called critical exponents [e.g., ν in (4)] are respectively identical for many apparently different physical systems. For example the Curie transition in magnets with strong uniaxial anisotropy, the liquid-gas transition at the critical point and some order-disorder transitions in alloys belong to the same *universality class*. It turned out that not the fine details but some general symmetry properties of the order parameter like its dimension, the dimension of the space in which the system is embedded, etc. determine these universality classes.

The singularities characterizing the phase transition points are mathematically possible in an infinite system only. One can take the view that the linear size L of the system considered is an additional "field" and the transition to the critical state happens for $1/L \to 0$. An important consequence of these considerations is the well-founded hypothesis of generalized homogeneity in systems with a critical point:

$$A\left(\frac{T - T_c, 1}{L, X_i}\right) = b^{y_A} A\left[(T - T_c)b^{y_T}, \left(\frac{1}{L}\right) b^{y_L}, X_i b^{y_{X_i}}\right], \qquad (5)$$

where A is the singular part of a characteristic quantity, b is an arbitrary factor, $X_i - s$ are fields and the $y_\alpha - s$ are critical exponents.

The far-from-equilibrium processes are connected to the physics of thermodynamic phase transitions in many ways. For example the grown patterns are often fractals,[4,5] i.e., [statistically] self-similar structures with non-trivial dimension—a situation analogous to the equilibrium critical state. Consequently the concepts of scale invariance can be applied and the methods of studying critical phenomena like renormalization group can be adapted. Furthermore, stationary stable patterns or morphologies can be considered as analogues to thermodynamic phases and the transitions between different morphologies resulting from the change of some control parameters correspond to the phase transitions.

3. Laplacian pattern formation

3.1. FORMALISM AND LINEAR STABILITY ANALYSIS

A broad class of pattern growth phenomena is related to the Laplace equation. The fundamental set of equations is[5-8]

$$\nabla^2 u(\mathbf{r}) = 0 \qquad (6a)$$

$$u(\Gamma) = f[\kappa(\Gamma), v_n(\Gamma), \dots] \qquad (6b)$$

$$v_n(\Gamma) = -B\nabla u(\Gamma)\mathbf{n}, \qquad (6c)$$

where u is a field, Γ represents the moving boundary, v_n is the normal velocity of the boundary, κ is the curvature and B a coefficient. The parameters of the problem are the capillary length d_0 which is proportional to the surface tension and the driving force usually represented by the value of u at some distant boundary. Diverse physical processes can be modelled by (6a)-(6c) like dendritic crystal growth, viscous fingering, dielectric breakdown or electrochemical deposition. The seemingly simple set of governing equations hide a considerable amount of complexity due to the nonlinearity brought in *via* the moving boundary conditions (6c). (We confine ourselves mainly to two-dimensional problems.)

As a basic example let us consider the Mullins-Sekerka instability[9] in a two-dimensional circular geometry as it is realized in a radial Hele-Shaw cell[10,11]. As a concrete form of f in (6b) we take

$$f = \Delta - d_0 \kappa \tag{7}$$

(we neglect velocity dependent terms). Here $u(\mathbf{r}) = [p(\mathbf{r}) - p(R)]/p(R)$ with p being the pressure which is kept constant at a circle of radius R, i.e., $u(R) = 0$. The gas is injected at the origin through a circular hole with a pressure $u(0) = \Delta$ and it is assumed that the pressure is constant within the gas. d_0 of (7) is a length determined by the material constants and the distance between the plates of the cell.

The solution with $\Gamma = \rho_0$ (circular interface) is

$$u_0(r) = \left(\Delta - \frac{d_0}{\rho_0} \right) \frac{\ln(r/R)}{\ln(\rho_0/R)}. \tag{8}$$

The corresponding interfacial velocity is determined using (6c):

$$v_0 = \frac{B(\Delta - d_0/\rho_0)}{\rho_0 \ln(R/\rho_0)}. \tag{9}$$

Here $v_0 > 0$ for $\Delta > d_0/\rho_0$. If ρ_0 is the radius of the injecting pipe, d_0/ρ_0 determines the critical pressure above which the pattern starts to grow.

If Δ is slightly above d_0/ρ_0, the interface is initially circular. Linear stability analysis shows the theoretical limit of the radius where the circle looses its stability. Let us assume a small perturbation of the form $(\delta\rho)_m(t) = \delta_m e^{\omega_m t} \cos(m\varphi)$, i.e.,

$$\Gamma_m = \rho_0 + (\delta\rho)_m, \tag{10}$$

with $m = 1, 2, 3 \ldots$. Up to first order in δ_m the curvature corresponding to (10) is

$$\kappa_m = \frac{1}{\rho_0} + \frac{m^2 - 1}{\rho_0^2} (\delta\rho)_m(t). \tag{11}$$

The new solution is supposed to have the following form:

$$u_m(\mathbf{r}) = u_0(\mathbf{r}) + A_m r^{-m} e^{\omega_m t} \cos(m\varphi), \tag{12}$$

where A_m is of the order of δ_m and we allow a slight violation of the boundary condition at R. A_m is calculated using (7) and (11):

$$A_m = -\delta_m \rho_0^{m-1} \left[\frac{d_0}{\rho_0} (m^2 - 1) + \frac{\Delta - d_0/\rho_0}{\ln(\rho_0/R)} \right]. \tag{13}$$

ω_m is obtained from (6c) using the (12) form of $u_m(\mathbf{r})$ and keeping terms up to linear order in δ_m:

$$\omega_m = (m-1)v_0 \left[1 - \frac{d_0 B}{v_0 \rho_0^2} m(m+1)\right]. \tag{14}$$

Since $v_0 > 0$ for $\rho_0 > d_0/\Delta$, the sign of ω_m depends on the difference in the brackets. Since $a = \ln(R/\rho_0)$ depends weakly on ρ_0 we consider it as a constant and obtain

$$\rho_m^* = \rho_0^* \left(1 + \frac{m(m+1)}{a}\right), \tag{15}$$

where $\rho_0^* = d_0/\Delta$ and ρ_m^* is the critical radius at which ω_m becomes positive. According to (15) the first instability occurs with $m = 2$ when the bubble is growing.

3.2. MODE SELECTION

For some ρ^* the the symmetric mode becomes unstable against infinitesimal perturbations. We have seen that even in the framework of the Landau theory nonlinear terms are needed to find the new stable phase. This is true for the considered far-from-equilibrium processes as well. One would think that for $\rho_2^* < \rho < \rho_3^*$ the amplitude of the $m = 2$ mode serves as a proper order parameter. Inspite of this picture suggested by the linear stability analysis the observed number of fingers is usually not less than $m = 5$ (Fig. 1). The reason for this discrepancy may be that some modes with low m indices appear via first order transitions. (It was shown[12] by generalizing the Landau theory to pattern formation that in the three-dimensional problem of crystal growth the simplest symmetry breaking mode occurs in a discontinuous transition.) Nonlinear and finite noise analysis is necessary to clarify this point. It is interesting to note that the number of main branches in diffusion-limited aggregation is also about 5 (see Refs. 13).

Fig. 1: Radial viscous fingering pattern which started to grow with the $m = 6$ mode.

What happens if the growth starts from a circular interface with some $\rho_0 \gg d_0/\Delta$, for which many of the $\omega_m - s$ are positive? The mode with the maximal velocity is a possible candidate to be selected.[10] From (14) we get

$$m^* = \left[\frac{1}{3\ln(R/\rho_0)} \left(\frac{\Delta\rho_0}{d_0} - 1 \right) + 1 \right]^{1/2}, \qquad (16)$$

which is somewhat modified[14,15] if an additional v_n-dependent kinetic term is included in (6c):

$$f = \Delta - d_0\kappa + \beta v_n^\gamma. \qquad (17)$$

The dependence of m^* on Δ and ρ_0 was roughly verified in Ref. 15.

When following the development of viscous fingering pattern, it turns out that the phases loose their stability during the growth. A sequence of tip splitting instabilities leads to a complex pattern. By now it is not clear whether the fluctuations during the growth play an essential role here or the initial perturbations are mainly responsible for the apparent randomness of the interface (see T. Vicsek's lecture in this volume[16]).

The absence of a stable symmetry breaking mode is typical for Laplacian growth without anisotropy. In the presence of strong enough anisotropy, however, the situation is different. It turns out that anisotropy occurring either in the surface tension or in the shape of the cell or in other physical parameters stabilizes the tips of the fingers (for an excellent review see Ref. 8). In this case the velocity of the tip is selected from a discrete set of solutions of a nonlinear eigenvalue problem called microscopic solvability condition. The fastest mode—which is the linearly stable one—is then realized. This mechanism was shown to work in both dendritic crystal growth and viscous fingering. Let us here mention the result on the velocity of a dendritic tip only[17]: it is proportional to $\Delta^4\alpha^{7/4}$ where Δ is the driving force (undercooling for dendritic growth) and α is the anisotropy in the surface tension. Very recently this result was experimentally supported[18].

3.3. DLA AND ITS MODIFICATIONS

Diffusion-limited aggregation (DLA) is the basic cluster model of Laplacian growth.[19] In DLA the growing object (aggregate) starts from a seed. A particle is released from a distant source and it makes a random walk. If it touches the aggregate it sticks to it and a new particle is released. The resulting object is a fractal, i.e., its mass M scales with the linear size ρ as

$$M \sim \rho^D, \qquad (17)$$

with $D \approx 1.7$ in two dimensions, thus a DLA cluster is reminiscent of the critical state.

The relation to Laplacian growth is clear since the probability of finding a random walker satisfies the Laplace equation and the sticking rule corresponds to (6c).

There are several modifications of the original DLA model corresponding to interesting theoretical and experimental problems. Let us first consider the following model[20]. The diffusing particle is attracted by the cluster and is characterized by drift with velocity w towards the origin. In other words it is slightly more probable to jump towards the cluster than in other directions. Therefore the dimension of the walk will be changed from $d_w = 2$ to $d_w = 1$. The aggregation problem with $d_w = 1$ is known to lead to homogeneous non-fractal clusters ($D = d$). However, if w is small, its effect on the cluster cannot be seen immediately and there is a crossover from fractal to homogeneous behaviour as a function of the size of the object. From dimension analysis the characteristic length is $\xi = B/w$ with B being the diffusion coefficient. A simple scaling *Ansatz* for the density $n(r, w)$ of the cluster can be proposed to describe the above mentioned crossover:

$$n(r, w) = r^{D-d} f\left(\frac{r}{\xi}\right), \qquad (18)$$

with

$$f(x) \to \text{const} \quad \text{if} \quad x \to 0$$

$$f(x) \to x^{d-D} \quad \text{if} \quad x \to \infty.$$

This *Ansatz* leads to $n(r \gg \xi) \sim w^{d-D}$.

A similar problem rises in multiparticle aggregation where particles with concentration c are distributed initially and their diffusive motion governs the aggregation process. Calculations[19] on the lattice version of this model are comparable with (18) where $\xi \sim c^{-\alpha}$ and $\alpha \approx d - D$.

The situation is just the opposite if the original DLA model is modified in such a way that the probability p of sticking of a particle to the aggregate is less than unity. This change does not affect D. Since $p < 1$ simulates in some sense the surface tension, the object resulting from this growth process has an open fractal structure above some lower, p-dependent cutoff which diverges as $p \to 0$.

Although we have not emphasized, most of the calculations have been carried out on lattices. However, since the anisotropy is relevant to Laplacian growth, it is not surprising that the underlying lattice affects the geometry of the growing cluster in a relevant way. At early stages of the growth (up to several thousand particles) the visual inspection does not tell significant differences between lattice and off lattice DLA clusters and the calculated fractal dimensions are also close to each other. Hence the early belief of universal behaviour in D. Very large scale calculations showed that the lattice axes represent directions of easy growth and e.g., on the square lattice the cluster becomes gradually cross-shaped.[22]

Why does the effect of anisotropy show up so slowly? The reason is that the fluctuations due to the random character of the walks act against the manifestation of anisotropy. The actual shape of lattice DLA-clusters emerges from an interplay between fluctuation and anisotropy.

The fluctuations in DLA can be controlled by the so called noise reduction method[23,24]: instead of adding the randomly walking particle to the cluster after it hits a growth site, one keeps counting the trajectories terminating at a given site. Only those sites are added to the cluster at which the counter reaches a prescribed value m. Large m corresponds to a low level of noise.

For fixed number of particles in the cluster an interesting morphological transition can be observed[23] as a function of m (Fig. 2). Not too large "noisy" DLA-clusters grown on the square lattice are open branching structures like off lattice aggregates. If the noise is reduced the anisotropy due to the grid breaks through and

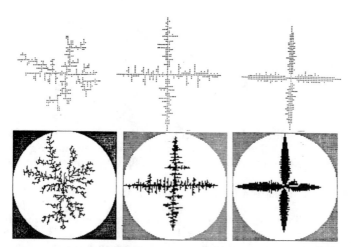

Fig. 2: The role of randomness in growth of Laplacian patterns on the square lattice. First row: noise-reduced DLA.[23] Second row: numerical solution of the problem of viscous fingering in a system of pipes with variable radii (from Ref. 29). Noise is decreasing from left to right.

the clusters become cross-shaped with stable tips reminiscent of dendritic crystals. For very large values of m even the sidebranches vanish and one recovers needles growing out of a center.

On the other hand, if m is fixed and the size of the cluster is increased, a transition from the needle to the dendrite with side branches occurs. This is due to the instability of the flat edge of the needle against fluctuations always present at finite m. Eckman et al.[25] have shown that the dependence of the size M^* at which the transition needle → dendrite takes place on m is $M^* \sim [\log(m)]^3$.

It is very interesting that a slight alteration of the boundary conditions which has practically no effect in the $m = 1$ case causes significant differences in the structure of the patterns if noise reduction is present[24]. The boundary condition used in Ref. 24 corresponds to the dielectric breakdown model[26]: the walking particle is allowed to jump onto the cluster and only then it sticks to it.

The dielectric breakdown model is generalized using an η parameter which changes (6c) to $v_n(\Gamma) = -B[\nabla u(\Gamma)\mathbf{n}]^\eta$. $\eta = 1$ corresponds to DLA ($D \approx 1.7$). D depends continuously on η (see Ref. 26) with the limits $D(\eta = 0) = 2$ and $D(\eta = \infty) = 1$.

4. Experimental results on Laplacian morphologies

4.1. TIP SPLITTING VERSUS STABLE TIPS

The structures generated by DLA and its modifications are closely related to the patterns observed in experiments on growth processes. For example two-dimensional crystal growth can be forced to obey tip splitting instead of dendritic stable tips if randomness is introduced by making the holder plates rough[27] (*Fig.* 3).

Viscous fingering is particularly well suited to investigate the effect of influencing factors on Laplacian growth. The effect of anisotropy was demonstrated by Ben-Jacob et al.[28] who achieved dendritic, snowflake-like viscous fingers by scratching a regular mesh of sixfold symmetry into one of the plates of the cell. The

Fig. 3: The crystalline anisotropy stabilizes the tip in this two-dimensional dendritic growth. Noise imposed onto the system by making the holder plate rough causes a transition to the tip splitting phase (from Ref. 27).

fluctuation induced transition from the phase with stable tips to the tip splitting structure was nicely demonstrated in an experiment with a square lattice consisting of pipes of random diameters.[29] The randomness can be characterized by the width of the distribution of the radii and for large randomness DLA-type fingers grew while small randomness resulted in dendritic shapes in full analogy with the noise reduced DLA simulations.

What happens if in a radial Hele-Shaw cell the viscous liquid is inherently anisotropic? Injecting air into a nematic liquid crystal is a realization of this situation.[30] The outward flow aligns the elongated molecules radially resulting in an viscosity with radial anisotropy. An interesting reentrant morphological transition can be observed as a function of the applied pressure of the injected gas. For low pressure the shape of the interface is very similar to viscous fingers in isotropic media. If higher pressure is applied the tips of the fingers are stable and the bubble becomes dendritic. The further increase of the pressure results again in a tip splitting phase. The probable explanation of this behaviour is that at low pressure the alignment is not sufficient while at too high pressure the orientation becomes chaotic.

Couder et al.[34] demonstrated that in viscous fingering with an isotropic liquid a small perturbation due to a tiny bubble at the tip of the finger is enough to produce the sufficient "anisotropy" for the stabilization of the tip. Furthermore, phases with splitting, stable and oscillating tips were observed as a function of the size and position of the tip.

Electrochemical deposition can also be considered under some circumstances as an experimental realization of Laplacian growth. As a function of the applied voltage a transition from the DLA-type open structure[31] to a dendritic one was observed.[32,33] Remarkably, using electron microscopy and X-ray diffraction technique, the former was shown to be polycrystalline while long range crystalline order was found in the latter case.[32] This underlines the role of anisotropy in dendritic growth.

4.2. DENSE BRANCHING MORPHOLOGY

In the absence of strong enough anisotropy the tips of the fingered parts of the growing interfaces are splitting. This may lead to a DLA-type, open, fractal structure but it is also possible that the number of branches increases with the radius and the dimension of the object will be equal to the embedding dimension. In fact, both situations can be found in experiments.

The first example could again be viscous fingering[14]. At high pressure the viscous fingering pattern is non-fractal and we do not know whether for large radii

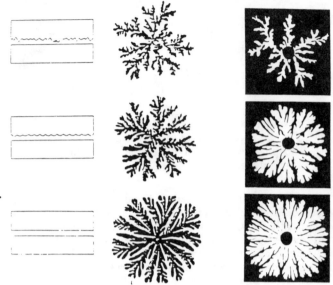

Fig. 4: Transition from open fractal to dense branching morphology. The series on the left hand side demonstrates the role of randomness in producing DLA-type patterns in an isotropic liquid (the prepared cell is also indicated) (from Ref. 36). On the right hand side a similar transition is shown in a smectic liquid crystal as a function of the pressure of the injected air.[37]

it is fractal or not. (The viscous fingers growing in a non-Newtonian liquid without surface tension are definitely fractal.[35]) It has been suggested that the number of branches as a function of the radius can be obtained from the m^* (see Ref. 16) calculated using the boundary condition (6b) with a kinetic term.[14,15]

It is clear that randomness results again in a transition towards DLA. This can be demonstrated by preparing the cell appropriately[36] (Fig. 4). Again, liquid crystals provide an example where the effect shows up without preparing the cell[37] (Fig. 4). In this case it is a smectic which exhibits the crossover from the DLA-type structure to the dense branching morphology as a function of the increasing pressure. At low pressure it is probably the domain structure in the smectic liquid crystal which causes the randomness while the high pressure destroys the domains.

The homogeneous morphology can also be observed in electrodeposition[32,33] (Fig. 5). This experiment demonstrates impressively the stability of the nearly circular envelope of the observed patterns. In the light of the linear stability analysis (c.f., Section 3.1.) this is again a theoretical challenge[38]. An apparent difference between the simulations mentioned in Section 3.3. exhibiting a crossover from DLA to a homogeneous phase is that the surface of the clusters become rough during the growth while this does not seem to be the case for the dense branching morphology. The development of dense branching morphology during crystallization is discussed in detail in other lectures at this school.[39]

4.3 COMPLEX MORPHOLOGICAL DIAGRAMS

We have seen that the discussed three morphologies (DLA-type fractal, dendritic with stable tips and dense branching morphology) can be found in the three main experimental realizations of Laplacian growth: viscous fingering, electrodeposition

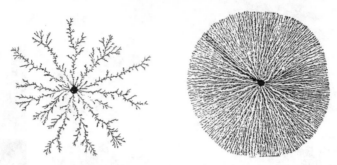

Fig. 5: Open fractal (low voltage) and dense branching (high voltage) morphologies in electrochemical deposition of Zn from aqueous solution of zinc sulphate (after Refs. 32, 33).

and solidification.[49] Of course, the different structures grow sometimes under under very different circumstances. For example the growth of a snowflake[40] is hardly comparable with crystallization in amorphous $GeSe_2$.[41] Therefore, it is of crucial interest whether transitions between different morphologies can be achieved in a system by changing some of its parameters. *Morphological diagrams* can be obtained in experiments if more than one variable parameters are changed in analogy with equilibrium phase diagrams.

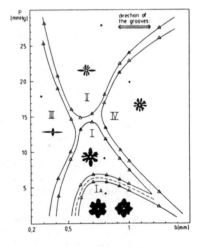

Fig. 6: Morphological diagram of radial viscous fingering patterns produced in a cell with imposed uniaxial anisotropy.[42]

In viscous fingering the natural parameters are the pressure of the injected gas and the distance of the plates of the cell. Figure 6 shows a morphological diagram measured in a radial Hele-Shaw cell with imposed uniaxial anisotropy[42]: A set of parallel grooves was etched on one of the plates. We see that virtually all possible combinations of stable and splitting tips in the directions parallel and perpendicular to the grooves occur, depending on the applied pressure p and the plate separation b. The reason is that the anisotropy of the cell is coupled to the driving force in a complicated way through the boundary condition (17). The overall tendency is that increasing b acts towards tip splitting while increasing p towards tip stabilization parallel to the grooves.

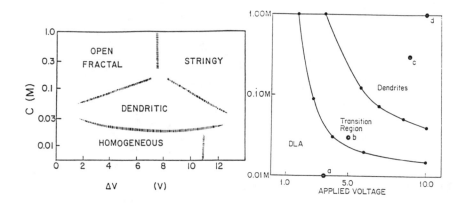

Fig. 7: Morphological diagrams electrodeposition patterns obtained under very similar circumstances–Ref. 32, (on the right) and Ref. 33 (on the left).

Interesting morphological diagrams were measured in electrodeposition cells. In this system the three main morphologies could be observed by changing well controllable parameters, the applied voltage and the concentration of the solution.[33] Surprisingly, different results were obtained in *almost* the same experiments![32,33] Both diagrams of Fig. 7 were measured by deposition of Zn from aqueous solution of zinc sulphate in a cell consisting of two parallel plates between which the solution was positioned. A metal ring was the anode and a fine wire put into the cell through a hole in the upper plate was the cathode. The differences are in the material of the electrodes, the plate separation and the radius of the cell. Obviously these seemingly minor effects influence the physics dramatically.

There is not enough space to discuss many further interesting clusters[43] or between DLA and Eden clusters.[44] However, as a final example I would like to show a so called Nakaya diagram of ice crystals taken from.[45] The natural parameters are here the excess vapour density and the supercooling (Fig. 8). The diagram is extremely complex! Ice crystals grow sometimes as needles and sometimes like plates and it is a relatively small region where the well known snowflakes can be observed (dendritic growth). Effort has been concentrated so far[46,47] mainly to reproduce the dendritic ice crystals and it would be useful to extend the calculations to the full three-dimensional problem. There are obviously important aspects which should be incorporated into a more general treatment of non-equilibrium growth in order to attack the general problem of morphological transitions.

5. Conclusion

As I tried to emphasize in this lecture, there are analogies between morphologies produced by pattern growth and thermodynamic phases as well as respective transitions among them. However, the situation is much more complicated for far-from-equilibrium processes than for most of the systems at or near to equilibrium.[2] One

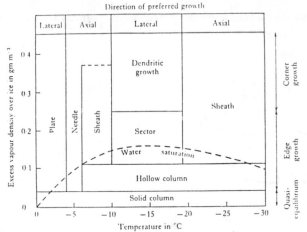

Fig. 8: Nakaya diagram of ice formation (from Ref. 45).

has often a feeling only that a specific morphology, e.g., a DLA-type pattern should be considered as a "phase." It is not clear what are the order parameters and what are the ordering principles. Sometimes many new modes become available at the instability points and new approaches are necessary to proceed here.[8] Multifractal analysis can also lead to a deeper insight to morphological transitions.[48]

We have seen that, in analogy with critical phenomena, rather different physical systems show similar behaviour. However, it is not clear whether the analogies are due to the similar underlying equations only and particular minor differences play essential role. We hope that some kind of universality concept will work also for pattern growth, although it seems that e.g., in the explanation of the dense branching morphology system-specific explanations have been put forward.[34,38,39]

Fluctuations may play an even more important role in Laplacian growth than in equilibrium transitions: it is possible that the strength of noise should also be considered as a parameter of the problem like the driving force, the anisotropy, the size of the pattern etc. What are the critical or transition values of these parameters? Are the transitions sharp and is it possible to define their order? Much more experimental, numerical and theoretical work is needed to solve these questions.

1. H. E. Stanley, *Introduction to Phase Transitions and Critical Phenomena* (Oxford University Press, 1971), for RG see C. Domb and M. S. Green (eds.), *Phase Transitions and Critical Phenomena*, Volume 6 (Academic, New York, 1976).

2. H. Haken, *Synergetics* (Springer, Berlin, 1978).

3. L. D. Landau and E. M. Lifshits, *Statistical Physics* (Pergamon, Oxford, 1977).

4. B. B. Mandelbrot, *The Fractal Geometry of Nature* (Freeman, San Francisco, 1982).

5. T. Vicsek, *Fractal Growth Phenomena* (World Scientific, Singapore, 1988); H. E. Stanley and N. Ostrowsky (eds.), *On Growth and Form* (M. Nijhoff, Dordrecht, 1986); L. Pietronero and E. Tossatti (eds.), *Fractals in Physics* (North Holland, Amsterdam, 1986)

6. J. S. Langer, *Rev. Mod. Phys.* 52, 1 (1980).

7. D. Bensimon, L. P. Kadanoff, S. Liang, B. I. Shraiman and C. Tang, *Rev. Mod. Phys.* 58, 977 (1986).

8. D. Kessler, J. Koplik and H. Levine, *Adv. Phys.*, (in press).

9. W. W. Mullins and R. F. Sekerka, *J. Appl. Phys.* 34, 323 (1963).
10. L. Paterson, *J. Fluid Mech.* 113, 513 (1981).
11. See T. Vicsek's short description in this volume.
12. Z. Rács and T. Tél, *Phys. Rev.* A26, 2986 (1982).
13. R. C. Ball, *Physica* 140A, 62 (1986), P. Meakin and T. Vicsek, J. Phys. A20, L171 (1987).
14. E. Ben-Jacob, G. Deutscher, P. Garik, N. Goldenfeld, and Y. Lareah, *Phys. Rev. Lett.* 57, 1903 (1986).
15. Á. Buka, P. Palffy-Muhoray and Z. Rács, Z, *Phys. Rev.* A36, 3984 (1987).
16. T. Vicsek, lecture at the Institute for Technical Physics, HAS.
17. A. Barbieri, D. C. Hong and J. S. Langer, *Phys. Rev.* A35, 1802 (1987).
18. H. C. Chou and H. Z. Cummins, *Phys. Rev. Lett.* 61, 173 (1988).
19. T. A. Witten and L. M. Sander, *Phys. Rev.* B27, 5686 (1983).
20. P. Meakin, *Phys. Rev.* A28, 5221 (1983); R. Jullien M. Kolb and R. Botet, *J. Physique* 45, 395 (1984); P. Meakin: J. Chem. Phys. (1984).
21. R. F. Voss, *J. Stat. Phys.* 36, 861 (1984); P. Meakin in *Time-Dependent Effects in Disordered Materials*, eds., R. Pynn and T. Ryste, (Plenum, New York, 1987), p. 45.
22. P. Meakin, R. C. Ball, P. Ramanlal and L. M. Sander, *Phys. Rev.* A35, 5233 (1987).
23. J. Kertész and T. Vicsek, *J. Phys.* A19, L257 (1986).
24. J. Nittmann and H. E. Stanley, *Nature* 321, 663 (1986).
25. J.-P. Eckmann, P. Meakin, I. Procaccia and R. Zeitak (preprint).
26. L. Niemeyer, L. Pietronero and H.J. Wiesmann, *Phys. Rev. Lett.* 57, 650 (1983).
27. H. Honjo, S. Ohta and M. Matsushita, *J. Phys. Soc. Japan* 55, 2487 (1986).
28. E. Ben-Jacob, Y. Godbey, N. Goldenfeld, J. Koplik, H. Levine, T. Mueller and L. M. Sander, *Phys. Rev. Lett.* 55, 1315 (1985).
29. J. C. Chen and D. Wilkinson, *Phys. Rev. Lett.* 55, 1982 (1985).
30. Á. Buka, J. Kertész and T. Vicsek, *Nature* 323, 424 (1986).
31. M. Matsushita, M. Sano, Y Hayakawa, H. Honjo and Y. Sawada, *Phys. Rev. Lett.* 53, 286 (1984).
32. D. Grier, E. Ben-Jacob, R. Clarke and L. M. Sander, *Phys. Rev. Lett.* 56, 1264 (1986).
33. Y. Sawada, A. Dougherty and J.P. Gollub, *Phys. Rev. Lett.* 56, 1260 (1986)
34. Y. Couder, O. Cardoso, D. Dupuy, P. Tavernier and W. Thom, *Europhys. Lett.* 2, 437 (1986); Y. Couder, lecture at the Institute for Technical Physics, HAS.
35. G. Daccord, J. Nittmann and H. E. Stanley, *Phys. Rev. Lett.* 56, 336 (1986).
36. J. Nittmann, G. Daccord and R. Lenormand, in *Fragmentation, Form and Flow in Fractured Media*, eds., R. Englman and Z. Jaeger (Ayalon Offset Ltd., Haifa, 1986), p. 556.
37. V. K. Horváth, J. Kertész and T. Vicsek, *Europhys. Lett.* 4, 1133 (1987).
38. T. C. Halsey, *Phys. Rev.* A36, 3512 (1987); D. G. Grier, D. A. Kessler and L. M. Sander, *Phys. Rev. Lett.* 59, 2315 (1987).
39. Y. Lereah and G. Deutscher, lectures at the Institute for Technical Physics, HAS.
40. U. Nakaya, *Snow Crystals: Natural and Artificial* (Harvard University Press, Cambridge, 1954).
41. G. Radnócsy, T. Vicsek, L. M. Sander, and D. Grier, *Phys. Rev.* A35, 4012 (1987).
42. V. K. Horváth, T. Vicsek and J. Kertész, *Phys. Rev.* A35, 2553 (1987).
43. R. Lenormand, *Physica* A140, 114 and lecture at the Institute for Technical Physics, HAS.
44. A. T. Skjeltorp, *Phys. Rev. Lett.* 58, 1444 (1987).
45. T. Kobayashy, *Phil. Mag.* 6, 1363 (1960).
46. F. Family, D. Platt and T. Vicsek, T. *J. Phys.* A20, L1177 (1987).
47. J. Nittmann and H. E. Stanley, *J. Phys.* A20, L1184 (1987).
48. C. Amitrano, L. de Arcangelis, A. Coniglio and J. Kertész, *J. Phys.* A21, L15 (1988).
49. T. Vicsek and J. Kertész, *Europhys. News* 19, 24 (1988).

VISUALIZATION AND CHARACTERIZATION OF MICROPARTICLE GROWTH PATTERNS

A. T. SKJELTORP* and G. HELGESEN†

*Institute for Energy Technology, N-2007 Kjeller, Norway
†Dept. of Physics, University of Oslo, N-0316 Oslo 3, Norway

ABSTRACT. Uniformly-sized polymer microparticles are used for two-dimensional modelling of aggregation and crystal growth. The results are compared with computer simulation models which are believed to apply to the physical realizations.

1. Introduction

The enormous upsurge in computational and experimental activities in the areas of aggregation and crystal growth were prompted by the development of Mandelbrot's concepts of fractal geometry[1] and Witten-Sander's diffusion-limited aggregation (DLA) model.[2] The surprising development of the DLA model is that the simulated patterns resemble experimental patterns found for a diverse range of apparently different physical processes like viscous fingering, dielectric breakdown and electrodeposition.[3] Because typical colloidal aggregation patterns found in experiments look quite different from the simulated DLA-structures, various modifications have been introduced to come closer to the physical realizations as discussed in recent review articles on this subject.[4,5]

The main purpose of this presentation is to show how uniformly sized microspheres have been "tailor-made" to model the growth into macroscopic objects ranging from ramified aggregates to single crystals. The experiments were partly aimed at probing the following questions: How do strength and range of the interparticle attractive interactions influence the growth patterns for diffusing particles? To what extent do nondiffusive particles produce similar patterns when the aggregation process is induced by some random external forces on the particles?

2. Experimental

A wide range of monosized polymer microparticles produced by the unique two-step swelling method[6] were used in the experiments. These included nonmagnetic spheres in the size range $1.1 - 25\mu m$ and 3.6 μm magnetizable[7] sulfonated spheres containing 30% (weight) iron oxide in the form of evenly distributed grains in a thin shell ($\sim 0.2\mu m$) near the surface. Figure 1 shows schematically the very simple setups in the experiments. For the diffusive aggregation studies, Fig. 1(a), the spheres were dispersed in water and confined to a monolayer between two planar glass plates. The separation between the plates could be adjusted evenly by using a small fraction of slightly larger spheres as spacers. In Fig. 1(b), spheres were dispersed onto the surface of saline water (0.1 N NaCl) either in dry form or dispersed in methanol for various nondiffusive aggregation experiments. A light microscope with a digital image analysis system was used for quantitative analysis of the resulting growth patterns.

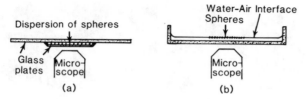

Fig. 1: Schematic experimental setup: (a) Dispersion of spheres confined to a monolayer between glass plates; (b) spheres trapped on the water-air interface.

3. Aggregation of nonmagnetic spheres

The purpose of the first part of this section is to summarize the change in growth patterns for *diffusing* spheres going from non-equilibrium to a true equilibrium situation. This is governed by the strength of the attractive interactions between the particles. For this purpose it is possible to vary the interparticle potential from strongly repulsive (Coulomb) to weakly attractive ("secondary minimum") and strongly attractive (van der Waals) by adding "counter-ions" (e.g., Na^+) in the host fluid to reduce the effects of the bound charges on the spheres.

Figure 2 summarizes a wide range of experimental[8,9] and simulated[10,11] growth patterns. Figure 2(a) represents a relatively fast diffusion-limited cluster aggregation (DLCA) process where particle-particle, particle-cluster and cluster-cluster aggregation takes place. The fractal dimension D for this was found to be $D = 1.49 \pm 0.06$ using the radius of gyration technique[2] and $D = 1.48 \pm 0.05$ using the box counting technique.[12] This is in very satisfactory agreement with the simulated DLCA result,[10] also allowing cluster rotations, Fig. 2(A), producing a fractal dimension $D = 1.485 \pm 0.015$. Figures 2(b)-2(d) show examples of successively slower particle-cluster growth by reducing the attractive interactions between the spheres as discussed above. This clearly demonstrates that if added particles are given more and more time to find a favourable place on the perimeter of the growing aggregate (reduced sticking probability, migration) the structures are becoming more and more regular. This trend is also nicely reproduced in recent computer simulations[11] shown in Figs. 2(B) - 2(D).

The remainder of this section will demonstrate aggregation of *non-diffusive* particles induced by some other random forces. In one series of experiments the sample was prepared by dropping $25\mu m$ spheres on the flat surface of water containing 0.1 N NaCl. The spheres were trapped on the water-air interface creating dimples around each sphere. This apparently produced attractive random surface tension forces between the spheres and clusters resulting in a relatively slow cluster-cluster aggregation.[13,14] Figure 3(a) shows typical aggregates after a few minutes. After several hours, the clustering process ended up in one large cluster, Fig. 3(b). The fractal dimension was found to be $D = 1.50 \pm 0.05$ for the isolated clusters in Fig. 3(a) using the radius of gyration technique. This value is very close to the DLCA value $D = 1.49 \pm 0.06$ found for the diffusing particles discussed earlier. Using the box counting technique, the fractal dimension was found to be $D = 1.78 \pm 0.04$ for the gelating cluster case in Fig. 3(b). This is quite close to the simulated value $D = 1.75 \pm 0.07$ found for high concentration cluster aggregation[15].

Another interfacial aggregation process was prepared as follows[16,17]: A small droplet of a suspension of $5\mu m$ polystyrene spheres stabilized in methanol was

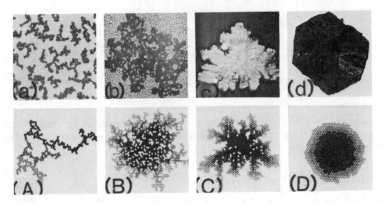

Fig. 2: Top row: Microscope pictures of aggregated spheres [4.7 μm in (a), 1.1 μm in (c)-(d)] for successively slower growth velocity v_g (defined as the average growth of radius of gyration): (a) ramified clusters ($v_g \simeq 10^{-3}\mu$m/sec); (b) rough single crystal with holes ($v_g = 3 \times 10^{-4}\mu$m/sec); (c) dendritic crystal ($v_g \simeq 10^{-5}\mu$m/sec); (d) faceted hexagonal crystal ($v_g \simeq 5 \times 10^{-6}\mu$m/sec). Bottom row: (A) Computer simulations for cluster-cluster aggregation including rotations; (B)-(D) DLA growth with increasingly effective "surface tension."

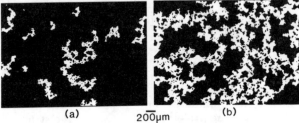

Fig. 3: Slow surface tension induced random cluster-cluster aggregation of 25μm spheres trapped on the water-air interface (a) below and (b) above the gel point.

carefully dispersed onto the surface of water with 0.1 N NaCl. This produced a highly turbulent spreading of the spheres and a very rapid (few seconds) aggregation into one cluster. Figure 4(a) shows an overall picture of the dendritic aggregate ($\sim 5 \times 10^7$ spheres) and a magnified portion in Fig. 4(b). It is interesting to note that this aggregate looks similar to that in Fig. 2(c) which has grown much slower for diffusing spheres.

4. Aggregation of Magnetic Spheres

Diffusing particles interacting via long-range dipolar forces produce aggregates with quite different morphology than the ones discussed so far. This is intuitively apparent as dipoles "like" to align and form chains. It is of interest to study how the patterns change for gradual increase in the strength of this type of interaction and a possible connection to DLCA.

 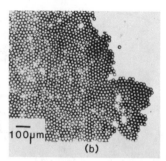

Fig. 4: Fast particle-cluster aggregation induced by convection of methanol-dispersed spheres on the interface between air and saline water as discussed in the text. Typical dendritic cluster (a) with a magnified portion (b).

The 3.6μm spheres used in the experiments (Sec. 2) could be magnetized to various levels of remanent magnetization $M_r = 0 - 1.40 \pm 0.13$emu/g, Fig. 5(A). The dispersion was stabilized so that the electrostatic and van der Waals interactions between the spheres were negligible compared to the magnetic forces. The magnetized spheres may be considered as point dipoles with magnetic moment $\mu = M_r(\pi d^3/6)$. The dimensionless parameter which determines the effective strength of the dipole-dipole interaction relative to the disruptive thermal energy is

$$K_{dd} = \frac{\mu^2}{d^3 k_B T}.$$ (1)

The experiments[18] were performed using samples with concentrations less than 10% with a random initial distribution of spheres confined to a monolayer [Fig. 2(a)]. To avoid the influence of the earth magnetic field and stray fields, the samples were enclosed in mumetal.

Figures 5(a)-5(c) show a series of aggregated structures formed after a few hours for various experimental conditions. It is seen that there is an increasing tendency to form chains and open loops as K_{dd} increases, reflecting the preference of alignment of the dipoles.

The fractal dimensions of these aggregates were determined from the usual log-log plot of the radius of gyration versus the number of particles in each cluster. Figure 6(a) shows the variation of D versus K_{dd} for no external fields. As may be seen, D becomes significantly lower as K_{dd} increases and agrees within experimental error with the simulated value $D = 1.23 \pm 0.12$ for large K_{dd}.[18] Also included in Fig. 6(a) is the value $D = 1.49 \pm 0.06$ found for cluster aggregation of nonmagnetic spheres (Sec. 3). As may be seen, this result appears to be close to the limiting value also for magnetized spheres as $K_{dd} \to 0$.

The temporal evolution of the cluster size distribution was analyzed using the following proposed scaling relation[19] for cluster-cluster aggregation:

$$n_s(t) \simeq S^{-2}(t) f\left(\frac{s}{S(t)}\right).$$ (2)

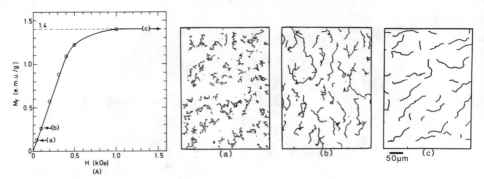

Fig. 5: (A) Remanent magnetization M_r versus external applied field H. Aggregates for increasing reduced magnetization: (a) $K_{dd} = 16$; (b) $K_{dd} = 100$; (c) $K_{dd} = 1360$.

Here, $n_s(t) = N_s(t)/N_0$ with $N_s(t)$ the number of clusters with s particles and N_0 the total number of particles. The mean cluster size is given by

$$S(t) = \frac{\Sigma n_s(t)s^2}{\Sigma n_s(t)s},$$ (3)

where the sums are taken over all clusters. It is expected that

$$S(t) \sim t^z,$$ (4)

for $t \to \infty$ with z a critical exponent.[19]

Figure 6(b) shows the experimental results for the clustering of the $K_{dd} = 1360$ particles for a time span of $t = 8 - 165$ min. As may be seen, there is fair data collapse with an exponent $z = 1.7 \pm 0.2$. For the less magnetized spheres, similar results were obtained. The fitted exponent z was thus found to decrease slightly with reduced K_{dd}, reaching a limiting value $z = 1.4 \pm 0.2$ for low values of K_{dd} which is the same as the earlier simulated result for nonmagnetic DLCA.[19]

5. Conclusions

The main objective with this presentation has been to show that aggregating microparticles in a plane produce a variety of patterns depending on the strength and range of the interparticle interaction. None of the results resemble the growth patterns seen in the original DLA model. This is not surprising as there is no reason to believe that various experimental conditions should be well represented by a simple model. However, various modified models including more of the physics are promising for further progress in this area. It is also obvious that the time has come to explore the use of multifractal dimensions in these processes.[4]

The research was supported in part by Dyno Industrier A/S and NAVF. The advice from Paul Meakin and Remi Jullien, assistance from Helmer Fjellvåg and samples supplied by John Ugelstad and collaborators at SINTEF are also greatfully acknowledged.

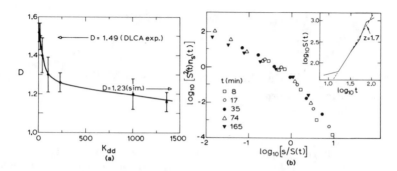

Fig. 6: (a) Fractal dimension D verus dipolar coupling constant K_{dd}. $D = 1.49 \pm 0.06$ is the DLCA value for nonmagnetic spheres. For large K_{dd}, D approaches the simulated value $D = 1.23 \pm 0.12$ as discussed in the text. The solid line is only a guide to the eye. (b) Scaling of the temporal evolution of the cluster size distribution ($K_{dd} = 1360$). The inset shows the fit of $z = 1.7 \pm 0.2$ to the log-log plot of average cluster size versus time [Eq. (4)].

1. B. B. Mandelbrot, The Fractal Geometry of Nature (Freeman, San Francisco, 1982); J. Feder, Fractals (Plenum, New York, 1988).

2. T. A. Witten, Jr. and L. M. Sander, Phys. Rev. Lett. 47, 1400 (1981).

3. T. Vicsek and J. Kertéss, Europhysics News 19, 24 (1988).

4. P. Meakin, in Phase Transitions and Critical Phenomena 12, 335 (1988).

5. P. Meakin, Adv. Colloid Interface Sci. 28, 249 (1988).

6. J. Ugelstad et al., Adv. Colloid Interface Sci. 13, 101 (1980).

7. J. Ugelstad et al., in Polymer Reaction Engineering, eds. K. H. Reichert and W. Geiseler (Hüthig & Wepf, Heidelberg, 1986), p. 77.

8. A. T. Skjeltorp, Phys. Rev. Lett. 58, 1444 (1987).

9. A. T. Skjeltorp, in Time Dependent Effects in Disordered Materials, eds. R. Pynn and T. Riste (Plenum, New York, 1987), p. 1.

10. P. Meakin and R. Jullien, J. Phys. (Paris) 46, 1543 (1985).

11. R. Botet, R. Jullien, and A. T. Skjeltorp, La Recherche 18, 1246 (1987).

12. R. Voss, in Scaling Phenomena in Disordered Systems, eds. R. Pynn and A. Skjeltorp (Plenum, New York, 1985), p. 1.

13. Similar experiments were reported earlier (Ref. 14) using much larger wax balls.

14. C. Allain and B. Jouhier, J. Phys. (Paris) Lett. 44, L421 (1983).

15. M. Kolb and H. Herrmann, J. Phys. A 18, L435 (1985).

16. This technique was used earlier (Ref. 17) for smaller silica spheres producing stringy aggregates.

17. A. J. Hurd and D. Schaefer, Phys. Rev. Lett. 54, 1043 (1985).

18. G. Helgesen, A. T. Skjeltorp, P. M. Mors, R. Botet, and R. Jullien Phys. Rev. Lett. (in press).

19. T. Vicsek and F. Family, Phys. Rev. Lett. 52, 1669 (1984).

ORIGIN OF FRACTAL ROUGHNESS IN SYNTHETIC AND NATURAL MATERIALS

DALE W. SCHAEFER, ALAN J. HURD and ANDREW M. GLINES

Sandia National Laboratories
Albuquerque, NM 87185-5800

ABSTRACT. Fractally rough surfaces have been observed in a variety of synthetic and natural materials. To explain these observations, we propose three kinetic models: ballistic polymerization, kinetic percolation, and reaction-limited dissolution. These models are applied to gas-phase growth of pyrogenous materials, solution polymerization of silicates, and the evolution of geologic materials.

1. Introduction

Since the observation by Bale and Schmidt[1] that fractal surfaces could be characterized by small angle scattering, fractal roughness has emerged as a persistent pattern in a variety of materials. Fractal roughness has been observed[1-3] in minerals such as coal, sandstone and limestone, as well as in synthetic materials such as fumed silica,[4] carbon black[5] and silicates[6] prepared by polymerization of $Si(OCH_3)_4$ and $Si(OC_2H_5)_4$.

Surface roughness has also been postulated to explain the dependence of Brunauer-Emmett-Teller (BET) surface areas on the molecular size of the absorbing species. From BET data, Pfeifer and Avnir presented evidence for fractally rough surfaces in a host of substances.[7] Although the BET data can be explained by other factors such as bottlenecks and surface energetics, Ross et al. have confirmed non-classical behavior for colloids where surface roughness had been established by scattering techniques.[8]

The repeated observation of fractal surfaces suggests that kinetic processes are active that generate roughness. The purpose of this article is to identify some of these processes. Our strategy is to seek likely sources of roughness in chemical and physical processes present in the precursors of rough materials and to map these processes onto kinetic models that produce surface roughness.[9]

2. Small Angle Scattering

The most practical way to demonstrate fractal roughness at submicroscopic length scales is by scattering techniques. Bale and Schmidt[1] demonstrated that fractally rough surfaces give scattering profiles of the form,

$$I(K) \sim K^{D_s - 6}, \tag{1}$$

where I is the scattered intensity as a function of momentum transfer K. The Porod slope $(D_s - 6)$ is equal to -4 for a smooth surface (i.e., $D_s = 2$) but lies between -3 and -4 for rough surfaces for which D_s is between 2 and 3. Figures 1 and 2 show scattering data taken by us on synthetic materials (Fig. 1) and by others[1-3] on natural substances (Fig. 2). All these materials are fractally rough on length scales between 1 and 100Å. Below, we trace roughness to gas-phase ballistic growth

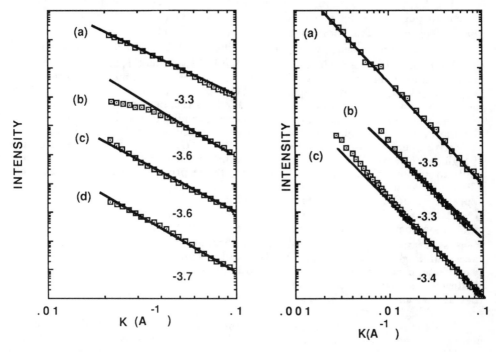

Fig. 1: SAS for synthetic materials.
(a) Fumed silica, (b) Black pearls
carbon black, (c)TMOS, (d) TEOS.

Fig. 2: SAS data for minerals.
(a) Lignite,[1] (b) Mudstone,[3]
(c) Limestone.[2]

for pyrogenous materials (carbon black and fumed silica) and to reaction-limited polymerization for solution silicates. For minerals, on the other hand, we believe roughness arises from reaction-limited dissolution.

It should be noted that, although Bale and Schmidt derived their formula for a self-similar surface, it applies to a self-affine surface as well.[10] No distinction is made between self-affinity and self-similarity in this study.

Occasionally, Porod slopes steeper than -4 are observed corresponding to D_s less than 2 in Eq. (1). These "subfractal" surfaces[10] are observed when interfacial layers are present such as a surfactant coating a colloidal particle.[5,10] Subfractal roughness means that surface mass density fluctuations decrease with distance in power-law fashion.

3. Ballistic Polymerization

The pyrogenous materials, fumed silica (Fig. 1a) and carbon black (Fig. 1b), are prepared by burning volatile precursors, silicon tetrachloride, and fossil fuel, respectively. These materials are similar in structure, consisting of 100Å colloidal particles aggregated into highly ramified clusters. The data in Fig. 1a,b show that the primary colloidal particles have rough surfaces.

Roughness in pyrogenous materials may arise because of the tendency for the system to polymerize by a ballistic monomer-cluster (MC) process in the initial phase of growth.[9] Transport is ballistic because of the long mean free path ($\sim$ 5000Å) of small molecules (e.g., monomeric silicic acid, $Si(OH)_4$ in the case of fumed silica) in flames. Once any cluster reaches about 100Å diameter, however, its motion becomes diffusional over similar distances. In the presence of many monomers, the large sluggish clusters grow by accretion of the speedy ballistic clusters. In addition, the sticking probability, based on van der Walls forces, decreases with particle size.

Ballistic deposition leads to self-affine surface roughness with $D_s = d - \alpha$ where d is the dimension of space and α is the length scaling exponent of the height of the active zone.[11] From simulations, $\alpha = 1/3$ so $D_s = 2.67$ for $d = 3$. Because of high flame temperatures, however, sintering competes with ballistic growth and tends to smoothen the surface. Therefore, flame processes should generate self-affine surfaces with $D < 2.7$, consistent with Fig. 1a and 1b. Smoother surfaces are expected at higher temperature and after longer residence times.

MC polymerization will cease when flame chemistry stops producing polymerizable monomers. In addition, when all the clusters become large compared to the mean separation of small molecules in the background gas ($\sim$ 50Å), the reacting species undergo a transition from ballistic to Brownian motion. As a result of these two effects, a crossover is expected from ballistic MC growth to diffusion-limited cluster-cluster aggregation. This crossover in growth habit accounts for the fact that almost all materials produced in flames are uniformly dense on short length scales and highly ramified on long scales.

Ballistic polymerization is probably more appropriate to fumed silica than carbon black. The primary carbon black particles often contain substantial crystalline graphite. Crystalline phases indicate growth near equilibrium. Consistent with this observation is the fact that rough carbon black seems to be less common than smooth varieties. Unfortunately, the differences depend on many proprietary processing steps. We are currently sorting out these complexities.

4. Kinetic Percolation

Ballistic polymerization clearly cannot explain the data for the solution polymerized silicates in Fig. 1c and 1d. For these systems, fractally rough colloidal particles are observed when alkoxides like $Si(OCH_3)_4$, TMOS, are polymerized under strongly basic conditions. Under basic conditions, polymerization is by monomer cluster growth because the most facile reaction pathway requires inversion[12] of one of the reacting partners, an operation which is difficult if both are tied into clusters. Furthermore, condensation,

$$2 - Si(OH) \rightarrow - Si - O - Si - + H_2O$$

must be preceded by hydrolysis,

$$- Si(OCH_3) + H)2O \rightarrow - Si(OH) + CH_3OH.$$

This requirement also favors MC growth when hydrolysis is rate limiting.[10] Thus base-catalyzed chemistry favors reaction-limited MC growth (Eden model).

In pure Eden growth, all sites are equally reactive corresponding to complete hydrolysis of TMOS to silicic acid, $Si(OH)_4$. In fact, when hydrolysis is rate-limiting, both the monomers and clusters have unreactive blocked sites corresponding to unhydrolyzed alkoxy groups. Fractal roughness results from these blocked sites.

For TMOS, all sites on monomers and clusters are equally reactive toward hydrolysis.[13] If hydrolysis is purely statistical, growth is limited by hydrolysis on the cluster, leading to a growth process that might be called kinetic percolation (KP) by analogy to an algorithm for generating percolation clusters developed by Leath.[14]

On a computer, percolation clusters[15] are formed by randomly filling sites on a lattice like rain drops falling on a surface. In Leath's algorithm, however, clusters are "grown" by starting from a seed site and randomly occupying sites neighboring the seed. Only a fraction P of sites are selected. Further generations are similarly added on an ever increasing perimeter. If this algorithm is restricted to a single cluster, the cluster will always die (i.e., at some generation no perimeter sites are selected) if P is below the percolation threshold P_c. Above P_c, since there is finite probability that every generation will survive, growth occasionally proceeds without limit. KP describes MC growth that is limited by hydrolysis on the cluster.

If hydrolysis is sufficiently aggressive, P is effectively above the percolation threshold and the geometry of the resulting cluster can be determined from percolation simulations. Above threshold, the largest clusters are compact with rough surfaces[16] accounting, at least qualitatively, for the observations of Fig. 1c.

If TEOS (Fig. 1d), the ethoxide, is used instead of TMOS, then fractally rough colloids are observed under substantially less basic conditions. For TEOS, steric effects limit hydrolysis on condensed clusters[17] so kinetic percolation is not appropriate. Now growth is limited by monomer hydrolysis leading to conventional Eden growth but with blocked sites. This poisoned Eden process[6] also produces fractally rough clusters and accounts for the TEOS data in Fig. 1d. It should be noted that kinetic percolation and poisoned Eden growth are identical if the blocked-site distribution is random.

5. Reaction-Limited Dissolution

It is unlikely that either of the above models is an appropriate explanation for fractally rough porosity in geologic materials. It seems reasonable, however, that naturally occurring roughness arises from dissolution. In a dissolving environment, surface tension is effectively negative in the sense that greater interfacial area is energetically favored. Enhanced surface area can develop by the inverse of Ostwald ripening (small pores appearing at the expense of large), a near equilibrium process. In addition, however, surface area can increase at constant porosity by roughening. Such a process could explain fractal roughness in minerals exposed to a dissolving environment.

To test the relationship between dissolution and fractal roughness, we followed the x-ray scattering profile as smooth colloidal silica dissolved. We chose Cab-O-Sil M5, a pyrogenous silica known to consist of smooth primary particles 100Å in diameter.[4] The upper curve of Fig. 3 confirms the smooth nature of the pristine material on length scales below 50Å.

Pristine Cab-O-Sil M5 was dissolved in a pH = 13 buffer for periods up to several hours. The lower curves in Fig. 3 show the data at 5 and 66 minutes. Over a one hour period, the solution pH dropped approximately one pH unit. At the observation times listed in Fig. 3, the solutions were quenched with HCl below pH = 5 where silica is extremely insoluble.

The data in Fig. 3 show that the surface roughens as dissolution proceeds. We confirmed this result through in situ scattering studies where scattering profiles were measured as dissolution proceeds. We found Porod slopes as large as -3.3 corresponding to a surface fractal dimension of 2.7.

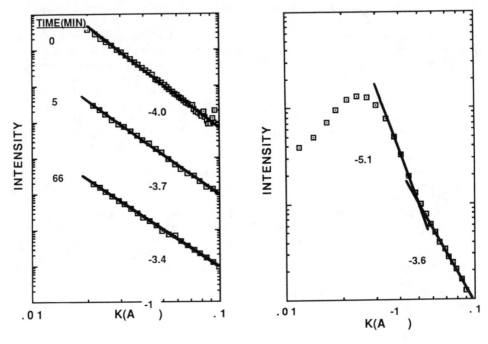

Fig. 3: Dissolution of Cab-O-Sil.   **Fig. 4:** Porous Vycor glass.

Dissolution then provides a reasonable explanation for naturally occurring surface roughness. It is interesting to note that Cohen[18] has postulated that the topology of the large scale porosity of oil-bearing sandstones can be explained in terms of 'negative surface tension'. If surface tension is positive, conventional Ostwald ripening leads to closed porosity just as it leads to smooth surfaces. In a dissolving environment, however, negative surface tension can lead to inverse ripening, connected porosity, and rough surfaces. Because the mechanism leading to connected porosity is a near-equilibrium phenomenon, it is not necessarily consistent with dissolution-induced roughening proposed here since the latter is a kinetic process that occurs far from equilibrium.

We also looked for evidence of the roughening effect in synthetic porous materials. Figure 4, for example, shows neutron scattering data for porous Vycor glass. This material is prepared by dissolving the boron rich phase out of a borosilicate glass. The scattering curve[19] shows a peak indicative of spinodal decomposition in the dense precursor glass. In the highest K region, however, a limiting slope of -3.6 is observed, consistent with short-scale surface roughness. Finally, in the region $K = 0.04 \text{Å}^{-1}$ the Porod slope is -5.1, indicating a subfractal interface.

We interpret the Vycor data to mean that a swollen interfacial gel layer[20] develops first. This swollen gel then dissolves leading to short length-scale roughness. The swollen gel layer must have a power-law density profile perpendicular to the surface to be consistent with the Porod slope less than -4. It must be recognized, however, that the conventional interpretation in terms of a uniformly broadened interface[21] is also consistent with the data.

6. Conclusion

We have demonstrated that simple kinetic models can explain the existence of self-similar or self-affine roughness in synthetic as well as geologic materials. In all cases, roughness arises when growth or dissolution takes place far from equilibrium.

1. D. Bale and P. W. Schmidt, Phys. Rev. Lett. 53, 596 (1984).
2. P.-Z. Wong, J. Howard, and J.-S. Lin, Phys. Rev. Lett. 57, 637 (1986).
3. D. F. R. Mildner, R. Resvani, P. L. Hall, and R. L. Borst, Appl. Phys. Lett. 48, 1314 (1986).
4. A. J. Hurd, D. W. Schaefer, and J. E. Martin, Phys. Rev. A35, 2361 (1987).
5. D. W. Schaefer and A. J. Hurd in *The Chemistry and Physics of Composite Media*, eds. M. Tomkiewics and P. N. Sen (Proc. Electrochem. Soc. 85-8, Pennington NJ, 1985), p. 54.
6. K. D. Keefer and D. W. Schaefer, Phys. Rev. Lett. 56, 2376 (1986).
7. P. Pfeifer and D. Avnir, J. Chem. Phys. 79, 3558 (1983); 80, 4573 (1984).
8. S. B. Ross, D. M. Smith, A. J. Hurd, and D. W. Schaefer, "Surface Roughness in Vapor-Phase Aggregates via Adsorption and Scattering Techniques," Langmuir 4, 977(1988).
9. D. W. Schaefer, Bull. Mat. Res. Soc. 13-2, 22 (1988).
10. D. W. Schaefer, J. P. Wilcoxon, K. D. Keefer, B. C. Bunker, R. K. Pearson, I. M. Thomas, and D. E. Miller, in *Physics and Chemistry of Porous Media II*, eds. J. R. Banavar, J. Koplik and K. W. Winkler (Am. Inst. of Phys. Conf. Proc. 154, New York, 1987), p. 63.
11. P. Meakin, P. Ramanlal, L. M. Sander, and R. C. Ball, Phys. Rev. A34, 5091 (1986).
12. K. D. Keefer in *Better Ceramics Through Chemistry*, eds. C. J. Brinker, D. E. Clark and D. R. Ulrich (Mat. Res. Soc. Symp. Proc. 32, Pittsburgh PA, 1986), p. 15.
13. R. A. Assink and B. D. Kay in *Better Ceramics Through Chemistry*, eds. C. J. Brinker, D. E. Clark and D. R. Ulrich (Mat. Res. Soc. Symp. Proc. 32, Pittsburgh PA, 1986), p. 301.
14. P. L. Leath, Phys. Rev. B 14, 5046 (1976).
15. D. Stauffer, *Introduction to Percolation Theory* (Taylor and Francis, London, 1985).
16. R. F. Voss, J. Phys. A 17, L373 (1984).
17. J. C. Pouxviel and J. P. Boilot in *Better Ceramics Through Chemistry III*, eds. C. J. Brinker, D. E. Clark and D. R. Ulrich (Mat. Res. Soc. Symp. Proc. 121, Pittsburgh PA, 1986), p. 39.
18. M. H. Cohen in *Physics and Chemistry of Porous Media II*, eds. J. R. Banavar, J. Koplik, and K. W. Winkler (Am. Inst. Phys. Conf. Proc. 154, New York, 1987), p. 3.
19. D. W. Schaefer, B. C. Bunker, and J. P. Wilcoxon, Phys. Rev. Lett. 58, 284 (1987).
20. B. C. Bunker, G. W. Arnold, D. E. Day, and P. J. Bray, J. Non-Cryst. Solids, 87, 226 (1986).
21. W. Ruland, J. Appl. Cryst. 4, 70 (1971).

ELECTRODEPOSITION

G. HELGESEN
Department of Physics, University of Oslo
Blindern, N-0316 Oslo 3, Norway

ABSTRACT. The purpose of this demonstration is to show a very simple method to grow fractal aggregates.

It is now recognized that many processes in nature give rise to structures having fractal patterns: aggregation, viscous fingering, dielectric breakdown and electrodeposition.[1] In many of these processes one needs specialized equipment like glass beads, pumps, tailormade cells etc. and well controlled conditions to obtain a good result. But using the simple electro-deposition technique described below it is quite easy to produce a lot of different types of deposits showing a wide range of forms from almost compact structures via dendritic crystalline forms to ramified fractal "trees".[2] Many of the forms which can be obtained have a striking resemblance with forms observed in living plants and trees.

To make the electrodeposition cell one uses an ordinary microscope slide cut to a length of less than 50 mm (to fit into a slide projector). The two electrodes can be made by painting stripes of a silver paint solution (Emetron Leitsilber 200) on each of the long sides of the slide. When the electrodes are dry, put a few droplets of water on the slide, gently put on a microscope cover glass and press lightly a piece of porous paper on the glass to remove excess water. The distance between the glass plates can be adjusted by a very low concentration of 10-25 μm polystyrene spheres as spacers, but this is usually not necessary because the thickness of the porous silver electrodes will keep the plates apart. Seal along the edge of the cover glass with epoxy. Then connect the electrodes to the poles of a 4.5 volt battery with thin wires, using Scotch tape to fix the wires to the electrodes. After about 30 minutes it should be possible to see small deposits growing on the negative electrode.

To make a closer inspection it is easy to place the slide (with wires connected to the battery) in a slide projector and observe the details of the growth. A competing process in many cases is the electrolysis of water producing gas bubbles in the cell, but usually this process will not be important before the deposition has almost finished. It is possible to replace the water in this cell by a metal-ion solution ($CuSO_4$, $ZnSO_4$) or use another metal in the electrodes, but by varying concentration and voltage one can obtain the same types of deposits as in the silver/water case.

1. F. Argoul, A. Arneodo, G. Graseau and H.L. Swinney, preprint
2. G. Helgesen, in Time-dependent effects in disordered materials (R. Pynn and T. Riste, eds) Plenum, NY 1987.

FLOW PATTERNS IN POROUS MEDIA

ROLAND LENORMAND and GERARD DACCORD
Dowell Schlumberger
B.P. 90 42003 Saint Etienne Cedex 1
France

ABSTRACT. This paper presents the physical background corresponding to *hands on* experiments of displacements of fluids though porous media shown during the conference.

1. Immiscible Displacements in 2-D Porous Media

A 2-D porous medium is a kind of Hele-Shaw cell (two transparent plates separated by a small gap) filled with a monolayer of granular material. In our case we use *micromodels* made of transparent resin cast on a photographically etched mold,[1] but similar results are obtained with glass beads between plates.[2]

1.1. CAPILLARY DISPLACEMENTS

In this case, viscous and gravity forces are assumed to be negligible and consequently, all the displacement mechanisms are linked to capillary forces and randomness due to the different sizes of pores.

Generally speaking, when one fluid (say oil) is slowly displacing another immiscible fluid (say water) in a capillary tube of diameter D_0, the fluid for which the contact angle θ (between the tube and the meniscus) is smaller than $\pi/2$ is called the *wetting fluid* (W); the other one is the *nonwetting fluid* (NW). The pressure in the NW fluids exceeds the pressure in the wetting fluid by a value P, called capillary pressure and linked to the interfacial tension γ by the Laplace law: $P = 4\gamma \cos\theta/D_o$. A displacement where the NW fluid is pushing the wetting fluid is called *drainage*, the reverse is *imbibition*.

1.1.1. *Physical Mechanisms.*

Experiments show that both fluids can flow simultaneously in the same duct with different velocities, the wetting fluid remaining in the extreme corners of the cross-section and roughness of the walls.

A displacement can be divided into three parts:

(i) flow of the injected fluid from the entrance towards the moving meniscus,
(ii) displacement of the meniscus,
(iii) flow of the displaced fluid towards the exit.

The displacement of the meniscus in the network has been presented in detail in previous papers.[3] The main result is the effect of the geometry of the pores which can introduce a selection of displacement mechanisms. In *drainage*, there is no interaction between adjacent menisci and the resulting pattern is governed by the smallest throats.

In *imbibition*, we found two main cases:

(i) *large pores*, when the size of the pore (site of the network) is large compared with duct diameters; the wetting fluid will preferentially collapse inside a duct.

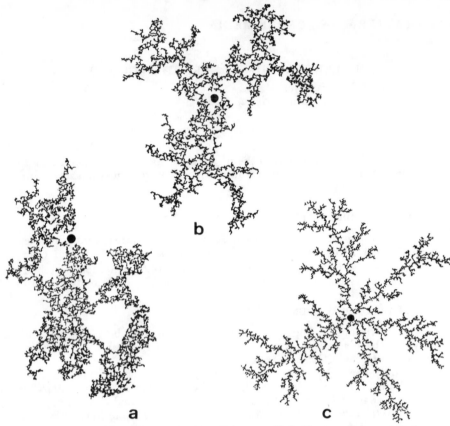

Fig. 1. Air displacing very viscous oil in a radial micromodel containing 250,000 ducts. a) very low flow rate: viscous forces are negligible and the pattern corresponds to invasion percolation; b) intermediate flow rate: cross-over between percolation and D.L.A.; c) high flow rate: the pattern is very similar to D.L.A.

(ii) *small pores*, in the opposite case, the meniscus collapses inside the pore. The former case leads to percolation type patterns, the later to flat interfaces at large scale.

The nonwetting fluid can flow only in the bulk of the ducts. So the flow occurs only if a continuous path of ducts or pores filled with this phase exists either towards the entrance (during drainage) or the exit (during imbibition) of the network. Otherwise the NW fluid is trapped.

We observe 3 kinds of flow of the wetting fluid:

(i) flow in the bulk of the ducts,
(ii) along the corners when the NW fluid fills the central part of the duct,
(iii) by *film* along the roughness of the walls, but only in case of strong wettability and very low flow rate.

1.1.2. *Drainage*. Capillary forces prevent the NW fluid from spontaneously entering a porous medium. It can only enter a duct (diameter D_o) when the pressure exceeds the capillary pressure. From a statistical point of view a duct with $D > D_o$ is an "active" or "conductive" bond and a duct with $D < D_o$ an inactive bond. The fraction p of active bonds can easily be deduced from the throat size distribution.

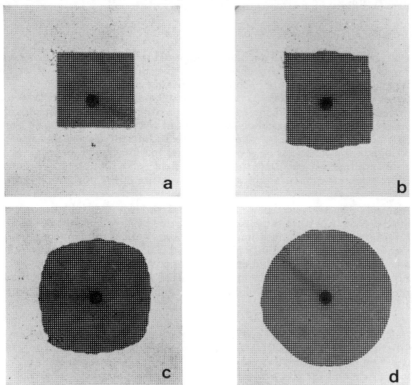

Fig. 2: Radial injection of oil displacing air in a regular square micromodel at different flow rates and viscosities: a) 0.18 cc/mn, 14 cP; b) 1.8 cc/mn, 14 cP; c) 18 cc/mn, 14 cP; d) 18 cc/mn, 100 cP.

At a given pressure P, the injected fluid invades all the percolation clusters connected to the injection point (Fig. 1a); this mechanism has been called invasion percolation[4]. During the displacement, the wetting phase is trapped in the network when the invading NW fluid breaks the continuous path toward the exit (in 2-D, the W fluid cannot escape by flowing via the corners).

1.1.3. *Imbibition.* The type of displacement depends upon two factors: the pore geometry (large or small pores) and the possibility of flow by film along the roughness.

In the case of *small pores*, without flow by film, and with a very narrow distribution of pore diameters, the meniscus instability inside a pore leads to filling the network line after line, without trapping. The result is a kind of *crystal growth* (Fig. 2a).

At a given stage of a capillary displacement, menisci in the pores do not "see" the exit because we are assuming a zero pressure drop in the fluids. Consequently, they are described by "local models," i.e., at each step, the interface between the two phases moves towards an adjacent site according to some local rules. The mechanism is different for viscous displacements.

1.2. VISCOUS DISPLACEMENTS

In this case, capillary forces are negligible compared to viscous forces. However, we need a small amount of capillary effects, otherwise the fluids mix together inside the ducts and the problem is more complex (miscible displacements).

Viscous displacements, either stable or unstable are governed by the pressure field between the entrance and the exit. Consequently, even in the case of a stable displacement, a local model based on some rules at the interface cannot be used for modeling viscous displacements. A model, called *Gradient Governed Growth* has been developed simultaneously by several authors[5] to solve this problem, using both a continuum approach to calculate the pressure field and a discrete displacement of the interface which accounts for the granular structure of the porous medium.

1.2.1. *Unstable displacement.*
If we assume that the injected fluid has negligible viscosity, the growing pattern can also be represented by a model known as *Diffusion Limited Aggregation.*[6] A computation based on a network of *random* conductances, leads to similar results. The validity of this model is demonstrated by the similarity between computer simulations and experimental patterns.

For instance, Fig. 1c shows the displacement of a very viscous oil by air in a radial geometry. The measured fractal dimension again is very close to the theoretical value $D = 1.70$.

1.2.2. *Stable displacement.*
The corresponding pattern can be described by an *anti-DLA* statistical model.[7] Anti-DLA consists in releasing particles near a compact aggregate. The particles move at random until it reaches an occupied site. In this case, the particle and the site are removed and another particle is released.

1.3. TRANSITION BETWEEN CAPILLARY AND VISCOUS DISPLACEMENTS

The above-described displacements are three extreme cases of more general displacements involving the simultaneous competition of viscous and capillary forces (gravity and inertial effects are always assumed to be negligible during the experiments).

In the general case we are dealing with three kinds of forces: viscous forces in the displaced fluid (labeled 1), viscous forces in the displacing fluid (labeled 2) and the capillary forces. Now, with 3 kinds of forces we can define two dimensionless numbers which characterize a given experiment:

(i) The viscosity ratio $M = \mu_2/\mu_1$.

(ii) The capillary number, ratio of the viscous forces in fluid 2 and capillary forces: $Ca = (q\mu_2)/(S\gamma)$, where q is the total voluminal flow rate, S the cross-sectional area of the sample and γ the interfacial tension.

Consequently, all the experiments performed with a given sample (micromodel) can be displayed in a diagram with axes representing the viscosity ratio M and the capillary number Ca.

1.3.1. *Drainage.* We have studied the transition between the three main types of displacement in the case of drainage[7]. All the points corresponding to the fractal displacements are located within two distinct domains in this diagram (Fig. 3): at low Ca for capillary fingering, and low M for viscous fingering. In addition, a domain at high Ca and M corresponds to the stable displacements (anti-DLA). As a consequence of this "phase-diagram", we are expecting a smooth transition between percolation and DLA patterns for a given, very low value of M, by varying the flow rate of the displacement, a result which has been observed experimentally (Fig. 1, from percolation to DLA, air displacing a viscous oil).

1.3.2. *Imbibition.* Fig. 2 shows the evolution of the pattern when viscous forces increase. In this experiment, oil is displacing air. At low rate (capillary forces are dominant) the pattern is a square (square network) and progressively it tends toward a circular shape due to stabilizing effects of viscous forces.

Fig. 3: "Phase-Diagram" for drainage in log-log scale.

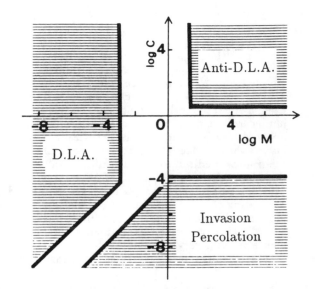

2. Reactive Flow

It has been known for a long time that injecting a reactive fluid into a soluble porous medium yields very ramified dissolution patterns. The physical phenomenon of the formation of dissolution patterns involves the flow of a liquid in a porous medium coupled with a chemical reaction.

We used the system plaster/water because its properties are well known and it is convenient for studying various geometries.[8,9] Good reproductions of the final dissolution patterns are obtained by injecting a low melting point alloy (Wood's metal) into the dried plaster. After cooling, the plaster is dissolved. Figure 4 shows the result in a two-dimensional radial cell, consisting of a thin disk of plaster (typically 1 mm thick and 200 mm diameter).

At first glance, this structure is very similar to other experimental growth patterns associated with the model of diffusion limited aggregation (DLA): dielectric breakdown, electrodeposited metal leaves, viscous fingering ... This analogy can be understood by comparison with the above-described viscous fingering.

Fig. 4: Photograph of a dissolution pattern obtained in a 2-D plaster cell.

The velocity u of a fluid of viscosity μ in a porous medium of permeability k is proportional to the pressure gradient: $u = -k/\mu$ grad P. Here k/μ is the mobility and P the pressure field. During the displacement of a viscous fluid by a non-viscous one, viscous fingering occurs due to the very sharp increase of mobility when a small perturbation appears at the interface between the fluids (the viscosity jumps from a finite value to a very small value). In the case of dissolution, the injected reactive fluid and the saturating non-reactive fluid have the same viscosity but, at the interface (the reactive front), the permeability jumps from a low value in the porous medium to a quasi-infinite value in the etched channels. Consequently, the effects on the mobility term are analogous to viscous fingering.

The two-dimensional patterns are quantified by digitizing the photographs and computing the density-density correlation functions. We obtained $D = 1.6 \pm 0.1$, in reasonable agreement with results from DLA.

We also performed experiments with 3-dimensional samples. All the experiments produce very ramified, tree-like structures and a technique, based on capillary properties, have been developed for measuring the 3-D fractal dimension.[10]

1. R. Lenormand and Zarcone, Proc. P.C.H. conf., Tel-Aviv, (1984), Phys. Chem. Hydrodyn. 6 (1985) 497.
2. K. J. Mäløy, J. Feder and T. Jøssang, Phys. Rev. Lett. 55 (1985) 1885.
3. R. Lenormand and C. Zarcone, Soc. Petrol. Eng. paper 13264 (1984).
4. R. Lenormand and C. Zarcone, Phys. Rev. Let. 54 (1985) 2226.
5. J. D. Sherwood and J. Nittmann, J. Physique 47 (1986) 15.
6. T. A. Witten and Sander, Phys. Rev. B 27 (1983) 5686.
7. R. Lenormand, E. Touboul and C. Zarcone, J. Fluid Mech. 189 (1988) 165-187.
8. G. Daccord and R. Lenormand, Nature 325, 41, (1987).
9. G. Daccord, Phys. Rev. Lett. 58, 479, (1987).
10. R. Lenormand, A. Soucemarianadin, E. Touboul and G. Daccord, Phys. Rev. A 36, 4, (1987).

VISCOUS FINGERING IN A CIRCULAR GEOMETRY

Y. COUDER

Groupe de Physique des Solides de l'Ecole Normale Supérieure
24 Rue Lhomond, 75231 Paris Cedex 05, France

1. Introduction

The instability giving rise to Saffman-Taylor viscous fingering occurs at the interface between two fluids moving between narrowly spaced solid plates (i.e., a Hele Shaw cell). The interface is unstable when it is the less viscous fluid which forces the more viscous fluid to recede. The flow of each fluid is dominated by the viscous dissipation on the plates and the mean velocity in the cell's plane is given by the Darcy law

$$V = -\frac{b^2}{12\mu}\nabla p, \qquad (1)$$

where b is the plates spacing and μ the viscosity of the fluid. The pressure p, because of the incompressibility of the fluid, obeys a Laplace law

$$\Delta p = 0. \qquad (2)$$

Surface tension has a stabilizing influence on the interface. It is taken into account by adding a boundary condition for the pressure jump at the interface

$$\delta p = T\kappa. \qquad (3)$$

where T is the surface tension and κ the local two dimensional curvature.

In the following we limit ourselves to the case where one of the fluids is a gas of negligible viscosity compared to the other fluid (i.e., a liquid). The linear analysis[1] of the stability of a plane interface moving at velocity V shows that the wavelength of maximum instability is

$$\ell_c = \pi b\sqrt{\frac{T}{\mu V}}. \qquad (4)$$

We will call ℓ_c the capillary length scale.

The Saffman-Taylor instability is one of the simplest pattern generating system. As was first noted by Paterson,[2] it can be compared to the theoretical model of diffusion limited aggregation (DLA). This model, introduced by Witten and Sander,[3,4] can be found in Refs. 5-7. In this model of aggregation of randomly walking particles, the probability P of visit of a site is given by $\Delta P = 0$ and the mean velocity of growth of a region of the interface is $V_n = (\nabla P)_n$. Thus the equations of growth of DLA are those of Saffman-Taylor fingering in the limit of zero surface tension (and vanishing thickness of the cell).

A large number of variants of both problems have been investigated; for instance viscous fingering can become dendritic with local perturbation of the finger tip,[8-10] dendritic or fractal in engraved Hele Shaw cells of periodically varying thickness[11,12] or with anisotropic fluids,[13] fractal with non-Newtonian shear thinning fluids.[14,15] On the other hand numerical simulations of DLA with modified sticking rules (see Refs. 16 and 5-7, and Refs. therein) simulated system with tunable noise and anisotropy, producing a wide variety of figures.

In this lecture we will limit ourselves to the simplest case. We present the results of experiments on the pure Saffman-Taylor instability in the axisymmetric geometry (where DLA aggregates are also usually grown). This geometrical configuration was first introduced for viscous fingering by Bataille[17] and more recently investigated by Paterson,[18] Ben Jacob et al.[19] and Rauseo et al.[20] The oil is contained in a cell formed of two circular glass plates and air penetrates either at the center (divergent flow) or at the periphery (convergent flow). In both cases the geometry creates unsteady states of the flow simply because the ratio of the perimeter of the interface to ℓ_c changes constantly.

In the divergent case the initial front is circular and its first destabilization usually gives rise to 5 to 7 initial fingers (Fig. 1). As the front moves on, they undergo tip splitting and patterns with 10 to 14 relatively regular fingers are observed. In part I we will investigate the selective action of capillarity on shape of the fingers of these patterns. The further evolution is dominated by side branching and complex patterns are obtained. In part II, we will characterize them and compare them to DLA. Finally in part III we will compare convergent fingering to convergent DLA.

2. The Selective Action of Surface Tension on the Initial Patterns

[The results presented in this part were obtained in collaboration with H. Thomé, M. Rabau, and V. Hakim.[21]]

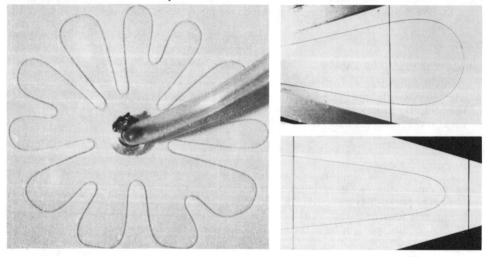

Figs. 1-2: Figure 1 is a pattern obtained at low velocity. Figures 2a and 2b show fingers in divergent and convergent cells.

The initial destabilization occurs when regions of the circular front stop and troughs are formed. These troughs become fjords as the fingers move on. We address ourselves to the shape of these fingers. Naively each of these fingers looks as if it was growing independently from its neighbor so we built Hele Shaw cells with a sector shaped geometry where a single finger moves between lateral walls forming an angle θ_0. It can move in the widening direction (divergent finger) or in the narrowing one (convergent finger).

For vanishing values of θ_0 these are the cells used by Saffman and Taylor[22] where the walls are parallel at a distance W. In these cells the classical results can be summarized by a unique control parameter[23]

$$\frac{1}{B} = \frac{12\mu V}{T}\left(\frac{W}{b}\right)^2 \approx 118\left(\frac{W}{\ell_c}\right)^2.$$

Saffman and Taylor found a family of analytical solution for the interface in the absence of surface tension. They are characterized by the ratio λ of the finger width to the channel width. In the experiment steady fingers are observed and it remained a puzzle during a long time that, at large value of $1/B$, λ tends towards the limit $\lambda_0 = 0.5$. Recent theoretical works[24-26] have shown that his selection is due to the action of surface tension. In reverse we showed[9] that this selection is lifted when local anisotropy is introduced at the tip.

In sector shaped cells, we plot the angular fraction occupied by the finger θ_f/θ_0 versus a local value of $1/B$ (where we replace W by the cell's width $\theta_0 r_0$ at the tip). In these conditions the λ versus $1/B$ plot is very similar to the classical one but with different asymptotes $\lambda_0(\theta_0)$ and different stability limits.

One single finger moving in a divergent cell has at first a spatial evolution of its width because $1/B$ grows with r_0. When it reaches the asymptotic value $\lambda_0(\theta_0)$ its width remains constant and its lateral sides can, in this range, be extrapolated at the apex of the cell (Figs 2a and 2b). Then ultimately it becomes unstable.

The dependence $\lambda_0(\theta_0)$ (Fig. 3) is approximately linear for all the convergent fingers (which are narrower) and for divergent fingers when $\theta_0 \leq 25°$ we find (θ_0 is in degrees)

$$\lambda_0(\theta_0) = 0.48 + (3.2 \pm 0.1)10^{-3}\theta_0.$$

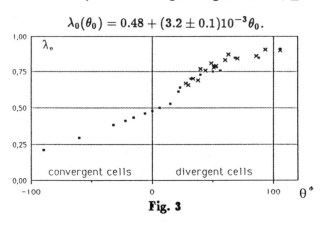

Fig. 3

For $\theta_0 \geq 25°$ the evolution is similar, but the finger becomes rapidly unstable towards tip splitting and the asymptotic width not completely reached.

Returning to the circular pattern shown on Fig. 1, each of its finger has the shape, width and stability that it could have in a cell that would have walls along the bisectors of the two neighboring fjords. These results give an interpretation of a statistical properties of these patterns. The proportion of air of the perimeter of a circle, is a decreasing function of the radius r of the circle. This is a result of the $\lambda_0(\theta_0)$ dependence given by Fig. 3; the mean values of θ_0 decreasing as the number of fingers grow, so does the mean value of λ_0.

Finally it is worth noting that the observed fjords have sides crossing at the center of the cell. This differs markedly from the constant width observed in previous theoretical[27] and numerical[28] works.

3. Complex Divergent Patterns

The statistical properties of complex divergent patterns have been specifically investigated by Ben Jacob *et al.*[19] and by Rauseo *et al.*[20] who reached different conclusions. The first group concludes that the pattern reaches a space filling shape that they call dense branching morphology (this term is also used to describe regimes of growth in annealing and electrodeposition.[19] The second group concludes to a fractal growth of the DLA type.

In order to obtain a complex pattern, it is necessary to have l_c as small as possible compared to the cells dimension. As most fluid have comparable values of surface tension, this means b small or μ and V large. In all cases, it results in high applied pressures, so that special care has to be taken to avoid plates flexions.

We work in cells with 1.5 cm thick glass plates and we apply suddenly a strong depression at the outer periphery (where the plates are held with spacers). Air at the atmospheric pressure moves in at the center. Photographs of the pattern are taken with a motorized camera. These photographs are then digitized with a resolution 512×512. By enhancement of the contrast we can then seek for the boundary of the pattern. We cover the field with grids of boxes of varying size ϵ. The partition function Z_q is defined by

$$Z_q = \sum_{i=1}^{N} p_i^q,$$

where p_i is the proportion of the cell i occupied by the boundary of the pattern. The scaling of Z_q gives the values of the Rényi[29] dimensions D_q

$$Z_q \propto \epsilon^{(q-1)D_q}.$$

Figure (5a) shows a graph of $\ln[Z_q/(q-1)]$ versus $\ln \epsilon$. The values of D_q as a function of ϵ are shown on Figure (5b). They appear to be nearly constant in the range $\ln(\epsilon) \in [-5.5, 0]$ with the following values:

q	0	1	2	4	10	15
D_q	1.76	1.77	1.77	1.74	1.70	1.69

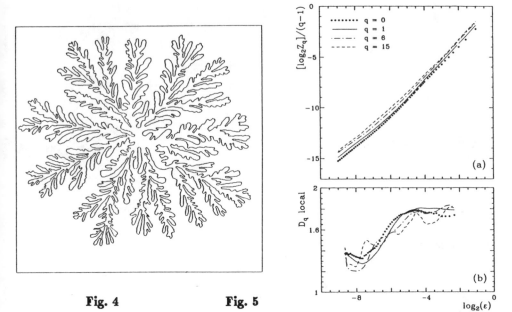

Fig. 4   **Fig. 5**

For $\ln \epsilon < -5.5$ the values of D_q drop towards 1. The cut-off corresponds to scales smaller than the capillary length ℓ_c where the boundary become smooth.

A similar investigation of DLA clusters, by Argoul *et al.*[31] point to a unique value $D_q = 1.6$ suggesting that these aggregates are self similar. Although there is a certain scatter in our results due to insufficiently branched patterns, our results is coherent with theirs. The fractal behaviour described above confirms the previous measurements of Rauseo *et al.*[20] but are not compatible with the dense branching morphology reported by Ben Jacob *et al.*[19] A probable interpretation is that, in spite of the cautions they took, there was plates flexion in their experiment which was performed with perspex plates. We performed an experiment to exaggerate the effect of a lift of the plates. Figure 6 shows the pattern obtained when the top glass plate has been replaced by a 2 mm thick perspex plate. The fingers now move in a gap of varying thickness, and this stabilizes the extremity of all the branches at a well determined position of the widening gap. This was first observed by Mc Ewan and Taylor[33] in the peeling of adhesive tape. In our case the flexions of the plates are axisymmetric so that they maintain a circular envelope to the finger. The flexions of the plates being axisymmetric, they maintain a circular envelope to the pattern.

In fact the viscous fingering in a widening gap is itself very interesting. In a recent work[35] we show that it is akin directional crystal growth.

4. Convergent Patterns

In order to obtain the reverse axisymmetric flow, we start the experiment injecting oil at the center of an empty horizontal circular cell, until it occupies a well centered circular zone (without reaching the spacers). Then we rapidly siphon the oil out by the center, so that air at atmospheric pressure penetrates at the periphery. A

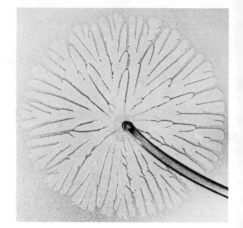

Fig. 6: Dense branching morphology.

typical pattern is shown on Fig. 7.

As we did not know of any DLA calculation in this geometry, we undertook a simulation where the random walkers are emitted at the center of a circle of finite radius, and stick at the periphery of the circle. We used the on-square lattice algorithm proposed by Ball and Brady.[34] A typical DLA aggregate of this type is shown on Fig. 8.

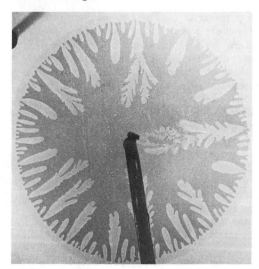

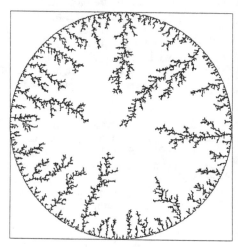

Fig. 7   **Fig. 8**

In both the fingering and the aggregate growth, the evolution of the pattern is different from the classical one. The region available for the growth has a shrinking perimeter instead of an expanding one. As a result, the competition between neighboring fingers dominate the dynamics, tip splitting and side branching tend to become secondary processes. (This trend of increased competition already existed when DLA aggregates were built up on a linear base instead of on a central seed). Another specific characteristic of this new configuration is that the growth ends at a finite time when a finger reaches the center. In other terms the growth stops as soon as a percolating aggregate is formed between the periphery and the center.

Finally this different dynamics does not seem to affect the fractal dimensions, and, to the resolution that we have obtained up to now, we find similar statistical properties in the convergent and divergent cases.

1. R. L. Chuoke, P. Van Meurs and C. Van der Pol, Tr. AIME 216, 188 (1959).
2. L. Paterson, Phys. Rev. Lett. 52, 1621 (1984).
3. T. Witten and L. M. Sander, Phys. Rev. Lett 47, 1400 (1981).
4. T. Witten and L. M. Sander, Phys. Rev. B 27, 5686 (1983).
5. *On Growth and Form: Fractal and Non-Fractal Patterns in Physics*, eds. H. E. Stanley and N. Ostrowsky (Martinus Nijhof, Dordrecht, 1983).
6. *Fractals in Physics*, eds. L. Pietronero and E. Tossatti (North Holland, Amsterdam, 1986).
7. *Statistical Physics*, ed. H. E. Stanley (North Holland, Amsterdam, 1986).
8. Y. Couder, O. Cardoso, D. Dupuis, P. Tavernier and W. Thom Europhys. Lett. 2, 437 (1986).
9. M. Rabaud, Y.Couder and N. Gerard, Phys. Rev. A 37, 935 (1988).
10. G. Zocchi, B. E. Shaw, A. Libchaber, and L. P. Kadanoff, Preprint.
11. E. Ben Jacob, R. Godbey, N. D. Goldenfeld, J. Koplik, H. Levine, T. Mueller and L. M. Sander, Phys. Rev. Lett. 55, 1315 (1985).
12. R. Lenormand and L. Zarcone, Phys. Rev. Lett. 54, 2226 (1985).
13. V. Horvath, T. Vicsek, and J. Kertész, Phys. Rev. A, 35, 2553 (1987).
14. G. Daccord, J. Nittmann and H. E. Stanley, Phys. Rev. Lett. 56, 336 (1986)
15. H. Van Danme, F. Obrecht, P.
16. J. Nittmann, G. Daccord and H. E. Stanley, Nature 314, 141 (1985).
17. J. Bataille, Revue Inst. Pétrole 23, 1349 (1968).
18. L. Paterson, J. Fluid Mech. 113, 513 (1981).
19. E. Ben Jacob, G. Deutscher, P. Garik, N. D. Goldenfeld and Y. Lereah, Phys. Rev. Lett. 57, 1903 (1986).
20. S. N. Rauseo, P. D. Barnes and J. V. Maher, Phys. Rev. A 35, 1245 (1987).
21. H. Thomé, M. Rabaud, V. Hakim and Y. Couder, "The Saffman-Taylor Instability: From the Linear to the Circular Geometry" (preprint).
22. P. G. Saffman and G. I. Taylor, Proc. Roy. Soc. London, Ser. A 245, 312 (1958).
23. G. Tryggvason and H. Aref, J. Fluid Mech. 154, 287 (1983).
24. R. Combescot, T. Dombre, V. Hakim, Y. Pomeau and A. Pumir, Phys. Rev. Lett. 56, 2036 (1986).
25. B. Shraiman, Phys. Rev. Lett. 56, 2028 (1986).
26. D. C. Hong and J. Langer, Phys. Rev. Lett. 56, 2032 (1986).
27. D. Bensimon and P. Pelcé, Phys. Rev. A 33, 4477 (1986).
28. L. M. Sander, P. Ramanlal, and E. Ben Jacob, Phys. Rev. A 32, 3160 (1985).
29. A. Rényi, *Probability theory* (North Holland, Amsterdam, 1970).
30. H. Froechling, J. P. Crutchfield, D. Farmer, N. H. Packard and R. Shaw, Physica D 3, 605 (1981).
31. F. Argoul, A. Arnéodo, G. Grasseau and H. L. Swinney, "Self Similarity of Diffusion Limited Aggregates and Electrodeposition Clusters" (preprint).
32. T. C. Halsey, M. H. Jensen, L. P. Kadanoff, I. Procaccia and B. I. Shraiman, Phys. Rev. A 33, 1141 (1986).
33. A. D. Mc Ewan and G. I. Taylor, J. Fluid Mech. 26, 1 (1966).
34. R. C. Ball and R. M. Brady, J. Phys. A 18, 809 (1985).
35. M. Rabaud, V. Hakim, H. Thomé and Y. Couder, "Directional Viscous Fingering" (in preparation).

CONSTRUCTION OF A RADIAL HELE-SHAW CELL

T. VICSEK

Institute for Technical Physics, HAS,
Budapest P.O. Box 76, 1325 Hungary

Here a few details are given for those who are interested in the construction of a Hele-Shaw cell. To build a versatile Hele-Shaw cell is not particularly troublesome but the application of a few tricks makes the experiments easier to carry out. A schematic picture of a possible arrangement is shown in Fig. 1. The related information is given below.

(a) The sizes of the upper and lower plates are 27×27 cm and 34×34 cm, respectively. They are made of good quality glass of width 5 mm.

(b) Air is injected into a viscous fluid through a hole of radius 3 mm drilled at the center of the upper place.

(c) In order to prevent the viscous fluid from flowing out, walls are attached to the plates in a manner shown in Fig. B.1.

(d) The distance between the plates is regulated by inserting between them thin metallic plates (used for checking spark plugs).

(e) At large pressures one needs to clamp the two plates together either by using screws or a heavy frame.

The above basic arrangement can be modified in a number of ways, and correspondingly, many types of viscous fingering patterns can be observed. Inserting a third glass plate (in between the two original ones) with a mesh etched on its surface is able to study the effects of anisotropy. If small balls are spread randomly on the surface of the third plate the structure of the interface becomes a random fractal, similar to the geometry of diffusion-limited aggregates.

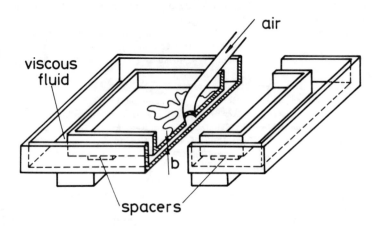

Fig. 1: Schematic picture of a radial Hele-Shaw cell. Its crossection is also indicated. The distance between the plates is denoted by b.

GROWTH AND VISCOUS FINGERS ON PERCOLATING POROUS MEDIA

A. AHARONY,[1,2] U. OXAAL,[2] M. MURAT,[1] Y. MEIR,[1]
F. BOGER,[2] J. FEDER[2] and T. JØSSANG[2]

[1] *School of Physics and Astronomy*
Beverly and Raymond Sackler Faculty of Exact Sciences
Tel Aviv University, Tel Aviv 69978, ISRAEL

[2] *Institute of Physics, University of Oslo*
Blindern, 0316 Oslo, NORWAY

ABSTRACT. Growth models and viscous fingers are studied on simple percolation models of porous media. Studies include computer and real experiments on square lattice models, at the percolation threshold, and exact calculations of deterministic flow on non-random fractal models. Crossover away from the threshold is also analyzed, using both computer simulations and scaling theory.

1. Introduction

Many growing fractal aggregates, like viscous fingers (VF), dielectric breakdown (DB), or diffusion limited aggregates (DLA), exhibit very similar branching structures (for reviews, see ref. 1), with very close fractal dimensionalities. Much of the existing research concentrates on real and computer experiments of growth on planar *non-random* substrate. In contrast, the practical application of these ideas aims to explain *fluid displacement in porous media*, e.g., using water to push oil out of a rock. Clearly, such a porous medium is *non homogeneous*. The aim of this present presentation is to discuss the effects of *randomness* on the growth.

1.1. PERCOLATION

The simplest model of a porous medium is the *percolation* model.[2,3] Sites on a lattice are randomly chosen to be open or blocked, with probabilities p and $(1-p)$. Below the threshold concentration p_c there exist only finite connected clusters, and there can be no flow through the lattice. Very far above p_c there exist only few blocked sites and the system looks homogeneous. However, *at* p_c the spanning connected cluster is *fractal*,[4] and its mass within length scale L behaves as

$$M(L) \sim L^D. \tag{1}$$

Most of the sites on this cluster belong to "dead ends," and the fluid in them cannot move. Flow is restricted only to the *backbone* of the cluster, which is also a fractal with a lower fractal dimensionality D_B.

1.2. OUTLINE

In the present paper we review work on viscous flow and growth on fractal structures, like the percolation cluster. In Secs. 2 and 3 we describe computer and real viscous fingers experiments on a model planar square percolating cluster, at the threshold $p_c = 0.593$. These fractal clusters cannot be treated analytically, because they are *random*. To obtain an *analytic* feel for flow on fractals, Sec. 4 describes *exact* calculations on *model fractals*.

All the above applies *at* p_c, when fluid is *completely* forbidden from flow in the "blocked" sites. In practice, one needs to know how the behavior *crosses over* to that on a *homogeneous* substrate, as either p grows from p_c to 1 or the conductivity of the "blocked" sites grows from 0 to 1. This *crossover* is discussed in Sec. 5.

2. Computer Experiments

We constructed specific realizations of a site diluted square lattice, of various sizes, at $p_c = 0.593$. On each of these, we simulated DLA's by sending random walkers from the outside and sticking them to the aggregate on touching.[5] We also used the stochastic DB rule,[6] solving Laplace's equation $\nabla^2 P = 0$ on the unoccupied backbone sites [i.e., writing $P(\vec{r}) = \sum_\delta P(\vec{r}+\vec{\delta})/4$, where sites $(\vec{r}+\vec{\delta})$ are neighbors of site $\vec{r}$ and requiring that $P(\vec{r}+\vec{\delta}) = P(\vec{r})$ if site $(\vec{r}+\vec{\delta})$ is blocked to ensure that $\vec{\nabla}P = 0$ on the boundary] and then adding the perimeter site i to the aggregate with probability $p_i \sim |(\vec{\nabla}P)_i|^\eta$. For $\eta = 1$ we find[7] both types of rules to yield practically the same fractal dimensionalities, $D \simeq 1.3$. However, the amplitude [mass $= A(length)^D$] of the DB was larger by a factor 1.3 from that of the DLA.

For $\eta = 0$ (equal growth probabilities on all perimeter sites with finite $\vec{\nabla}P$), we find[8] $D \simeq 1.5$. Unlike Eden model growth in uniform space, where $D = d$ (the Euclidean dimension), we have $D(\eta = 0) < D_B$. Most of the backbone mass remains "trapped." However, the $\eta = 0$ rule does allow the extraction of much more oil than the $\eta = 1$ rule.

3. Viscous Fingers Experiments

We constructed an epoxy cast porous model, identical to one of the spanning clusters (of size 147×147) used for the simulations. The model was filled with glycerol, and air was pushed through the central "seed" site.[8] At *fast* flow rates (30 pores/sec), the resulting fingers practically overlapped those from the computer experiments with $\eta = 1$, i.e., had $D \simeq 1.3$. At *slow* flow rates (0.33 pores/sec) the fingers reproduced the computer simulations with $\eta = 0$, having $D \simeq 1.5$. Slow flow is thus dominated by *local capillary effects*. Had the pores been *random* in size, one might end up with *invasion percolation*,[9] having yet another fractal dimensionality. Since our model had more or less equal pores, the capillary effects were the same and we end up with the $\eta = 0$ description.

4. Exact Results on Model Fractals

The simplest fractal model that imitates the percolation backbone was proposed by Mandelbrot and Given.[10] As shown in Fig. 1, each bond on this structure is replaced iteratively by L_1 smaller singly connected bonds, in series with a "blob" of L_2 and L_3 bonds in parallel.

If b is the length rescale factor, then the singly connected bonds scale as $L_1 = b^{1/\nu}$, the minimal path scales as $L_1 + L_2 = b^{d_{min}}$ and $L_1 + L_2 + L_3 = b^{D_B}$. We now solve a deterministic flow on this structure. Within one singly connected chain, the interface between the invading non-viscous fluid and the pushed viscous one obeys the equation $(d\ell_i/dt) = -K(\Delta P/\ell_i)^\eta$, where ΔP is the pressure difference

Fig. 1: Model fractal.

between the (advancing) non-viscous fluid and the end of the chain, a distance ℓ_i apart. Clearly, the singly connected chain, L_1, will be fully invaded. On the other hand, in the two parallel chains one has $(d\ell_2/d\ell_3) = (\ell_3/\ell_2)^\eta$, hence $\ell_3^{\eta+1} - \ell_2^{\eta+1} = L_3^{\eta+1} - L_2^{\eta+1}$. For $\eta > -1$, this implies that when $\ell_2 = 0$ the mass of the invading fluid is

$$ M = L_1 + L_2 + L_3 - \left(L_3^{\eta+1} - L_2^{\eta+1} \right)^{1/(\eta+1)}. \qquad (2) $$

For $\eta \leq -1$, $\ell_2 = 0$ implies $\ell_3 = 0$, hence $M = L_1 + L_2 + L_3$, and the invader fills the backbone.

The structure in Fig. 1 is not self similar on length scales which are not integer powers of b. However, when the number of iterations increases, the structure becomes finely divided and the growing aggregate becomes fractal. Using (2) in that limit, writing $M = b^{D(\eta)}$ and taking $b = 1 + \epsilon$ yield[11]

$$ D(\eta) = D_B - \left[(D_B - d_{\min})^{\eta+1} - (d_{\min} - 1/\nu)^{\eta+1} \right]^{1/(\eta+1)}, \qquad (3) $$

for $\eta > -1$. For two-dimensional percolation, this yields $D(1) \simeq 1.32$ and $D(0) = 2d_{\min} - 1/\eta \simeq 1.51$ in excellent agreement with both computer and real experiments.

For finite L_i's and a finite number of iterations, the stochastic DLA model would yield different dimensions than those quoted above for the deterministic flow. However, in the self similar, finely divided limit, the two models turn out to give exactly the same structures.[12]

5. Crossover to Homogeneous Behavior

Consider now the percolation structure, in which a fraction p_c of the bonds have a high conductivity (equal to 1), while the others have a low conductivity, equal to R. Flow can now leave the spanning infinite cluster, and move to other clusters. Flow is also possible for $p < p_c$. Two-dimensional simulations of both DLA and DB, with $\eta = 1$, showed a crossover from the fractal behavior described above, with $D(1) \simeq 1.3$, for short length scales, to that on uniform substrate, with $D \simeq 1.7$. The crossover length scales as $L_R \sim R^{-a}$, with $a \simeq 0.25$.[13] This result can be interpreted using a scaling theory, in which

$$ M(L, R) = L^{D_2} f(L/L_R), \qquad (4) $$

with $D_2 \simeq 1.3$, and with

$$F(x) \sim \begin{cases} \text{const,} & x \ll 1 \\ x^{D_1 - D_2}, & x \gg 1 \end{cases}, \tag{5}$$

where $D_1 \simeq 1.7$. This also implies that when $L \gg L_R$, one expects

$$M(L) \sim R^{a(D_1 - D_2)} L^{D_1}. \tag{6}$$

Similar scaling applies when $R = 0$ but $p > p_c$.[13]

6. Conclusions

At the percolation threshold, we see that the restricted geometry of the backbone of the spanning cluster effectively determines that of the growing aggregates. The differences between the different values of η, related to the speed of the flow, are then understood: larger η's prefer non-branching fast-running flows, while small η's tend to fill more of the backbone.

Our exact calculations on non-random fractals allow analytic comparisons between processes. For example, they proved the universality of DLA and DB, and explained deviations for finite sizes. This tool should certainly be applied to more growth problems.

In real situations, one is never exactly at the percolation threshold. Our crossover description, Sec. 5, is only the beginning of a systematic study of scenarios leading away from our idealized picture. An obvious necessary generalization is to include a distribution of conductivities of the different pores. This remains to be studied in the future.

Acknowledgements: Research at Tel Aviv University was supported by grants from the U.S.–Israel Binational Science Foundation and the Israel Academy of Sciences. Work in Oslo was supported by VISTA, a research cooperation between the Norwegian Academy of Science and Letters and STATOIL and NAVF.

1. H. E. Stanley and N. Ostrowsky, eds., *On Growth and Form* (Martinus Nijhoff, Dordrecht, 1986).
2. D. Stauffer, *Introduction to Percolation Theory* (Taylor and Francis, London, 1985).
3. A. Aharony, in *Directions in Condensed Matter Physics*, eds G. Grinstein and G. Mazenko (World Scientific, Singapore, 1986), pp. 1-50.
4. B. B. Mandelbrot, *The Fractal Geometry of Nature* (Freemen, San Francisco, 1982).
5. T. A. Witten and L. M. Sander, Phys. Rev. Lett. **47**, 1400 (1975).
6. L. Niemeijer, L. Pietronero and H. J. Wiesmann, Phys. Rev. Lett. **52**, 1033 (1984).
7. M. Murat and A. Aharony, Phys. Rev. Lett. **57**, 1875 (1986).
8. U. Oxaal, M. Murat, F. Boger, A. Aharony, J. Feder and T. Jøssang, Nature **329**, 32 (1987).
9. D. Wilkinson and J. F. Willemsen, J. Phys. A **16**, 3365 (1983).
10. B. B. Mandelbrot and J. Given, Phys. Rev. Lett. **52**, 1853 (1984).
11. Y. Meir and A. Aharony (unpublished).
12. E. Arian and A. Aharony (unpublished).
13. P. Meakin, M. Murat, A. Aharony, J. Feder and T. Jøssang (unpublished).

STRUCTURE OF MISCIBLE AND IMMISCIBLE DISPLACEMENT FRONTS IN POUROUS MEDIA

T. JØSSANG, U. OXAAL, J. FEDER, K. J. MÅLØY and F. BOGER

Department of Physics
University of Oslo
Box 1048, 0316 Oslo 3, NORWAY

ABSTRACT. We demonstrate displacement processes of fluids by other fluids in porous and non-porous Hele-Shaw cells. We also demonstrate a new result[9] showing the fractal nature of the dispersion front of a tracer when it is abruptly added at the injection site in an experiment where a viscous fluid is pumped into a two dimensional porous medium. The concentration contours of the tracer are self-affine fractal curves with a (local) fractal dimension $D \simeq 1.42 \pm 0.05$. The dispersion front may, on the average, be described by the hydrodynamic dispersion with a longitudinal dispersion coefficient $D_{\parallel} = U d_{\parallel}$, where U is the average flow velocity and $d_{\parallel}$ is a characteristic length of the order of a pore diameter. This result is valid for dispersion at high Péclet numbers $Pe = U d / D_m$, where D_m is the molecular diffusion coefficient of the dye.

Displacement processes of fluids by other fluids are demonstrated in porous and non-porous Hele-Shaw cells. A Hele-Shaw cell consists of two parallel transparent plates separated a distance b. The equation for the flow velocity $\mathbf{U}$, derived from the Navier-Stokes equations governing flow in the Hele-Shaw cell is $\mathbf{U} = -k/\mu(p + \rho g z)$ where p is the pressure, ρ the fluid density, g the component of the acceleration of gravity along the z-coordinate of the cell, k the permeability and μ the fluid viscosity. For the Hele-Shaw cell $k = b^2/12$. A more complete discussion of this problem is given in Ref. 1. Note that $\mathbf{U}$ is the average velocity over the thickness of the cell, and that $g \approx 0$ for a horizontal cell. When $g \approx 0$ the equation of continuity for incompressible fluids gives $\nabla \mathbf{U} = -\nabla^2 \cdot p = 0$. This is the Laplace equation, and it is characteristic of many potential problems.

We will demonstrate the situation when an invading fluid (index i) displaces a defending fluid (index d) starting at the center of a radial Hele-Shaw cell. Two cases arise: The viscosity μ_d of the defending fluid may be larger than or smaller than the viscosity μ_i of the invading fluid.

First we demonstrate the well-known situation where $\mu_i \gg \mu_d$, giving a stable displacement front. In the other case of $\mu_i \ll \mu_d$ we demonstrate the viscous fingering instability (VF) of Fig. 1. Viscous fingers form at large displacement rates, i.e., when long range viscous forces dominate over local (capillary) forces. The relative strength of these forces is measured by the capillary number $C_a \sim \mu_d \mathbf{U}/\sigma \sim$ viscous forces/capillary forces, where σ is the interfacial tension between the two fluids. The fingering pattern shown in figure 1 is obtained at a rather high $C_a{}^1$.

The flow of a fluid in a *porous* medium is controlled by the same equations as the flow in the Hele-Shaw cell, except that k now is the actual mean permeability of the medium. However, experiments show[2] that the fingering dynamics is strikingly different. In Fig. 2 we show the results obtained for $\mu_d \gg \mu_i$ at moderately high capillary numbers $C_a \sim 0.05$. We show viscous fingering in two 2-dimensional porous models, consisting of one regular (Fig. 2a)[1] and one random (Fig. 2b)[2] single layer array of blockings (to the fluid flow) between the two glass plates of the Hele-Shaw cell.

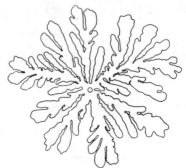

Fig. 1: Radial viscous fingering in a circular Hele-Shaw cell.[1] Air displacing glycerol at $C_a = 0.1$.

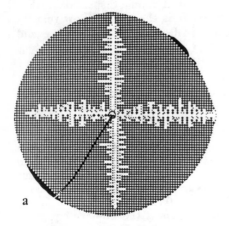

a   b

Fig. 2: Viscous fingering in 2-D model porous media confined between the two plates of a Hele-Shaw cell. (a) Air displacing glycerol at $C_a = 0.05$ in a medium consisting of a regular square array of cylinders 1 mm in diameter. (b) Air displacing liquid epoxy at $C_a = 0.04$ in a medium consisting of a random monolayer array of 1.6 mm glass spheres.

The fractal nature of *immiscible* displacement in a porous medium has been studied experimentally[2-5] and is found to be well described by the simulations in many situations[4,6].

The *dynamics* of the displacement process in random models has recently been studied both experimentally[2,7,8] and by simulations[7,8] and a good agreement is generally found.

In the random porous 2-dimensional model we also demonstrate a recent result[9] showing the *fractal structure* of the displacement front between a fluid and the same—but colored fluid.

The two fluid phases are perfectly miscible and they have the same density and viscosity. The tracer (colored) molecules disperse because of molecular diffusion and convection as shown in Fig. 3. An example of a digitized picture obtained using logarithmic response is shown in figure 3. The color scale used to encode the tracer particle concentration is shown in the insert and is chosen to emphasize the structure of the dispersion front. The figure illustrates the strong fluctuations of the fractal dispersion front caused by geometrical dispersion, which generates concentration 'spikes' in the radial direction. In effect the structure results because the transverse dispersion in the tangential direction is not sufficient to even out the large longitudinal fluctuations in the velocity field for the flow through the porous medium.

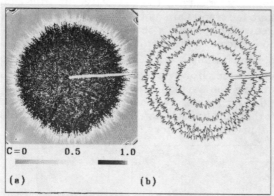

Fig. 3: The dispersion obtained by injecting glycerol colored by Negrosine into glycerol of the same viscosity and density in a two-dimensional porous model. (a) The observed transmitted light intensity color coded as shown in the insert. Time $t = 160s$. (b) The time development of the fractal contour where the concentration is $1/2$ the injected tracer concentration. The four contours correspond to increasing times: $t = 36s$; $t = 77s$; $t = 122s$; $t = 160s$.

The results are described quite well by the solution of the convective diffusion equation which can be written on the following form[10-15] $\partial C/\partial t = \nabla \cdot (\mathbf{D} \cdot \nabla C - \mathbf{U}C)$. Here $C(\mathbf{r}, t)$ is the tracer concentration as a function of position $\mathbf{r}$ and time t. The dispersion tensor $\mathbf{D}(\mathbf{U})$ depends in general on the imposed hydrodynamic (average) flow velocity $\mathbf{U}$. For homogeneous porous media this dispersion tensor is characterized by only two independent components: the longitudinal, $D_{\parallel}$, and the transverse, $D_{\perp}$, dispersivities. For poorly connected porous media one observes *anomalous* diffusion.

Dispersion of tracers in a stationary fluid ($\mathbf{U} = \mathbf{0}$) is due to ordinary molecular diffusion and $D_{\parallel} = D_{\perp} = D_m$, where D_m is the diffusion constant of tracers in the fluid. The relative importance of convective to diffusive transport is charactered by the Péclet number $Pe = Ud/D_m$, where d is a typical pore size. The question of the dependence of $D_{\parallel}$ on Pe in a (homogenous) porous medium is a delicate problem. For large Péclet numbers the position of a tracer (relative to the position expected from the average flow) can be described by a random walk process with an effective diffusion constant given by the Einstein relation: $D_{\parallel} = \frac{1}{2}d^2/\tau$, where the average time between 'steps' of length d is $\tau \sim d/U$. Therefore one finds[10] $D_{\parallel} \sim Ud$ i.e., the longitudinal dispersion is proportional to the flow velocity U for $Ud \gg D_m$. When stagnation effects are included the expression is modified by logarithmic terms $D_{\perp} \sim Ud\ln(U\ell/D_m)$, where ℓ is a length scale for diffusion out of stagnant regions[10,12,14,15]. At low Péclet numbers molecular diffusion across stream lines leads to *Taylor dispersion*[18] with a longitudinal dispersion given by $D_{\perp} = D_m + a^2U^2/48D_m$. This result was derived for flow in capillaries of radius a. Taylor dispersion dominates if the transverse mixing is large enough, i.e., if the time needed for diffusion across the capillary a^2/D_m is short enough compared with the convective time d/U.

The simplest geometry for solving the convective diffusion equation introduced above, is the case where the tracers are added as a step-function in a linear geometry at $x = 0$. The concentration profile is then given by $C(x,t) = 1/2(1 - erf[(x - R_0)/\lambda])$. Here $erf(x)$ is the error function, $R_0 = Ut$ is the position of the front and the width of the dispersion front is $\lambda = 2\sqrt{D_{\parallel}t}$. The solution for the radial geometry used here is more complicated. However, for the present purposes sufficient accuracy is obtained by estimating the average concentration as $C(r,t) = \sum c(r_i,t)/2\pi r_i$, where the sum extends over the observed concentrations, $c(r_i)$, in the pixels labeled by i. Figure 4 shows that this 'erf solution' (full curves) approximates the observed average concentration profile satisfactory.

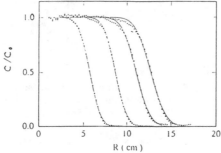

Fig. 4: The average concentration $C(r,t)/C_0$ as a function of r for different time stages t. The fitted functions $C/C_0 = \frac{1}{2}(1 - erf[(r - R_0)/\lambda])$ are shown as solid lines in the figure. a: $t = 36s$, $R_0 = 5.71 \pm 0.01$ cm, $\lambda = 1.44 \pm 0.02$ cm. b: $t = 77s$, $R_0 = 8.67 \pm 0.01$ cm, $\lambda = 1.55 \pm 0.02$ cm. c: $t = 122s$, $R_0 = 12.72 \pm 0.02$ cm, $\lambda = 1.95 \pm 0.02$ cm. d: $t = 160s$, $R_0 = 11.06 \pm 0.02$ cm, $\lambda = 1.83 \pm 0.04$ cm.

The *new feature* presented here appears when we analyze the dispersion front itself: Contours of constant concentration are *self-affine fractal curves*[19,1] confined by the width of the dispersion front. This fractal structure does *not* arise in dispersion of fluid flow in Hele-Shaw cells or capillaries, it is characteristic of dispersion in porous media. We also stress that our results have been obtained in homogenous models. In figure 5 we show the result of using the 'box-counting' method to estimate the fractal dimension of the front. The number of boxes $N(\delta)$ needed to cover the front with a box size δ (in pixels) is[19,1] $N(\delta) \sim \delta^{-D}$. We find that the dispersion front is fractal with $D = 1.42 \pm 0.05$. It is important to note that even when this structure is fractal it is well described by the average theory of the convective diffusion equation. In figure 5 we also see that for $\delta > \lambda$ there is a crossover to a lower fractal dimension. In fact this is to be expected since the front on the average is a circle (with $D = 1$) and therefore is a self-affine[19,1] curve. The fractal dimension is different for dispersion fronts in fractally porous media, such as dispersion in models that are near the percolation threshold.[8,16]

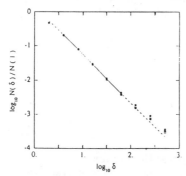

Fig. 5: The number of 'boxes' $N(\delta)$ needed to cover the contours where the tracer concentration equals $\frac{1}{2}$ the injected concentration as a function of the box size δ given in pixels. The line is a fit of the relation $N(\delta) = aN(1)\delta^D$, where $N(1)$ is the total number of pixels in the contour, a is parameter of the fit related to the crossover at small length scales. The fractal dimension is $D = 1.42 \pm 0.01$ is estimated from the fit in the region drawn with a full curve.

Acknowledgements: This work was supported by VISTA, a research cooperation between the Norwegian Academy of Science and Letters and by NAVF.

1. J. Feder, *Fractals*, (Plenum, New York), 1988.

2. K. J. Måløy, J. Feder, and T. Jøssang, Phys. Rev. Lett. 55, 2688 (1985a).

3. R. Lenormand and C. Zarcone, Phys, Rev. Lett. 54, 2226 (1985)

4. J. D. Chen and D Wilkinson, Phys Rev. Lett. 55, 1892 (1985).

5. T. M. Shaw, Phys. Rev. Lett. 59, 1671 (1987).

6. P. Meakin, in *Phase Transitions and Critical Phenomena* 12, editors C. Domb and J. L. Lebowitz, (Academic Press, New York), p. 336-489, (1987).

7. K. J. Måløy, F. Boger, J. Feder, T. Jøssang, and P. Meakin, Phys. Rev. A 36, 318 (1987).

8. U. Oxaal, M. Murat, F. Boger, A. Aharony, J. Feder, and T. Jøssang, Nature 329, 32 (1987).

9. K. J. Måløy, J. Feder, F. Boger, and T. Jøssang, submitted to Phys. Rev. Lett. (1988).

10. P. G. Saffman, J. Fluid Mech. 6, 321 and J. Fluid Mech. 7, 194 (1959).

11. D. F. Koch and J. F. Brady, J. Fluid Mech. 154, 399 (1985).

12. C. Baudet, E. Charlaix, E. Clment, E. Guyon, F. P. Hulin, and Leroy, in *Scaling Phenomena In Disordered Systems*, eds. R. Pynn and T. Riste, (Plenum, 1985), p. 399.

13. L. de Arcangelis, J. Koplik, S. Redner, and D. Wilkinson, Phys. Rev. B 31, 4725 (1986).

14. E. Charlaix, J. P. Hulin, and T. J. Plona, Phys. Fluids 30, 1690 (1987).

15. E. Charlaix, Thèse de doctorat de l'Université Paris 6 (1987).

16. E. Charlaix, J. P. Hulin, and T. J. Plona, Phys. Fluids 30, 1690 (1987).

17. Y. Gefen, A. Aharony, and S. Alexander, Phys. Rev. Lett. 50, 77 (1983).

18. G. I. Taylor, Pros. Roy. Soc. A 219, 186 (1953).

19. B. B. Mandelbrot, *The Fractal Geometry of Nature* (W. H. Freeman, New York), 1982.

DYNAMICS OF INVASION PERCOLATION

L. FURUBERG,[1] J. FEDER,[1] A. AHARONY,[1,2] and T. JØSSANG[1]

[1] *Institute of Physics, University of Oslo*
Box 1048, Blindern, 0316 Oslo 3, NORWAY

[2] *School of Physics and Astronomy*
Beverly and Raymond Sackler Faculty of Exact Sciences
Tel Aviv University, Tel Aviv 69978, ISRAEL

Invasion percolation[1] models the process of one fluid slowly displacing another in a porous medium. The displacement is governed by local capillary forces of the pores. Experiments on slow fluid displacements[2] show that the fluid-fluid front moves by invading local areas in bursts.

In computer simulations[1] of this process the porous medium is represented by a lattice where every lattice site is given a random number representing pore size. The model of the porous medium is initially filled with the displaced fluid. The invading fluid is injected at a set of source sites and proceeds into the medium by invading the smallest pore neighbouring the fluid front at every time step. The process is stopped when the invading fluid reaches the sink sites. We consider the displaced fluid to be incompressible so that if some region of the lattice is surrounded by the invading fluid, then no more growth can take place into that region. The resulting invader structure has a fractal dimension $D = 1.82$.[1]

We measure the space and time correlations of the displacement by considering the distribution $N(r, t)$, giving the probability that a site a distance r from a reference site is invaded a time t later than that (reference) site.

Using 15 simulations of size 300×600 (source and sink on opposite short sides) and 7 of size 500×500 (source at center, sink on edges), we found that

$$N(r, t) = r^{-x} f(r^z / t), \tag{1}$$

with $x = 1, z = D$. The dynamic scaling function $f(u)$ behaves as $u^{1.4} (u \ll 1)$ and $u^{-0.6} (u \gg 1)$, peaking at $u \sim 1$, i.e. $r_t \sim t^{1/D}$. The result $x = 1$ follows from $\int_0^\infty N(r, t) dr = 1$, while $z = D$ results from $\int_0^\infty N(r, t) r^{-(d-1)} dt = G(r) \sim r^{D-d}$, where $G(r)$ is the correlation function of the resulting structure.[3]

1. D. Wilkinson and J. F. Willemsen, J. Phys. A **16**, 3365 (1983).

2. K. J. Måløy, J. Feder and T. Jøssang, unpublished observations, 1985.

3. A. Aharony, in *Directions of Condensed Matter Physics*, eds. G. Grinstein and G. Mazenko (World Scientific Pub., Singapore, 1986), pp. 1-50.

DIRECTIONAL SOLIDIFICATION OF LIQUID CRYSTALS

J. BECHHOEFER,* A. SIMON,* A. LIBCHABER,* and P. OSWALD†

*The James Franck and Enrico Fermi Institutes
The University of Chicago
Chicago, IL 60637

†Laboratoire de Physique des Solides
Univ. de Paris—Sud
91405 Orsay, France

ABSTRACT. We describe an experiment on directional solidification of liquid crystals, in which we test predictions of the Mullins-Sekerka linear stability theory and begin to explore nonlinear, secondary instabilities.

1. Introduction

In recent years, physicists have begun to study the problem of pattern formation in non-equilibrium systems. Generally, such work is rooted in the physics of a particular situation—for example, convection, flame fronts, or electrodeposition—but the goal is to find pattern-formation mechanisms that are universal, or generic, in that they apply to many different situations. In this spirit, our discussion of directional solidification begins with a summary of the physics underlying the basic equations of motion and ends with a sketch of secondary instabilities.

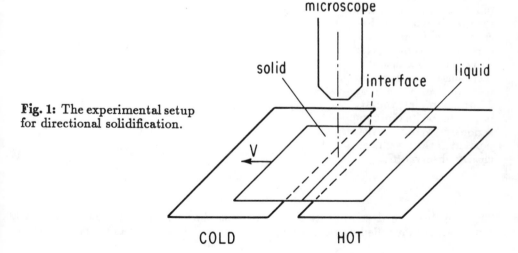

Fig. 1: The experimental setup for directional solidification.

Fig. 2: Phase diagram for a first-order transition in the presence of a small amount of impurity. N refers to the nematic phase, I the isotropic phase, and $N+I$ denotes the coexistence region.

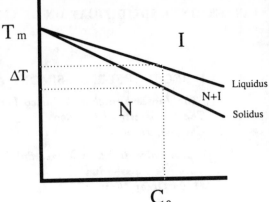

Fig. 3: Concentration profile of the zero-order, flat-interface solution to the equations of motion. The linear stability analysis of Mullins and Sekerka is a perturbation expansion about this solution.

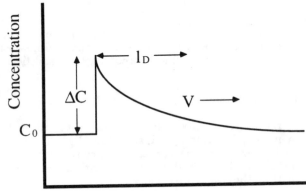

Fig. 4: Theoretical stability curve fit to experimental data points. Because we have not measured two parameters (k and d_0) and because the meniscus in the third dimension is important for 25 μm. samples, the fit is not to be taken too seriously. What is important is that we have clearly observed $G_{\max}$.

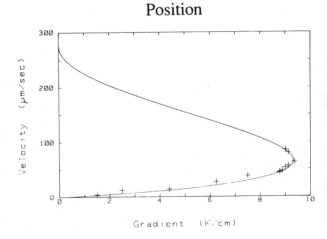

The directional solidification apparatus is shown in Fig. 1.[1] There are two temperature-controlled ovens, separated by a gap of 4 mm. A sample consisting of

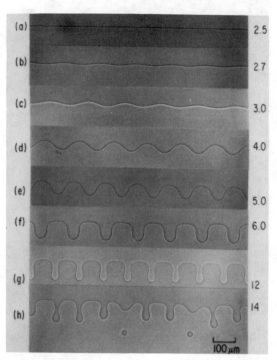

Fig. 5: The nematic-isotropic interface, showing steady-state shapes. The velocities in μm./sec. are marked by each picture. The nematic phase is below the interface, which is moving upwards into the isotropic phase.

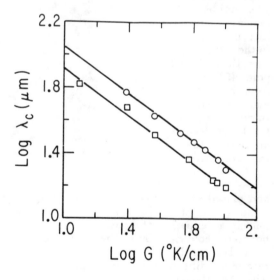

Fig. 6: Wavelength λ vs. gradient G at $v = v_c$, for $G \ll G_{max}$. The two curves are for differing impurity concentrations.

two 1 mm. thick glass plates with a 25 μm. spacer is filled with a liquid crystal

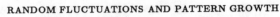

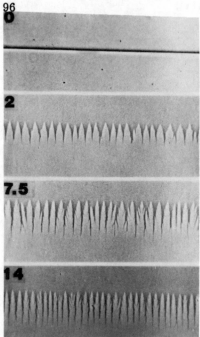

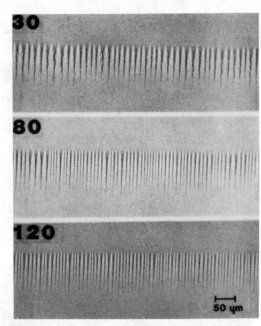

Fig. 7: Fingers of smectic B phase growing into the smectic A phase. The numbers indicate velocities, in μm./sec. In the last picture, the molecules point 45° away from the thermal gradient. In the other photos, the molecules are normal to the gradient.

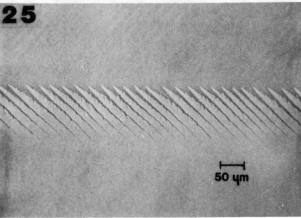

and placed atop the copper blocks comprising the ovens. The glass plates impose a linear temperature gradient upon the sample and an interface from in the gap. During an experimental run, the sample is pushed at constant velocity v from the hot to the cold side. Since the ovens are fixed in space, the front freezes at a velocity $-v$ to keep at the same average temperature. We thus have a way of controlling the rate of solidification and of putting ourselves in the reference frame moving along with the interface.

Our reasons for studying liquid-crystal transitions will emerge from the discussion below. For now, we note that the liquid crystal phases, which are intermediate

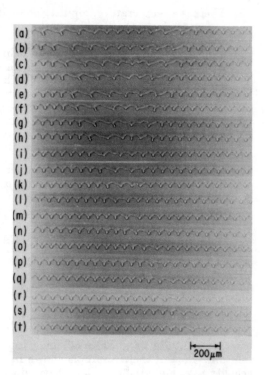

Fig. 8: Two solitons collide, one emerges. $v = 40\mu\text{m/sec}$, $G = 18$ K/cm.

between solid and liquid phases, have proportionately smaller latent heats of transition. For the transitions we study, the nematic-smectic A transition has a typical latent heat of less than 10^{-3} kcal/mol, the nematic-isotropic one of 10^{-1} kcal/mol, and the smectic B-smectic A one of 0.5 kcal/mol. By contrast, solid-liquid transitions range from 0.5 to 7 kcal/mol. The liquid crystals we used were 8CB and 4O8,[2] which were chosen for their convenient transition temperatures.

2. The Physics of Directional Solidification

The liquid crystal samples are doped with an impurity (e.g., 1% hexachloroethane). The instabilities of the moving interface we discuss below are due directly to the impurities. In a pure sample (operationally, one where the concentration of impurities is less than about 0.1%), we observe that the interface is always a straight line. The release of latent heat could lead to an unstable interface, but the latent heat is small and what little of it there is gets conducted away by the glass plates. The equations of motion are determined by three physical effects.[3,4]

(1) Impurities diffuse. In a solid-liquid sample, the ratio of the chemical diffusion constant of the low-temperature phase to that in the high-temperature phase is usually less than 10^{-5}. In the liquid-crystal experiments described here, it is 1/3 to 1/2. If the interface is moving at a velocity v, we can define a diffusion length $\ell_D = D/v$, where D is the diffusion constant of the high-temperature phase and v is the solidification velocity.

(2) There is a temperature-impurity phase diagram, illustrated in Fig. 2. For small concentrations, the liquidus and solidus curves separating the liquid and the solid phase regions from the coexistence region are straight lines. The diagram is generic to all first-order phase transitions with impurities and applies equally well to the liquid-crystal transitions considered here. One important difference, though, is that the ratio of slopes of liquidus and solidus lines, k, tends to 1 as the latent heat of transition vanishes. k is called the partition coefficient and is also equal to the ratio of the concentration of impurities on the high-temperature side of the interface to that on the low-temperature side. One can define a thermal length $\ell_T = \Delta T / G$, where ΔT is the temperature difference between liquidus and solidus lines and G is the thermal gradient.

(3) Curvature alters the interface temperature. The Gibbs-Thomson law gives the change in the melting temperature of a curved interface $\delta T = T_m d_0 \kappa$, where T_m is the transition temperature of the pure material, $d_0 = \gamma / L$ is the bare capillary length, γ is surface tension, L the latent heat, and κ the interface curvature. In the presence of impurities, the effective capillary length is $\ell_C = d_0 T_m / \Delta T$. d_0 is typically 1-2Å, but $T_m / \Delta T \approx 10^3$, so that ℓ_C is 0.1 to 1. μm.

Given three length scales ℓ_D, ℓ_T, and ℓ_C, we can define two control parameters, the dimensionless velocity $v \equiv \ell_C / \ell_D$ and temperature gradient $G \equiv \ell_C / \ell_T$.

An elementary solution of the equations of motion is a moving flat interface, shown in Fig. 3. The concentration jump at the interface is $\Delta c = c_0 (1 - k)/k$. The impurities diffuse away from the interface, with an exponential falloff of decay length ℓ_D. Mullins and Sekerka[3] analyzed the linear stability of the flat interface solution. Part of their results is shown in Fig. 4. Note that there is a maximum velocity v_{max} and gradient G_{max}, beyond which the interface is always stable. v_{max} and G_{max} depend strongly on k. For a solid-liquid transition, $k \approx 0.1$ or less, leading to $v_{max} \approx 10^2$ cm./sec. and $G_{max} \approx 10^7$ K/cm., at least. These are beyond the reach of a controlled directional-solidification experiment. For the nematic-isotropic interface, $k \approx 0.9$ and we have $v_{max} \approx 500 \mu m/sec$, $G_{max} \approx 10$ K/cm, which are more easily accessible experimentally. Physically, we can understand the shape of the stability curve in Fig. 4 by noting that in region 1, $\ell_D \sim \ell_T$, while $\ell_C \to 0$. The destabilizing diffusion term is checked by the thermal gradient, and surface tension is unimportant. In region 2, $\ell_D \sim \ell_C$, while $\ell_T \to \infty$. Here, diffusion is stabilized by surface tension, and the gradient is unimportant. Region 3 is at the crossover between stabilizing mechanisms, where $\ell_T \sim \ell_C$. The wavelength of the instability is about $\sqrt{\ell_D \ell_C} \approx 50$ to 150 μm.

3. Experimental Results

We have focused on three cases: the smectic A-nematic interface, the nematic-isotropic interface, and the smectic B-smectic A interface. In the isotropic phase, the molecules have no positional or orientational order. In the nematic phase, the molecules point in the same direction but have no positional order. In the smectic A phase, the molecules are further constrained to lie in layers but there is no positional order within each layer. In the smectic B phase, the molecules have long-range hexagonal ordering within each layer. The first two transitions were explored using 8CB; for the third, we used 4O8.

3.1. SMECTIC A-NEMATIC

In this case,[5] the transition is very nearly second order. Both v_{max} and G_{max} are so small that we always observe a flat interface. This case reminds us that the Mullins-Sekerka instability is associated with first-order transitions. For example, one can show that for small latent heat L, $v_{max} \sim L$.[6]

3.2. NEMATIC-ISOTROPIC

This is the case[5] most amenable to quantitative study. We work with a homeotropic sample, one where the molecules in the nematic phase point perpendicular to the glass plates of the sample. We have observed G_{max}, but v_{max} remains just out of reach. We also observe that the bifurcation from flat to unstable interface is supercritical, or second order. This agrees with a calculation by Caroli et al, which predicts that for $k > 0.45$, the bifurcation is normal, for small G.[7] A mark of the supercritical bifurcation is that the amplitude grows smoothly from zero as the velocity v is increased above the threshold v_c, and there is no hysteresis (see Fig. 5). Note that the shape just above v_c is nearly sinusoidal. As v is further increased, higher harmonics couple to the disturbance, leading to a squared-off shape for the interface. In Fig. 5h, we see that the shape is so distorted that droplets of isotropic fluid break off and are left behind the moving interface. This is a first example of a secondary instability of the regular cellular patterns shown in Figs. 5(b)-5(g).

Because the Mullins-Sekerka instability comes as a normal bifurcation, we can test quantitatively predictions of linear stability theory. For example, Fig. 6 shows the wavelength of the instability λ at v_c as a function of the gradient G. Theory predicts $\lambda \sim G^{-2/3}$ for small G. We measure $\lambda \sim G^{-0.72\pm0.03}$.

Recently, D. Kessler and H. Levine[8] have suggested that the cell amplitude should go to λ as at large v $k \to 1$, which our observations confirm. (The amplitude diverges for $k \to 0$, as has been observed in experiments on solid-liquid systems.[9])

3.3. SMECTIC B-SMECTIC A

This is a case[10] intermediate between the N-I and the solid-liquid transition. We study a planar sample, where the molecules in each phase are parallel to the glass plates of the sample. Here, the bifurcation from a flat interface is subcritical and the cells have the deep grooves typical of low-k impurities. Since the latent heat is moderate, the small k is not a surprise. Figure 7 shows a sequence of steady-state pictures of the moving planar interface. Note that one occasionally observes sidebranches, as one does with a solid-liquid interface.[11,12] Figure 7a-g shows the interface when the molecules are normal to the thermal gradient, while the smectic layers are along the gradient. The last picture shows the layers inclined about 45° to the thermal gradient. Here, one sees the effect of the anisotropy due to the planar anchoring. Physically, the surface tension varies with respect to the angle between the smectic layer and the thermal gradient. The anistropy is twofold, here, while in the solid-liquid transition it is fourfold or sixfold.

4. Advantages of Liquid Crystals

We have seen that one advantage of using liquid crystals to study directional solidification is that the parameters (particularly k) allow one to explore (nearly) the whole bifurcation diagram in velocity-gradient parameter space. Another advantage is that the typical transient times are short. The liquid-crystal samples need only a few seconds to come to steady state after a change in velocity; the solid-liquid samples can require hours. Also, the quantity of impurities transported by each run is so small that it is unnecessary to let the sample diffuse back to equilibrium. One can make many runs per day on a liquid-crystal sample, but only one run a day on a sample that has an interface between solid and liquid phases.

We note in passing that we have observed many features that seem to be traceable to the liquid-crystal aspects of our samples. These include the nature of the meniscus that forms between the two plates, nucleation near the interface, the motion of topological defect lines in the nematic phase, and a kind of buckling instability of the nematic-isotropic interface that is different from the Mullins-Sekerka instability. All these effects have been or will be discussed elsewhere.[6]

5. Towards the Future

We recently have begun to investigate secondary instabilities at the nematic-isotropic interface. One example is the formation of droplets, as discussed above. Another is the appearance, at velocities slightly lower than those needed to form droplets, of a soliton defect in the cellular interface pattern.[13] Figure 8 shows two solitons colliding. The soliton moves by stretching the leading cell and compressing the trailing one. Curiously, whatever the wavelength of the leading cell, the trailing one is always compressed to the same λ. Solitons can select the cell wavelength! When solitons of length m and n collide, they (partially) annihilate, and a soliton of length $m - n$ propagates in the direction of the longer initial soliton.

To conclude, liquid-crystal interfaces — particularly the nematic-isotropic — are well suited to the study of the Mullins-Sekerka instability. We have made some quantitative tests of the linear theory, and are exploring the intricate world of secondary instabilities.

This work was partially funded under NSF DMR 87-09502. J. B. has been supported by an AT&T Bell Labs Ph.D. Fellowship.

1. J. D. Hunt, K. A. Jackson, and H. Brown, Rev. Sci. Inst. 37, 805 (1966).
2. These are abbreviations for 4-n-octylcyanobiphenyl and butyloxybenzilidene, respectively.
3. W. W. Mullins and R. F. Sekerka, J. Appl. Phys. 35, 444 (1964).
4. J. S. Langer, Rev. Mod. Phys. 52, 1 (1980).
5. P. Oswald, J. Bechhoefer, and A. Libchaber, Phys. Rev. Lett. 58, 2318 (1987).
6. J. Bechhoefer, Ph. D. thesis, in preparation.
7. B. Caroli, C. Caroli, and B. Roulet, J. Physique 43, 1767 (1982).
8. D. Kessler and H. Levine, "Cellular Solutions for Highly Non-Equilibrium Directional Solidification," preprint.
9. See, for example, S. de Cheveign'e, C. Guthman, and M. M. Lebrun, J. Physique 47, 2095 (1986).
10. J. Bechhoefer, P. Oswald, and A. Libchaber, Phys. Rev. A 37, 1691 (1988).
11. R. Trivedi, Metall. Trans. 15A, 977 (1984).
12. H. Esaka and W. Kurs, J. Crystal Growth 72, 578 (1985).
13. A. Simon, J. Bechhoefer, and A. Libchaber, "Solitons and the Eckhaus Instability in Directional Solidification," preprint.

INTERFACIAL INSTABILITIES OF CONDENSED PHASE DOMAINS IN LIPID MONOLAYERS

H. MÖHWALD,* H. D. GÖBEL,† M. FLÖRSHEIMER,†
and A. MILLER‡

*Univ. Mainz, Inst. Phys. Chem.
Welder Weg 15, D6500 Mainz, FRG

†TU München, Physics Department E22
D8046 Garching, FRG

‡Konsortium f. Electrochem. Ind.
Zielstattstr. 80, D8000 München, FRG

1. Introduction

This is an unusual lecture of experimentalists who prefer to be taught instead of to teach. Yet I hope that not only we will be profiting but that the questions raised will be stimulating beyond the scope of this specific work. Thus I will introduce into our systems and into conceptually simple analytical techniques and then describe experimental results. It turns out that, depending system parameters, the observations can be explained by different concepts. These concepts will be described only qualitatively since at present the experimental data are available also mostly in a qualitative way. Nevertheless I presume that the discussion will contribute also to a better quantitative understanding of these systems between material- and bio-science.

2. Systems and Experimental Technique

Spreading an amphiphile, i.e., a molecule with a hydrophilic and a hydrophobic moiety, at the air/water interface one can prepare monolayers with hydrophilic parts pointing towards water and hydrophobic parts towards air. These monolayers can exist in different phases and phase transitions can be induced by varying environmental parameters. In this work we will consider the pressure induced transition from a fluid to a more ordered condensed phase. Initially we had called the latter phase solid.[1] However, recent diffraction data indicate that for some systems this phase may be a conventional two-dimensional solid whereas for others there exist smectic mesophases.[2] Yet the detailed nature of the more condensed phase is of minor relevance for this work. What matters is the existence of a first order phase transition. It is detected from a break in the slope of the isotherm measuring the surface pressure on continuously varying the two-dimensional molecular density by compression of a barrier.[3] Concerning molecular diffusion the systems are truly two-dimensional, i.e., development of the ordered phase is a two-dimensional process.

    The latter development can be directly observed by fluorescence microscopy using dye probes that are less soluble in the ordered compared to the fluid phase.[4] Hence the development of the ordered phase can be followed by inspection of the

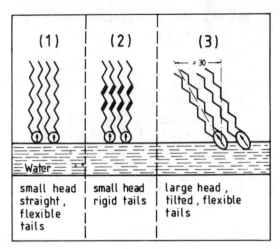

Fig. 1: Sketch of the three classes of systems considered. The circles or ellipses represent hydrophilic head groups with arrows indicating a preferential alignment of molecular dipoles at the air/water interface.

dark areas in the fluorescence micrographs and the dye distribution is measured via quantitative image analysis.

The types of molecules studied in this work are depicted in Fig. 1. Class (1) is represented by phospholipids like L-α-dimyristoylphosphatidylethanolamine (DMPE), possessing a small head group and a saturated hydrocarbon chain. For this molecule dense packing is determined by the projection of the hydrocarbon cross section ($\sim 20\text{Å}^2$/chain) on the surface plane. The tails are nearly vertically oriented and electron diffraction patterns of the ordered phase are hexagonally symmetric but exhibit only first order and rather broad spots.[5] Class (2) represents a molecule with an additional rigid group. In the example presented we had diacetylene groups introduced but have also examined systems with planar π-electron systems and their ordered phases exhibit also sharp high order electron diffraction spots with a pattern of low symmetry.[6] Class (3) represents systems where the packing is determined by the bulky but mobile head groups, an example being L-α-dipalmitoylphosphatidylcholine (DPPC). This gives rise to an especially soft ordered phase as can be observed on distorting a domain shape and looking for re-establishment of an equilibrium type structure. An additional complication in understanding these systems then results from the fact that to optimize van-der-Waals contact the tails tilt uniformly.[7] Since each molecule is associated with a dipole moment (indicated by arrows) a net in-plane dipole moment results. This gives rise to anisotropic and long range electrostatic forces adding to the dipolar forces due to partial alignment of perpendicular dipole components.

3. Experimental Results and Discussion

3.1. CLASS (1) SYSTEMS

Figure 2 shows a time sequence of fluorescence micrographs following the first observation of a domain for systems of class (1). The images were taken after quickly compressing the barrier and then stopping it. Following this the pressure exceeds that corresponding to the onset of the phase transition (π_c) and then relaxes to an equilibrium value due to formation of the new phase. In this case structure formation can be understood and modeled within the concept of constitutional supercooling[8]: Impurities expelled from the ordered phase enrich near the phase

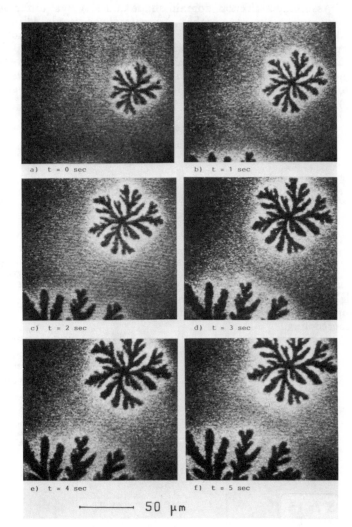

Fig. 2: Time development of a condensed phase DMPE domain following a fast barrier compression. $t = 0$ sec marks the time of first observation of a domain.

a) t = 0 sec

b) t = 1 sec

c) t = 2 sec

d) t = 3 sec

e) t = 4 sec

f) t = 5 sec

50 μm

boundary (Fig. 3) thereby increasing the melting pressure and impeding domain growth. For growth to continue these impurities have to diffuse from the interface, and therefore the process is diffusion limited. Taking the dye as typical surface active impurity the growth kinetics can be assessed via many measurable parameters: (i) The impurity distribution near the interface as a function of time from a densitometric analysis of the images presented in Fig. 2; (ii) The increase of domain diameter and area with time; (iii) The relaxation of pressure with time.

Assuming a circular domain shape and a typical diffusion coefficient the time dependence of the measurable parameters was simulated in reasonable agreement with the experiment.[8] However, there is one discrepancy not only between experiment and simulation but also within the experiment. The pressure relaxes about a factor of 3 faster than the domain area increases. This suggests a faster aggregation process preceding the formation of visible structures. It was confirmed by energy transfer experiments responding to local concentration changes on dimensions of 100Å.[9]

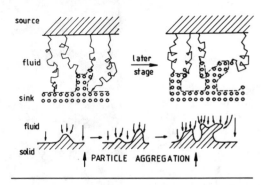

Fig. 3: Sketch of the DLA model (top) in comparison to constitutional supercooling.

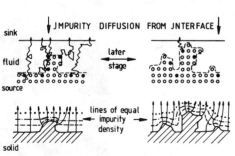

Whereas growth kinetics can be described almost quantitatively within classical models a complication arises considering the analysis of the irregular shapes. Qualitatively one can understand these also as due to constitutional supercooling favouring the development of an unstable growth front: Impurities near a tip can diffuse faster into the fluid monolayer phase than near a smooth boundary.[10] As the shapes resemble those of viscous fingers the analysis procedure of Nittmann, Daccord and Stanley[11] was applied (Fig. 4).[1] The circumference U was measured with rulers of length $n \cdot \ell_0$ (n = integer, ℓ_0 = unit length). It was tested for scaling according to

$$U(n) = \text{const } xn^{-d}(n \cdot \ell_0). \tag{1}$$

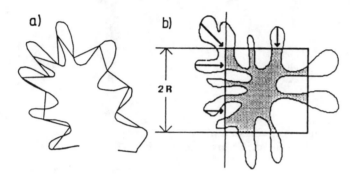

Fig. 4: Sketch of the procedures to measure U (Fig. 4a) and F (Fig. 4b). (a) A ruler of fixed length is started from a point at the boundary line to cut the boundary line in clockwise direction. There the next ruler is started, and the number N of steps to return to the starting point is counted. (b) The area F within a square of side length $2R$ is measured.

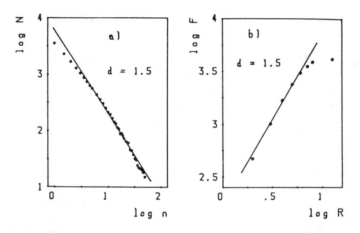

Fig. 5: Log/log plot of N in units of $n \cdot \ell_0$ ($\ell = 0.8\mu m$) versus n (Fig. 5a) and of F versus R (R in units of $8\mu m$) for a DMPE domain at an early development stage.

Following the original model the same fractal dimension d should also appear in the measurement of the condensed phase area fraction F within a square of length $2R^{11}$:

$$F(R) = \text{const}' \, x R^d. \tag{2}$$

An analysis according to Eqs. (1) and (2) was performed for about 10 domains of sufficiently large size but at an early development stage. A typical result is given in Fig. 5. The corresponding plots can be tested varying dimensions by a factor of 50 for U and by a factor of 5 for F. Considering the result for U (Fig. 5a) it is not

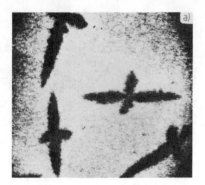

Fig. 6: (a) Solid phase domains of a diacetylene lipid. (b) Sketch of dendritic growth.

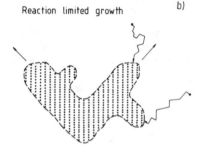

Reaction limited growth

exactly linear. At the lower end self-similarity is not fulfilled due to the influence
of line tension which at later stages of development affects the shape in making it
completely circular.[8] At the other extreme of dimensions the analysis is limited by
the overall domain size. The latter also concerns the analysis in Fig. 5b. Although
there the range of data considered is too small to test any scaling concept one notes
that a plot with the fractal dimension $D = 1.5 \pm 0.1$ also best fits the data of
Fig. 5b. This value is in accordance with numerical results using a DLA model.[11]
On the other hand we should stress that we had expected discrepancies between
experiment and simulation for the following reasons:

- There is long-range electrostatic repulsion between molecules within a domain[12] and this favors establishment of elongated structures.[13,14] This may
  be difficult to simulate and reduce d.

- The process observed very probably occurs in two stages: Growth of clusters
  and directional cluster aggregation[9] starting from a nucleation center. We are
  not aware that this has ever been simulated.

We should also remark that our simple model has been tested to some extent by
numerical simulations.[15,16] These have stimulated new experiments to vary the diffusion coefficients, e.g., by incorporating proteins and then to measure the influence
on the fractal dimension. Experiments have also been reported where the authors
claim that their system is pure enough and therefore domain growth is not limited
by impurity diffusion.[17] The shapes strongly resemble ours suggesting that they are
determined by DLA.

3.2 CLASS (2) SYSTEMS

Irregularities at the phase boundaries of class (2) systems do not develop since

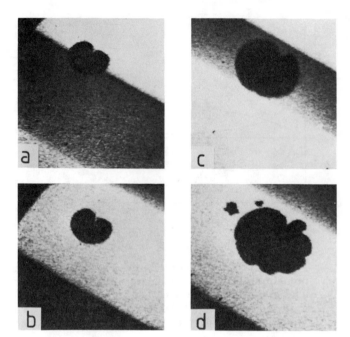

Fig. 7: Growth stages of a DPPC domain fixed under an electrode.

domain growth is not limited by mass transport but by the interfacial reaction. This leads to a dendritic and compact structure (Figs. 6a,b). In that case one can show from a combination of fluorescence polarization and electron diffraction experiments that the axes of fast growth correspond to diagonals of the unit cell.[18] In accordance with this the growth velocity cannot be controlled by incorporation of non-crystallizable impurities, even if these are added in molar ratios of 1/3. In these systems the nucleation rate is very low leading to large domain sizes and it is possible to over-compress the film such that the surface pressure is up to 9 mN/m above the value pc corresponding to the onset of the phase transition. The growth velocity largely varies for different domains and experiments at nominally identical conditions.

3.3 CLASS (3) SYSTEMS

If the condensed phase is not very densely packed initially formed domains quickly change shapes as required by local equilibrium. Then usually line tension minimizes the boundary energy and hence boundary length. This eliminates fractal or dendritic shapes kinetically determined and allows for smooth boundary lines. However, there is one interesting and specific effect leading to shape instabilities, and this is presented in Fig. 7. In the corresponding experiment a domain is fixed in an external field and the very slow growth of single domains is investigated over

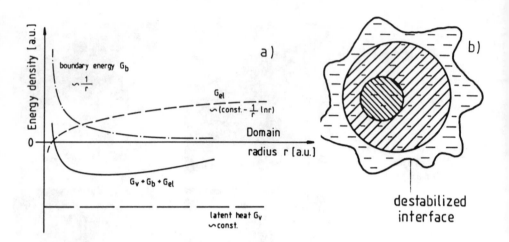

Fig. 8: (a) Sketch of contributions to the energy density of a circular lipid domain and of the total energy density as a function of domain radius. The graph shall indicate that the total energy assumes a minimum for finite domain size. (b) Sketch of growth stages indicating that energy minimization produces an unstable boundary line.

times of hours. One observes smooth boundary lines for small domain sizes. Exceeding a certain size, however, the boundary develops extrusions which appear to exhibit some periodicity. Although the specific system may be rather complicated due to an in-plane anisotropy expressed also in a non-circular domain shape for small domain size, the principal features can be explained assuming cylindrical symmetry (see Fig. 8b). For small domains line energy determines the shape which therefore is circular. Increasing the size this term looses importance but, due to its long range, electrostatic repulsion of molecules within a domain increases. This in turn favours elongated more irregular structures where for a given domain area the average intermolecular distance increases. The corresponding contributions to the free energy density assumes a minimum. Exceeding this size one may expect shape instabilities or transitions to non-compact or non-circular shapes as recently suggested[19,20] We propose that this is reflected in the findings of Fig. 7.

4. Concluding Remarks

We have demonstrated the existence of different types of instabilities of the fluid condensed-phase boundary lines in lipid monolayers. These systems exhibit two-dimensional mass transport in contrast to transport of latent heat which also occurs into the third dimension and are therefore suitable candidates to investigate low-dimensional processes. On the other hand we have to admit that most of the information obtained hitherto is of qualitative nature and that we raised many questions and answered merely few. Hopefully this could stimulate theoretical work that might also be useful to understand the technological aspects of structuring thin films as well as biophysical questions on the control of membrane inhomogeneity.

1. A. Miller, W. Knoll and H. Möhwald, Phys. Rev. Lett. 56, 2633 (1986).
2. I. R. Peterson, J. Mol. Electronics 3, 103 (1987).
3. O. Albrecht, H. Gruler and E. Sackmann, J. Phys (Paris) 39, 301 (1978).
4. M. Lösche, E. Sackmann and H. Möhwald, Ber. Bunsenges. Phys. Chem. 87, 848 (1983); R. M. Weis and H. M. McConnell, Proc. Natl. Acad. Sci. US 89, 4453 (1985).
5. A. Fischer and E. Sackmann, J. Phys. (Paris) 5, 517 (1984).
6. H. D. Göbel and H. Möhwald, Thin Solid Films 159, 63 (1988).
7. C. A. Helm, H. Möhwald, K. Kjaer and J. Als-Nielsen, Europhys. Lett. 4, 697 (1987).
8. A. Miller and H. Möhwald, J. Chem. Phys. 86, 4258 (1987).
9. M. Flörsheimer and H. Möhwald, Thin Solid Films 159,115 (1988).
10. J. S. Langer, Rev. Mod. Phys. 52, 1 (1980).
11. J. Nittmann, G. Daccord and H. E. Stanley, Nature 312, 141 (1985).
12. A. Fischer, M. Lösche, H. Möhwald and E. Sackmann, J. Physique Lett. 45, L785 (1984).
13. R. M. Weis and H. M. McConnell, Nature 10, 5972 (1984).
14. D. Andelman, F. Brochard and J. F. Joanny, J. Chem. Phys. 86, 3673 (1987).
15. H. C. Fogedby, E. Schwartz Sørensen and O. G. Mouritsen, J. Chem. Phys. 87, 6706 (1987).
16. O. G. Mouritsen, H. C. Fogedby, E. Schwartz Sørensen and M. J. Zuckermann in Time-Dependent Effects in Disordered Materials, eds. R. Pynn and T. Riste (Plenum Press, 1987).
17. K. A. Suresh, J. Nittmann and F. Rondelez, Europhys. Lett. (in press).
18. H. D. Göbel, H. E. Gaub and H. Möhwald, Chem. Phys. Lett. 138, 441 (1987).
19. D. J. Keller, J. P. Korb and H. M. McConnell, J. Phys. Chem. 91, 6417 (1987).
20. H. M. McConnell and V. T. Moy, J. Phys. Chem. 92,4520 (1988).

ONE "HANDS-ON" EXPERIMENT

PATTERN FORMATION OF MOLECULES ADSORBING ON LIPID MONOLAYERS

H. MÖHWALD, H. HAAS, AND S. KIRSTEIN

Univ. Mainz, Inst. Phys. Chem.
Welder Weg 15, D6500 Mainz, FRG

1. Introduction

For many technical applications of thin films it is highly desirable to control the lateral distribution of active species on a microscopic and submicroscopic level. We have shown previously[1] that this can be achieved for lipid monolayers at the air/water interface in the coexistence range of a fluid and an ordered phase where the specific lateral distribution of a dye probe was established. To make use of this principle we ask the following questions:

- Can the heterogeneous surface layer be used as a template to form an adjacent layer with similar molecular distribution via adsorption form the bulk?

- Can desired distributions in monolayers also be achieved for functional molecules like proteins incorporating into the monolayer after adsorption from the water phase?

Trying to answer these questions we are studying a complex sequence of events and we will show that the structure is determined by twodimensional as well as by three-dimensional processes and that the contributions of these processes can be varied in a predictable way.

2. Analytical Technique and Preparation

The experiments are performed using a Langmuir film balance with integrated fluorescence microscope.[2] The subphase contains the water soluble dye or protein and buffer in low concentration (10^{-6}M in the experiments reported) before the lipid is spread at the air/water interface. The film is then compressed into the fluid phase or into the fluid/ordered phase coexistence range and adsorption of the bulk component is observed. Maintaining the barrier at fixed position, adsorption of the dye effects a decrease of lateral pressure towards zero, since the Coulombic repulsion between the charged lipids is reduced by binding of the oppositely charged dye molecules. Adsorption of the protein in turn causes a surface pressure increase, because the protein is incorporated into the surface film. Dye aggregation following adsorption is visualized by fluorescence microscopy, since the aggregate exhibits a strong emission with maximum at 620 nm whereas the monomeric dye in water solution does not fluoresce. Emission of the fluorescently labeled protein is collected from a volume of thickness corresponding to the focal depth ($\sim 1\mu m$) of the microscope. Yet the strong enrichment at the interface enables measurement of sharp images.

3. Results and Discussion

3.1. DYE ADSORPTION

Figure 1 sketches system and interactions of the dye adsorption experiments. A

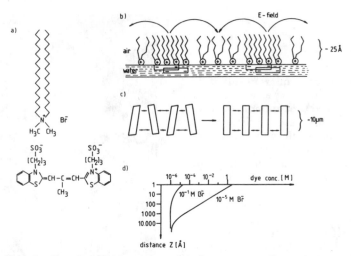

Fig. 1: Sketch of molecular structure (a), interactions (b and c) and distributions (d) in the dye adsorption experiments. Water soluble dyes (rectangles in b) are electrostatically bound to the charged lipid domains with ordered chains. Rectangularly shaped aggregates may be finally formed (c) that order due to electrostatic repulsion. The expected order of magnitude of dye concentration as a function of distance z from the interface is sketched for two different Br^- concentrations in the subphase (d).

positively charged ammonium dye forms a monolayer in the fluid/ordered phase coexistence range with nearly circular domains of the ordered phase[3] as previously observed for phospholipids.[4] The oppositely charged cyanine dye (Fig. 1a) is enriched near the surface and crystallizes to form J-aggregates.[5] Either due to a higher charge density or to more favorable molecular arrangement of the ordered domains aggregation may occur preferentially under the more condensed phase (Fig. 1b). If the aggregate binds to the surface layer an excess dipole density may result giving rise to long range electrostatic repulsion. The field is not screened in the adjacent air space (Fig. 1b), and this may allow establishment of periodic superstructures.[6] To minimize the electrostatic forces rod-like domains would tend to align parallel (Fig. 1c). Since dye aggregation is linked to the surface potential of the surface layer it should be possible to control it via the surface potential. One way of doing this would be to vary the subphase ionic strength. As sketched in Fig. 1d one expects that increasing the ionic strength reduces the surface concentration of the dye and reduces the depth of the volume of local enrichment. This in turn should affect nucleation conditions as well as the dimensionality of transport processes leading to structure formation. Whereas for large Debye lengths aggregating molecules stem from the half space below the surface layer, for small screening lengths they basically result from a small volume near the interface. In the latter case we therefore expect a two-dimensional process.

Figures 2, 3 and 4 basically verify the above expectations. Compressing a monolayer above the dye solution one observes rather uniformly fluorescing areas of dimensions near 100μm (Fig. 2a). The structure tends to recrystallize to form

Fig. 2: Fluorescence micrograph of J-aggregate emission (a-c) after compression of the surfactant monolayer into the ordered/fluid phase coexistence range (a) and after maintaining a fixed barrier position for 30 minutes (b). (c): Micrograph after expansion and recompression. (a-c): $T = 20°C$ subphase dye concentration: $10^{-6}M$ no other ions added, magnification according to the left bar, (d): Micrograph of the monolayer of (a) after transfer on glass slides, viewed through crossed polarizers, with scale given by the right bar.

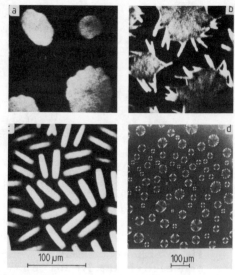

rectangle shaped domains with bright fluorescence polarized along their long axes (Fig. 2b). On expansion of the surface film the aggregate emission disappears, but on recompression it reappears with a dominance of the brightly fluorescing domains. The elongated structures tend to align parallel although they do not touch (Fig. 2c). Hence they exhibit a long range repulsion. The structures formed can be transferred on a solid support. Viewed through crossed polarizers one realizes that also the initially formed domains exhibit structure (Fig. 2d). They exist of radially aligned needles. Figure 3 displays the variation of textures on the surface, and this can basically be explained assuming that after expansion nuclei for aggregation persist. These are concentrated under areas where a domain existed before. This then yields a large number of domains within a circular area on recompression and less and hence larger domains at the periphery. Thus the domain size is determined by nucleation kinetics whereas the shape appears to be an equilibrium feature. As expected by Wulf's theorem the ratios of the two sides of the rectangle ($\sim 5 : 1$) is independent of size and the domain is compact. This holds for the area of high domain density where the aggregates are compact (right side of Fig. 4). If the domains are less compact the ratio decreases towards one, and in the other extreme the domains are nearly circular (left side of Fig. 4).

Performing a comparable experiment at high ionic strengths the structure appears qualitatively different (Fig. 4). Elongated domains grow radially without possessing solely one nucleation center. The structure is not compact, but the tips tend to round off on maintaining the film at a fixed surface area. Performing the processes of compression and stopping in distinct steps the shapes of Fig. 5a can be observed where each broadening of the radial extrusion corresponds to a barrier stop of approximately one minute. Maintaining the film at fixed conditions for a longer time the periphery becomes smooth and stronger fluorescing (Fig. 5b). In conclusion we have shown that aggregation during adsorption can be controlled via the surface layer. It is assisted by electrostatic attraction between positively charged surface and negatively charged dye. The finding that aggregation predominates under the ordered lipid domain, is, however, probably not due to the increased charge density but to steric interactions. This can be deduced since dye aggregation is

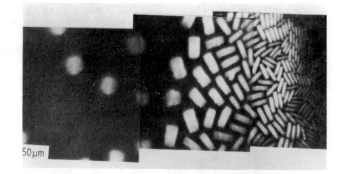

Fig. 3: Extended view on the surface at conditions of Fig. 2c.

also observed at high ionic strength although the dye concentration near the interface is many orders of magnitude smaller than for low ionic strength (see Fig. 1d). The observed images suggest that aggregation occurs highly anisotropically and is diffusion limited at low and reaction limited at high ionic strengths.

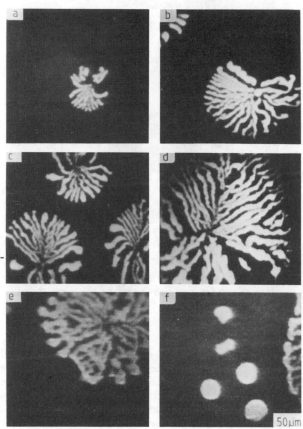

Fig. 4: Fluorescence micrograph of J-aggregate emission on compression of the monolayer from a-f. $T = 20°C$, 10^{-6}M dye concentration, 10^{-1}M NaBr. After compression to take Fig. 4d the barrier was stopped for some minutes to observe rounding of the tips (Fig. 4e). At very high pressure (Fig. 4f) additional circular domains form.

At later development times the domain shapes are uniform, corresponding to local equilibrium, and the sizes are determined by nucleation kinetics. Due to electrostatic interactions superlattices of aggregates can be formed. Although the surface layer determines nucleation conditions and the final nature of the superlattice, a 1 : 1 correlation between surface and aggregate does not exist.

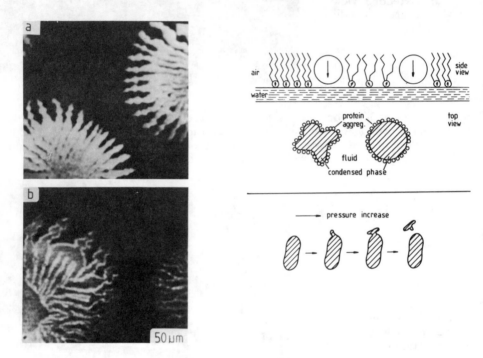

Figure 5   **Figure 6**

Fig. 5: Fluorescence micrograph at subphase conditions of Fig. 4 after stepwise barrier compression (a) and after maintaining the barrier fixed for a long period (b).

Fig. 6: Sketch of the arrangement of Con A patches at the fluid/condensed boundary (upper part) and of the Con A influence on shape changes.

3.2. PROTEIN ADSORPTION AT MONOLAYERS

It is an important biophysical and biotechnological step to achieve a defined lateral arrangement of proteins as functional molecules near interfaces. This can be achieved, since polypeptides penetrating into the monolayer prefer the fluid phase compared to the ordered one.[9] A very severe problem, however, results from the fact that proteins at interfaces tend to denature, and this leads to aggregation. Using fluorescently labeled proteins these aggregates can be observed as bright patches in the micrographs.[10] They then may arrange near the fluid/condensed

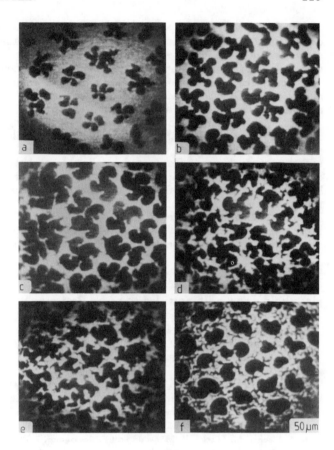

Fig. 7: Fluorescence micrographs of a monolayer of L-α-dipalmitoylphosphatidylcholine (DPPC) containing 3 mole % of the glyco lipid and 10^{-8}M Con A in the subphase $T = 20°C$, pH 7 buffer. $T = 20°C$, 0.3M NaCl, 10^{-4}M Ca^{2+} and Mn^{2+}, pH = 7. The surface pressure (above π_c) was increased from a-f.

phase boundaries as sketched at the top of Fig. 6. This arrangement can be understood considering the balance between steric and electrostatic forces. If the dipole density of the protein covered surface is smaller or of opposite sign compared to that of the condensed phase (indicated by arrows in Fig. 6) the latter phase attracts the protein.[10] However, due to packing restraints, it does not incorporate it. The above conditions and observations have been verified using the lectin Concanavalin A (Con A) as model protein. To get rid of patching we introduced a small fraction (between 1 and 3 mole %) of the lipid 1,2-distearoyl-3-maltosylglycerol, a glycolipid, which Con A specifically binds. Then Con A patches at least very little and still can be laterally arranged in a peculiar way. This is demonstrated by means of Fig. 7 displaying observations on elevating the surface pressure from a-f. Increasing the lateral pressure above a critical value π_c, condensed phase (dark) domains start to grow from individual nucleation centers into distinct directions (Fig. 7a). Further increase of the pressure results in compact growth of domains (Fig. 7b) but outer extrusions are developed (Fig. 7c-e) which finally disjoin from the parent domain (Fig. 7f). The sequence of events is sketched at the bottom of Fig. 6.

The findings can be understood as follows: The fluorescently labeled protein attached to the glyco lipid head group creates a membrane distortion and thus is enriched under the fluid lipid phase. It is attracted to the phase boundary thus reducing line tension, i.e., the energy to create this boundary. This allows for a highly carved boundary line and also enables development of the boundary instability as discussed above for a competition between line tension and electrostatic energy.[1] The structures are thus more compact if no Con A is present and also if Con A but no glyco lipid is used. The latter is then understood since Con A patches formed exhibit a lower boundary/area ratio and are therefore less edge active.

The final question to be answered is: What distinguishes Con A from any edge active impurity to cause the distinguished domain shapes? Compared to surface active dyes as impurities[11] Con A is strongly enriched at the ordered/fluid interface and thus very efficient. Compared to cholesterol, that was shown to induce spiral or lamellar shapes,[12,13] the protein may be more isotropic and therefore reduce the line tension of all domain faces in a similar way. Thus domains do not exhibit any pronounced anisotropy.

In conclusion this part of the work has shown that functional molecules like proteins can be arranged at interfaces similarly as dyes, but in addition the influence of these molecules on nucleation and edge tension has to be considered.

The work on dye adsorption was started in collaboration with Professor M. Shimomura, Tokyo to whom we are also grateful for supply of dye and surfactant. The glyco lipid was kindly provided to us by Professor Hinz, Regensburg.

1. H. Möhwald, H. D. Göbel, M. Flörsheimer and A. Miller, preceding article.
2. M. Lösche and H. Möhwald, Rev. Sci. Instr. 55, 1968 (1984).
3. M. Shimomura, K. Fujii, P. Karg, W. Frey, E. Sackmann, P. Meller and H. Ringsdorf, Jap. J. Appl. Phys. (submitted).
4. M. Lösche and H. Möhwald, Europ. Biophys. J. 11, 35 (1984).
5. M. Shimomura and T. Kunitake, J. Amer. Chem. Soc. 109, 9175 (1987).
6. D. Andelman, F. Brochard and J. F. Joanny, J. Chem. Phys. 86, 2673 (1987).
7. D. J. Keller, J. P. Korb and H. M. McConnell, J. Phys. Chem. 91, 6417 (1987).
8. A. Fischer, M. Lösche, H. Möhwald and E. Sackmann, J. Physique Lett. 45, 785 (1984).
9. W. Heckl, M. Lösche, H. Scheer and H. Möhwald, Biochim. Biophys. Acta 810, 73 (1985).
10. H. Haas, Diploma thesis, Univ. Mains 1988.
11. A. Miller and H. Möhwald, J. Chem. Phys. 86, 4258 (1986).
12. R. M. Weis and H. M. McConnell, J. Phys. Chem. 89, 4453 (1985).
13. W. M. Heckl and H. Möhwald, Ber. Bunsenges. Phys. Chem. 90, 1159 (1986).

LUCA PELITI

INTRODUCTION TO DENSE BRANCHING MORPHOLOGY

GUY DEUTSCHER

School of Physics and Astronomy
Raymond and Beverly Sackler Faculty of Exact Science
Tel Aviv University
Ramat Aviv, Tel Aviv, Israel

ABSTRACT. This chapter serves as an introduction to the two following ones on the same subject. It includes the necessary metallurgical background for the case where this morphology results from the crystallization of amorphous alloys. Other types of morphologies that can be obtained from the same chemical system are briefly discussed. The main features of the DBM growth process are described, with emphasis on the respective roles played by nucleation and diffusion as well as surface tension.

1. Introduction

The aim of this lecture is to give the necessary background material and ideas in order to help the reader understand more easily the following two lectures—one on the experiments by Y. Lereah and one on the theory of DBM by R. Hilfer.

DBM was clearly identified in 1986 (1) as a morphology distinctly different from other ones such as dendritic growth, DLA, etc., that have been much studied for many years. It was simultaneously discovered in different systems: recrystallized amorphous alloys (2), Hele-shaw cell experiments (3), electrolytic metal growth (4). The characteristic features observed in all these experiments are as follows:

   a. Growth starts from a central point and is two dimensional.

   b. The growing colony has a circular envelope.

   c. The colony has a branched structure but—contrary to DLA—it is not fractal on the scale of the colony. In other terms, its branched structure presents a characteristic length scale.

The remarkable feature of DBM is that the overall envelope has no front instability—it is in all cases neatly circular. But it contains branches that are unstable and split as they grow.

The lecture is structured as follows. In section 2 we briefly review different kinds of experiments that produce the DBM morphology. In section 3 we present the metallurgical background of amorphous annealing, and mention morphologies other than DBM that can also occur during such annealing. Section 4 focuses on the fundamental processes that govern DBM growth, particularly during its early stages.

2. Experimental Observations of DBM

Experimental evidence for a distinct Dense Branching Morphology was first stated by E. Ben-Yacob et al. (1). They noticed that the same pattern consisting of a circular envelope and leading branch tips emerged in two very different experiments that they performed, where interfacial dynamics is diffusion controlled.

2.1. THE HELE-SHAW DBM

The first is a Hele-Shaw cell experiment where oxygen gas is injected into a viscous fluid (glycerol) maintained between two plates less than 1 mm apart. Air penetrates in the liquid by growing invading branches that split as growth goes on. Contrary to what had been observed in previous simulations and experiments (see ref.1 for detailed references), branch splitting does not give rise here to a fractal structure. The dense branched structure has a well defined length scale that does not change during growth. In other terms, the number of leading branches at a given distance x from the center of the pattern does not depend on x, at least at not too short distances. A finite surface tension is seen to be necessary for DBM growth—DLA would be the limiting case of DBM for the case of zero surface tension.

2.2. AMORPHOUS ANNEALING

The authors remarked that a similar structure—although on scale 10,000 times smaller—appears during the annealing of an amorphous Al-Ge alloy. The similarity between the structures is indeed striking, considering the apparent difference between the two physical systems. These annealing experiments have been since then pursued in a very detailed way and are analyzed at length below and in the following two lectures. It suffices to say here that there are indeed three fundamental physical features that are common to these two systems: their growth is diffusion controlled; the surface tension is small; their is no significant anisotropy factor that could lead to dendritic growth.

2.3. ELECTROLYTIC METAL DEPOSITS

Similar patterns were reported at the same time in electrolytic metal deposits (5). Here a Zinc sulfate solution is confined between two plexiglass disks. A circular Zn electrode surrounds the fluid. The growth experimented can be performed with two separate control parameters: the applied potential and the concentration of the Zinc sulfate solution. Dendritic growth is observed in a wide range of potentials and at intermediate concentrations. But at the lowest concentrations the deposits are ramified and remain circularly symmetric. The transition between DBM and dendritic growth is seen to take place as the growth speed is increased (through an increase in concentration). At low speeds, tip splitting instability is understood to prevent the formation of regular dendritic patterns.

3. Metallurgical Processes in Amorphous Annealing

3.1. ELEMENTARY BACKGROUND

Basically, we consider an MS binary system whose equilibrium phase diagram is a simple eutectic one. The only stable phases are then: the liquid alloy MS; the solid solution of M in S, Sm; the solid solution of S in M, Ms. At the eutectic temperature Te and for a specific concentration Ce called the eutectic concentration, these three phases are in thermal equilibrium.

Above Te, and below the melting points of M and S, the solid solutions can coexist with the liquid phase. Below Te, the free energy of MS is at all concentrations higher than that of Ms and Sm. depending on the concentration of the starting liquid, we shall then have either one of the solid solutions, or a mixture of both. In particular, at the eutectic concentration we have what is called a "eutectic alloy", which often presents a lamellar Ms/Sm/Sm... structure whose period decreases for high velocities of the solidification front (high cooling rates).

Front instability and dendritic growth can occur during solidification of the alloy, through a process that is known as constitutional supercooling. This process has some relation with the instabilities observed in amorphous annealing, and it is worthwhile to explain it here in a few words. According to the phase diagram, at the liquidus temperature T1(C) the liquid is in equilibrium with a solid solution having a smaller solute concentration than the liquid. this means that when we cool the melt, through a wall for instance, the solid that forms rejects a certain amount of solute which tends to accumulate in the liquid near the solidification front. If the front velocity is high enough, the liquid far from the interface remains at its original concentration and is in a sense supercooled because at equilibrium its concentration should be higher than that of the solid while it in fact remains the same. The front can then become unstable because if a bulge forms, the solute can diffuse away laterally thus reducing the local supercooling on the sides of the bulge, while the tip remains strongly supercooled. The tip will then tend to advance faster than the lateral sides, and we can have dendritic growth—without thermal supercooling!

3.2. THE DIFFERENT MORPHOLOGIES OF METAL-SEMICONDUCTOR EUTECTIC ALLOYS

Let us now go back to the actual system that we wish to describe. Our starting material is not a liquid, but an amorphous solid alloy, composed of a metal M and a semiconductor S. The reason why it is amorphous is that it has been quenched directly form the vapor unto a room temperature substrate. When prepared under similar conditions, pure semiconductors such as Si and Ge are amorphous, and this is also the case for alloys such as our Al-Ge film when the Ge content is high enough (typically above 50%).

3.2.1. *The granular morphology.*

At lower Ge concentrations, the film comes down with a granular structure: the metal forms crystallized grains, surrounded by thin amorphous Ge coatings.

The amorphous phase and the granular morphology are clearly metastable. In the amorphous phase, neither phase separation nor crystallization have been achieved. In the granular morphology, the elements have been separated, bit only the metal has crystallized. Its existence show that the metal crystallizes more easily than the semiconductor. This is reasonably explained by the fact that the semiconductor, being a covalent solid, can reach a fairly low free energy just by local arrangements of the atoms (tetrahedra) without long range order. On the contrary, metallic bonding requires long range order to effectively lower the free energy. The stronger tendency of the metal to crystallize will be playing an important role when we discuss the early stages of DBM growth later on. The granular morphology is remarkable in many respects. Phase separation is achieved in a very specific arrangement where the semiconductor forms an amorphous coating of atomic thickness around the metallic grains. This is locally stable down to very small grain sizes—sometimes less that 20A. The very small semiconductor thickness allows sufficient electron tunneling form grain to grain for the granular film to be metallic as a whole,. It is tempting to conjecture that this metallic character stabilizes the granular morphology. Unfortunately, it has not attracted much theoretical attention until now.

3.2.2. *The Random Morphology.* The random morphology is obtained when the alloy is deposited from the vapor unto a substrate heated up to such a temperature that both constituents crystallize. They then nucleate at random locations independently and form clusters that are fractal up to a certain scale—beyond which the concentration is uniform. These films have been much studied as model systems of percolation in a random continuum(6). The random morphology is the stable form of our alloy. Notice that its realization implies that the surface tension between crystalline Al and Ge is not too large. This point will be important later on.

3.2.3. *Experimental conditions for DBM observation.* We now start from an amorphous Al-Ge alloy film with a Ge concentration of the order of 50deposited on a room temperature substrate as described above. This sample is introduced in a Transmission Electron Microscope where it is heated uniformly with the help of a small furnace. At a certain temperature we observe the apparition of a few nucleation centers. The temperature is then fixed, and we watch the growth of the crystallizing areas. They grow into colonies whose general morphological features do not depend critically on parameters such as concentration and annealing temperature. At maturity, they consist of a polycrystalline and branched Ge core surrounded by a circular Al envelope. Their detailed features are analyzed in the following two lectures. What we want to do in the next section is to describe what we think we understand about the early stages of growth—we might say the genesis of the colony.

4. Early Stages of DBM Growth

4.1. NUCLEATION

Growth starts from the nucleation of an Aluminum crystallite. We know this not so much from direct observation at the nucleation stage—which is very difficult to catch—but from the fact that at maturity, all the Al crystallites in the colony have the same crystallographic orientation. So they must all have a "common" ancestor. On the contrary, Ge is polycrystalline.

After it has formed, this Al crystallite is in local equilibrium with the amorphous Al-Ge matrix that surrounds it. This situation is different from regular eutectic solidification for two reasons: first, only one of the constituents has crystallized; second, the crystallite is in presence of an amorphous matrix rather than a liquid, and much below the eutectic temperature. We shall assume that in terms of local equilibrium, this amorphous matrix is equivalent to a liquid that would have been "frozen". The Ge concentration inside the Al crystallite must then be much higher than if the Al was in equilibrium with Ge crystallites (Fig. 1). That concentration should be roughly given by the continuation of the solidus line below the eutectic temperature. As discussed in the next lecture, this is indeed observed experimentally and gives us some confidence that our approach is correct.

4.2. Ge PRECIPITATION

What happens next? As the Al crystallite keeps growing with this high Ge solute concentration, Ge crystallites will necessarily at some point precipitate out from this Al matrix. This will occur because the interior of the Al crystal does not "know" that its border is in local equilibrium with the Al-Ge matrix. This precipitation is also a necessary condition for further growth of the Al crystal. Indeed, if precipitation did not occur, a thick layer of expelled Ge would build up in front of the Al crystal—as happens in the solidification process described in section 3. Growth would then slow down and eventually practically stop, the Al crystal being in contact with an almost pure Ge barrier.

Precipitation of Ge, when it occurs, will lower the Ge concentration inside the Al nucleus, which will in turn allow more Ge to diffuse in—thus avoiding the Ge build up in front of the growing nucleus and allowing its further growth.

Our conjecture is confirmed by two key experimental observations on large colonies (see next lecture): the concentration of Ge in solution in the Al is the same everywhere in the colony; there is no detectable build up of Ge ahead of the crystallization front.

We have thus reached some simple but important conclusions. Precipitation of the Ge inside the Al crystal is a necessary condition for the growth of the colony. The rate of precipitation will then necessarily play an important role in the growth kinetics. As precipitation occurs, Ge in solution will be immediately resupplied by a process described below, so as to keep the precipitation rate constant.

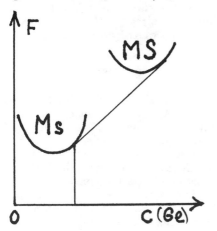

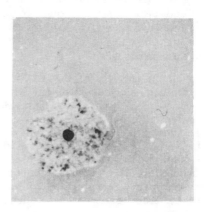

Fig. 1: Free energy diagram   Fig. 2: Early stage of colony growth

4.3. FURTHER GROWTH - DIFFUSION

We first remark that in the granular morphology, growth of the Al crystallites stops precisely when a continuous Ge coating forms around the grains. So this is indeed one possible solution: no further growth. It will occur when many crystallites of Al nucleate within a short time, that is so short that it does not allow further growth of the colony by the diffusion process described below.

Next, we note that the simplest morphology for the growing colony consists of a central Ge core surrounded by an Al rim. This solution allows continuous growth of the colony by epitaxial growth of each of the constituents on itself. Growth kinetics is dominated by diffusion of the Ge through the Al. In order to keep the Al concentration constant inside the colony, and equal to that of the surrounding matrix, the thickness L of the Al rim has to grow as the radius of the colony, R. The velocity of the diffusion front, equal to D/L where D is the coefficient of diffusion, would then vary as $1/R$: growth would slow as the colony grows.

The two features of this simple model: a single crystal core and a growth velocity that varies as $1/R$, do not correspond to experimental observations. It appears that the inner Ge/Al boundary is unstable, contrary to the outer Al/Al-Ge boundary which is in general stable. Instability of the inner boundary is not unexpected since crystalline Ge is in contact with a highly supersaturated solution of Ge in Al. At the temperature of the experiment, the equilibrium concentration of Ge in Al is of the order of 1%—while the actual concentration is of the order of 10%. This large difference is the driving force for the Ge growth—the analog of the amount of supercooling in pure dendritic growth, constitutional supercooling in the case of alloy solidification, or pressure gradient in the Hele-Shaw cell experiments. The analogy with constitutional supercooling is particularly interesting, since in both cases an anomalous concentration profile is the driving force for the instability.

The mechanism for the instability of this inner boundary is rather clear. Growth of the Ge core occurs through diffusion of Ge through the Al rim. Growth velocity is limited by D/L, as discussed above. Rapid local growth of the core feeds upon itself because it reduces the local value of L. As usual, surface tension must be sufficiently small for the instability to occur—and indeed we already noted that between Al and Ge it is small enough to allow the random morphology under appropriate circumstances.

5. Conclusion

The main features of the DBM growth process that we have described involve preferential nucleation of one of the constituents, M; precipitation of the other constituent S from an oversaturated solid solution; diffusion of S through an M barrier. Another feature that we have not discussed but which is essential during the growth of the mature colony is the polycrystalline way in which S grows on itself—but this will be discussed at length in the following lectures. The fact that S does not nucleate easily inside M is also important. DBM may only exist if the surface tension between the constituents falls in a narrow range—a moderate surface tension seems indeed to be a common feature of all the experiments reviewed here briefly. Additional work on other systems is clearly needed to establish this point more precisely.

It is a pleasure to acknowledge E. Ben-Yacob for his early interest and many conversations, S. Alexander for continuous illuminating discussions and Y. Lereah for a most fruitful collaboration. This work was partially supported by the US-Israel Binational Science Foundation and by the Oren Family Chair for Experimental Solid State Physics.

1. E. Ben-Yacob, G. Deutsher, P. Garik, Nigel D. Goldenfeld and Y. Lereah, Phys. Rev. Lett. 57, 1903 (1986).

2. G. Deutscher and Y. Leareah, Physica (Amsterdam), 140A, 191 (1986).

3. For earlier experiments on Hele-Shaw cells, see for instance E. Ben-Yacob, R. Godbey, N. D. Goldenfeld, J. Koplik, H. Levine, T. Mueller and L. M. Sander, Phys. Rev. Lett. 55, 1315 (1985).

4. Y. Sawada, A. Dougherty and J. P. Gollub, Phys. Rev. Lett. 56, 1260 (1986).

5. D. Grier, E. Ben-Yacob, Roy Clarke and L. M. Sander, Phys. Rev. Lett. 56, 1264 (1986).

6. For a review see for instance G. Deutscher, A. Kapitulnik and M. Rappaport in *Percolation, Structures and Processes*, eds. G. Deutscher, R. Zallen and J. Adler (Adam Hilger, 1983), p. 207.

MATERIAL FACTORS LEADING TO DENSE BRANCHING MORPHOLOGY IN Al:Ge THIN FILMS

YOSSI LEREAH

Department of Electronic Devices and Materials
Faculty of Engineering
Tel Aviv University, Tel Aviv 69978, Israel

1. Introduction

In the field of materials science the Laplacian pattern is known in crystals that exhibit dendritic shape during a liquid-solid transition. This shape is formed due to a gradient of either temperature or concentration, while surface tension and anisotropy are also important[1,2] The DLA shape is relative rare and was found to occur during crystallization of amorphous materials.[3-6] This shape is not consistent with materials nature, as materials do have finite surface tension, and some degree of anisotropy depending on the crystalline structure. Probably this is the reason that this structure is so rare and from my experience, not well controlled. (I do not relate here to the electro-deposition experiments). Structures like needles, which present anisotropy but the surface tension is not dominant, are also known to occur during solidification (e.g., Ref. 7).

A case in material science of isotropic Laplacian pattern in which surface tension plays an important role, was found during crystallization of Al:Ge thin films.[8,9] As will be shown here, this structure is stable and once the proper amorphous Al:Ge films are prepared under the right conditions, it becomes easy to control the patterns which are formed during their crystallization.

Here I will examine the different factors leading to this pattern.

2. Films Crystallization

Amorphous films of $Al_x Ge_{(1-x)}$ $(0.4 < x < 0.5)$, 550Å thick were prepared by simultaneous evaporation of Al and Ge from two sources. The films were crystallized in the transmission electron microscope by heating them to 250-350 C with a special furnace. The amorphous phase, which is an atomic scale mixture, crystallized to form colonies tens of microns in size, which contain Ge and Al crystals of hundred to thousands Å in size. The core of the colony is polycrystalline Ge having dense branching morphology which is surrounded by Al. The following process was detected[8,9]:

Nucleation of Al occurs once in a colony. The Al crystal grows at the colony boundary in a compact shape, keeping its original orientation. The Al is supersaturated by 10% of Ge. Ge crystals nucleate and grow inside the Al crystal. Preferred sites for nucleation of Ge crystals are the existing Ge-Al interface.

Theoretical considerations[10] show that the colony growth velocity V depends on Ge reaction rate (B) and its diffusion rate through the Al (D), the ratio of $V \propto \sqrt{BD}$ was found (for more details see R. Hilfer's chapter). Experimentally this ratio was verified by assuming that the Ge nucleation rate has similar dependence on temperature as the Ge reaction rate.

Let us see what are the material factors leading to the dense branching morphology and their influence on the structure details.

3. The Nucleation and Growth of Al

These factors govern the colonies number and their size. The colonies should be big enough to enable the developing of the D.B.M. Ge core inside. This means that Al nucleation should be rare, but still occur at temperatures lower than required for nucleation of Ge in the amorphous phase. The (almost) single crystalline nature of the Al seems to be of not much importance, as we have seen that Al rich films crystallize in similar morphology but with polycrystalline Al. These films are now being intensively studied because they provide a good opportunity to follow the nucleation stage of the colony.

4. The Ge Diffusion into the Al

The role of diffusion is shown in Fig. 1 by the screening effect. It is also reflected by the equality $V = \alpha D/\xi$ in different temperatures,[8] ξ is Al rim width, and is determined in Ref. 9. α was calculated by S. Alexander et al.[10] and experimentally depends on the exact determination of ξ (see also the following chapter by R. Hilfer).

Fig. 1: The typical Al:Ge colony taken at different times of its growth. Note the arrowed branch, which at Fig. 1a, is very little behind its neighbors' branches. This small retarding is sufficient for avoiding its further growth as can be seen in Fig. 1b, because of the screening effect.

5. The Al-Ge Interface

The Al-Ge interface energy is high enough to prevent the Ge from random crystallization, but on an existing Al-Ge interface. Yet, it should be low enough to allow the ramified structure. This structure is essential for the constant growth velocity of the colony, which was found experimentally.[8] If the Ge core had a round shape inside the Al, the growth velocity would decrease, as the Al rim width would increase, until it would practically vanish. It seems, as the Ge 'prefer' to 'suffer' the excess of surface energy for the benefit of the later phase transition which is energetically favored.

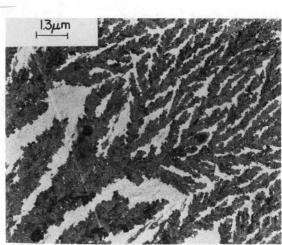

Fig. 2: Influence of crystallization temperature on the branch width. The left part was crystallized at lower temperature than the right part. The colony growth velocity of the left part was also lower than that of the right part.

6. Nucleation and Growth of Ge

These factors are responsible for the polycrystallinity of the core, i.e., its isotropic nature, and they determine the Ge crystal size. Experimentally, it was found[9] that the biggest Ge crystals determined the width of the core branches. This means that the higher the nucleation rate the smaller the Ge crystal size and the narrower the core branches. Qualitatively this can be seen in Fig. 2, where the left part was crystallized at a lower temperature than the right side. It was found that the average branch width λ varies with colony growth velocity V as $V^{-1/4}$.

This is consistent with the nucleation rate N which varies as $V^{1.36}$ according to experiments[8] or as $V^{1.5}$ according to theory[10] because we can write $NS = V$, where S is Ge crystal area which is equal to λ^2. This result leads to the conclusion that λ^2 varies with temperature as $\sqrt{D/N}$, i.e., it is material dependent.

7. The Two Dimensionality of the Samples

As one can see, the creation of the above structure involves many processes. All of them should occur through the whole thickness of the samples, simultaneously. Therefore, the sample should be thin enough, and the component concentrate should be uniform with respect to the thickness. Figure 3 shows a case in which fluctuations in evaporation rate causes some Al rich layers. In this film the Al was crystallized in a DLA shape losing its connection to the colony.

Acknowledgements: I would like to thank G. Deutscher for constructive discussions during my studies on 'crystallization of the amorphous thin films of Al:Ge' described above.

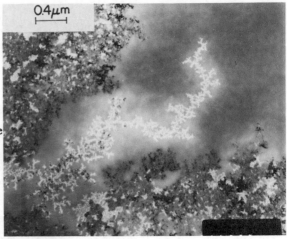

Fig. 3: Al with DLA shape grown at a velocity of 4Å/sec when nonhomogeneous amorphous Al:Ge films are crystallized.

1. W. Kurz and D. J. Fisher, Fundamentals of Solidification (Trans. Tech. Publications, 1986).

2. J. S. Langer, Reviews of Modern Physics 52, 1 (1980).

3. Y. Lereah, Proc. of the 43rd Annual Conference of Electron Microscopy Society of America (1985), p. 346.

4. G. Deutscher and Y. Lereah, Physica 104, 197 (1986).

5. M. Ozenbas and H. Kalebozan, J. Crys. Growth 78, 523 (1986).

6. G. Y. Radnocsi, T. Vicsek, L. M. Sander and D. Grier, Phys. Rev. A 35, 4012 (1987).

7. S. K. Chan, H. H. Reimer and M. Kahlweit, J. Cryst. Growth 32, 303 (1976).

8. G. Deutscher and Y. Lereah, Phys. Rev. Lett. 60, 1510 (1988).

9. Y. Lereah and G. Deutscher (to be published).

10. S. Alexander, R. Bruinsma, R. Hilfer, G. Deutscher and Y. Lereah, Phys. Rev. Lett. 60, 1514 (1988).

LIV FURUBERG

THEORETICAL ASPECTS OF POLYCRYSTALLINE PATTERN GROWTH IN Al/Ge FILMS

RUDOLF HILFER

Department of Physics
University of California, Los Angeles
Los Angeles, CA 90024, USA

ABSTRACT. These notes discuss recent theoretical approaches to polycrystalline fingering during annealing of amorphous Al/Ge thin films, and compare them to experiment.

Annealing of amorphous Al/Ge films can give rise to a highly branched polycrystalline pattern[1]. This demonstrates the existence of a remarkable new morphology, both for phase separation, and for pattern growth. As a pattern growth mode it is intermediate between diffusion limited aggregation (no crystallization, no nucleation) and dendritic single crystal growth (crystallization, but no nucleation). As a phase transformation morphology it is controlled by long range diffusion, not by interfacial mobility, but exhibits a linear time dependence for the radius of the precipitate ("colony")[1,2]. A typical precipitate ("colony") consists of a branched polycrystalline Ge core embedded in monocrystalline Al. The colonies are roughly circular and grow into the surrounding amorphous phase of $Al_x Ge_{1-x}$ with $x \sim 0.4$. See the contributions of G. Deutscher and Y. Lereah in this volume for more details on the morphology and the experiment.

My objective here will be to address the following three questions:

(a) Why is the phase boundary between crystalline Al and the amorphous phase stable although the colonies grow into the metastable amorphous phase?

(b) Why does the radius of a colony increase linearly with time, and which factors determine its velocity?

(c) What determines the strongly temperature dependent length scale of a colony?

1. Stability

During ordinary solidification from a melt the excess foreign atoms have to diffuse away from the solid/liquid interface. Because this diffusion process has to occur into the metastable phase and over large distances it will be more effective if the interfacial area is increased. This leads to the Mullins-Sekerka (MS) instability and a dendritic morphology for the growing crystal[3]. For the case of Al/Ge phase separation, the crystalline colonies also grow into the metastable amorphous phase, and one might expect to find the same MS-instability for the interface between the colony and the amorphous region. Instead, in the experiment, the shape of the colonies remains nearly circular during the growth process[1,2].

Growth of Germanium inside the colonies requires Ge-transfer from the amorphous region across the Al-rim to the polycrystalline Ge-aggregate. This can occur by diffusion of atomic Ge through regions of crystalline Al. Assuming that this is indeed the dominant diffusion process which limits the growth of the colony one realizes an important difference to the case of the MS-instability: The concentration gradient of the (Ge) diffusion field, and the growth velocity (of the Al/amorphous interface) have opposite orientation.

Let me consider a perturbation of the circular colony shape. Behind the most advanced regions of the perturbed interface the concentration gradient of atomic Ge will be smallest. Remember that Ge diffuses backwards from the interface (opposite to the growth direction.) On the other hand the gradient is high close to those regions of the interface that lag behind, in contrast to the ordinary case where the gradient is largest close to advanced tips. For the Al/Ge colony the most advanced tips are therefore slowed down relative to the rest of the interface, and the perturbation will be damped out. This explains one of the reasons for the stability of the circular colony shape observed in the experiment.

2. Linear Growth and Velocity Selection

The discussion above and experimental evidence[1,2] suggest the following central features for the colony growth:

1. The dominant diffusion is that of atomic Ge backward from the Al/amorphous interface into the crystalline Al.
2. Atomic diffusion in the amorphous phase is very slow compared with the crystalline phase.
3. Nucleation and growth of Ge crystallites occurs only at the interface between Al and Ge.
4. Nucleation of Al crystals in the amorphous phase is much more frequent than that of Ge but still rare; it controls the initiation of new colonies.

The first of these assumptions is central to the following treatment. It results from the observation made in section 1 that the Al/Ge interface is separated from the amorphous phase by a continuous Al rim and can only grow if Ge atoms diffuse across this rim. It will be seen that the interplay between the Al/amorphous boundary and the Al/Ge boundary which act respectively as source and sink for Ge atoms gives rise to the linear growth law.

To approach the problem more formally replace the local concentration of atomic Ge by its angular average $c(r)$ and that of crystalline Ge by its angular average $\rho(r)$. Then one finds

$$\frac{\partial c}{\partial t} = D_L \nabla^2 c - \frac{\partial \rho}{\partial t} \tag{1a}$$

on the "left" inside the Al and

$$\frac{\partial c}{\partial t} = D_R \nabla^2 c \tag{1b}$$

in the amorphous phase on the "right" of the interface. D_L resp. D_R is the diffusion constant on the left resp. right. At the interface one has $c = c_L$ on the Al side ($r = R_-$) and $c = c_R$ on the amorphous side ($r = R_+$). The growth process is described in its simplest form through

$$\frac{\partial \rho}{\partial t} = Bc\rho. \tag{2}$$

The phenomenological rate constant B describes the growth of the branched Ge structure and thus incorporates nucleation and growth of Ge crystallites. At the Al/amorphous interface, $r = R(t)$, the diffusion field must obey mass conservation:

$$\frac{d}{dt}\left[R(t)\left(\Delta c + \rho(R)\right)\right] = D_L \frac{\partial c}{\partial r}\bigg|_{r=R_-} - D_R \frac{\partial c}{\partial r}\bigg|_{r=R_+}. \tag{3}$$

Here $\rho(R)$ is a small seed concentration of crystalline Ge at the boundary. $\Delta c = c_R - c_L$ denotes the discontinuity in the concentration across the interface (miscibility gap).

The ramified Al/Ge boundary close to the Al/amorphous boundary acts as a sink for the diffusing Ge and from Eq. (3) this implies a finite concentration gradient and thus a finite velocity for the moving front. Transforming into the moving frame one writes

$$c(r,t) = c_0 f(z) \qquad \text{resp.} \qquad \rho(r,t) = c_0 g(z), \tag{4}$$

where $R = vt$, $z = (r - R)/\xi$, c_0 is the concentration of Ge in the amorphous phase and $\xi = D_L/v$ is the basic length scale in the problem. For sufficiently long times $(v^2 t/D_L \gg 1)$ the curvature of the interface can be neglected and one obtains a closed nonlinear equation for f [4]

$$f' + f'' = \beta f(1 - f - f'), \tag{5}$$

with the boundary conditions

$$f(-\infty) = f'(-\infty) = 0 \tag{6a}$$

$$f(0) = \frac{c_L}{c_0} \tag{6b}$$

$$f'(0) = 1 - \frac{c_L}{c_0} - \epsilon, \tag{6c}$$

where $\beta = c_0 B D_L/v^2$ is a dimensionless control parameter and $\epsilon = g(0)$ is the small seed concentration at the interface introduced in Eq. (3). The solutions to Eq. (5) are displayed in Fig. 1 in $f - f'$-space. Trajectories fulfilling the boundary conditions at $z = -\infty$ emerge from the origin with a slope $f'/f = 1/\zeta = [-1 + (1 + 4\beta)^{1/2}]/2$. The straight line $1 - f - f' = 0$ is a separatrix. The boundary condition Eq. (6c) determines a straight line parallel to the separatrix. First choose a value for β, then follow the associated flow line starting from (0,0) until it intercepts this straight line and read off the corresponding value $f(0) = c_L/c_0$. This determines β and thus v as a function of c_L (see inset of Fig. 1).

3. Length Scales

Before discussing the last question it is instructive to compare the theoretical consequences with experiment. The growth velocity is found to be constant as predicted. For slow velocities, i.e, $\beta \gg 1$, one derives the relation $v \approx (c_L/c_0)d_L/\xi\zeta$ [4]. It has been checked experimentally by comparing the temperature dependence of $D/\xi\zeta$ with that of v and was found to be in good agreement [2]. If the transformation

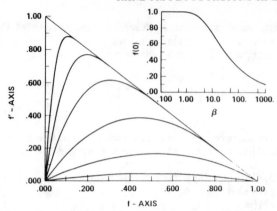

Fig. 1: Trajectories fulfilling the boundary conditions at $z = -\infty$ for selected values of β ($\beta = 0.2, 1, 5, 25, 100, 500$). Inset: Dependence of $f(0) = c_L/c_0$ on β for trajectories fulfilling all boundary conditions with $\epsilon = 0.001$.

were interface-controlled one might expect to find the same activated behaviour for v as for D_L and consequently a temperature independent length scale. This, however, is not borne out by experiment [2], and indeed the theory above predicts $v \propto (BD_L)^{1/2}$ showing that the temperature dependence of $\xi\zeta$ depends on that of B. Here lies a problem, because B is not a material parameter. Experimentally the activation energy of the nucleation rate has been measured [2], and it was found that the relation $v \propto (BD_L)^{1/2}$ is obeyed if B has the same activated behaviour as the nucleation rate. One also observes a broad distribution of crystallite sizes where the size of the largest Ge-crystallites is comparable to the width of a finger. These observations plus a theoretical consistency argument [4] suggest that the nucleation rate for Ge-grains at the Al/Ge interface plays a central role for the understanding of B, and for the length scale (resp. velocity) selection in the experiment.

As a first step let me pose the simpler question how nucleation alone determines a characteristic length for a competitive growth process. To be more specific consider the following highly idealized model for the growth of an isolated Ge-finger in the colony.

A polycrystalline Germanium finger is idealized as a rectangular shape of length L and width W, with $W < L$. To represent the grain structure consider the rectangular shape being filled with a hexagonal tiling. A grain corresponds to a connected region of elementary hexagons. Each grain is identified by a unique number. The rectangular shape grows unidirectionally keeping W constant. This idealizes the fact that most of the growth of a Ge-finger occurs close to its tip, while lateral growth is negligible. To simplify the growth dynamics assume that the rectangle grows stepwise through the addition of one layer of elementary hexagons at a time. The elementary hexagons may be thought of as critical nuclei which can either be incorporated into the existing grains touching the surface or nucleate a new crystallite. For simplicity assume that each of the elementary hexagons in the new layer can nucleate a new crystallite with probability p_{nu}. If an elementary hexagon does not nucleate a new grain, it is added to an existing grain in the previous layer according to the following rules: If the two hexagons in the previous layer which it touches (let me call them predecessors) belong to the same grain, then the new hexagon will also be added to that grain, i.e., it acquires the same number. If the two predecessors have different numbers, then the number of the new hexagon is chosen with probability 1/2 from the numbers of its predecessors.

In this model the growth of a single grain starts with a single hexagon and ends when that grain ceases to touch the surface. This can be caused by the nucleation of other grains on its surface or by screening through neighboring grains. The grain growth is competitive in that each grain can only grow at the expense of its neighbours because the overall available surface area W is kept constant.

Before presenting some of the results for this model let me point out that they seem to be relatively insensitive to the artificial rectangular geometry of the model above. This is seen from the fact that qualitatively similar results are obtained in a model where analogous nucleation rules have been introduced into diffusion limited aggregation with surface tension [5]. In that case a branched polycrystalline structure with finite finger width is produced. Other variants of the model have taken preferential nucleation at already existing grain boundaries into account [5,6].

To extract a length scale from this nucleation and growth model consider the grain length $X_\parallel$ along the growth direction. $X_\parallel$ is a random variable defined as the maximum linear extension of a grain in the direction perpendicular to to the growth surface. An approximate probability density $P(X_\parallel)$ for this quantity can be obtained in closed form in the limit of small p_{nu}[6]. Its asymptotic expansion is found to have the scaling form

$$P(X_\parallel) \sim X_\parallel^{-3/2} \exp\left(\frac{-X_\parallel}{\ell_\parallel}\right), \tag{7}$$

where

$$\ell_\parallel = \left(\frac{1}{2} - [\nu(1-\nu)]^{1/2}\right)^{-1}. \tag{8}$$

Here ν is a monotone function of the nucleation probability p_{nu} with $\nu(0) = 1/2$. The power law with exponent 3/2 is cutoff with an exponential function, and for $p_{nu} \to 0$ the decay length is divergent.

The form for $P(X_\parallel)$ can be easily checked by simulation. In Fig. 2 the density $P(X_\parallel)$ is plotted on a log-log scale for a system of width $W = 3000$ with the values $p_{nu} = 1\%$ (circles) and $p_{nu} = 0.1\%$ (crosses) for the nucleation probability. In the simulations $L = 10^4$ layers of size $W = 3000$ were added. The initial layer consisted of a few hundred different grains, and periodic boundary conditions were imposed identifying the first and the last element in each layer. In Fig. 2 a straight line of slope -3/2 has been drawn for comparison. Despite the approximations implicit in Eq. (7) and the relatively small scale of the simulation (3×10^7 hexagons) theory and simulation are in good agreement.

These results show that nucleation enters only into the cutoff function of the grain length distribution, and does not influence the exponent which is determined purely by the competitive growth process. Although the model does not involve the characteristic interplay of the two interfaces discussed in section 2, it shows that the nucleation parameter ν alone determines a characteristic length for the resulting grain structure. This in itself may be important for other growth models[6]. The model does however not give a better understanding of the parameter B in the previous section. For that a more detailed model of the nucleation process at the

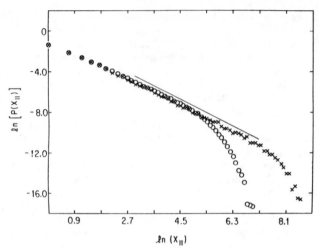

Fig. 2: Probability density $P(X_\parallel)$ of grain length for $p_{nu} = 0.001$ (crosses) and $p_{nu} = 0.01$ (circles). The straight line has slope -3/2.

Al/Ge interface is needed. It must be pointed out that the model above ignores all subsequent grain growth resulting from the reduction of the interfacial energy in the initial grain structure (Ostwald ripening), and the resulting grain size distributions can therefore not be expected to agree with those in the experiment.

In summary these notes have discussed some theoretical aspects of polycrystalline fingering in Al/Ge. The stability question was answered on the basis of the classical Mullins Sekerka analysis. The colony growth has been described in an averaged fashion. Within this phenomenological approach the linear growth law and the velocity selection were seen to follow from the boundary conditions. The theoretical predictions have been compared against experiment. This comparison indicated that nucleation of Ge at the Al/Ge interface plays a central role. Subsequently a simple model for a single polycrystalline finger involving nucleation and competitive growth has been introduced. It was studied analytically and by simulation. The distribution of grain lengths for this model was found to follow a scaling form with an exponential cutoff function whose decay rate depends only on the nucleation rate. The theoretical expression and the simulation results for the distribution appear to be in good agreement.

Acknowledgments: The work described here was carried out in collaboration with Professors S. Alexander, R. Bruinsma, P. Meakin and H. E. Stanley. I gratefully acknowledge financial support from the Deutsche Forschungsgemeinschaft.

1. G. Deutscher and Y. Lereah, Physica A 140A 191 (1986).

2. G. Deutscher and Y. Lereah, Phys. Rev. Lett. 60 1510 (1988).

3. J. S. Langer, Rev. Mod. Phys. 52 1 (1980).

4. S. Alexander, R. Bruinsma, R. Hilfer, G. Deutscher and Y. Lereah, Phys. Rev. Lett. 60 1514 (1988).

5. R. Hilfer, P. Meakin and H. E. Stanley, to be published.

6. R. Hilfer, to be published.

PATTERN FORMATION IN DENDRITIC SOLIDIFICATION

A. DOUGHERTY

Department of Physics, Haverford College
Haverford, PA 19041 USA

ABSTRACT. The growth of dendrites of NH_4Br from supersaturated aqueous solution is studied in order to test recent theories of dendritic growth. The stability constant σ^* is found to be 0.081 ± 0.02, in agreement with theoretical predictions. The mean initial sidebranch spacing λ is found to be proportional to the tip radius ρ, with $\lambda/\rho = 5.2 \pm 0.8$, compared to the theoretical prediction of $\lambda/\rho \approx 3.7$. These measurements are consistent with the hypothesis that steady state dendritic properties are determined by the surface tension anisotropy.

1. Introduction

A rich variety of patterns form in non-linear, non-equilibrium growth processes, depending on both the external conditions and the material properties. Many common materials, including a number of metals and metal alloys, grow with a non-faceted interface. These materials develop shapes that are macroscopically curved and depend sensitively on the growth conditions. In many cases, the system reaches a final steady state, known as dendritic growth, in which the tip of the solid phase advances with a constant shape and speed into the surrounding fluid.[1-3] An example for the dendritic growth of *ammonium bromide* is shown in Fig. 1, where the contours of a growing dendrite are shown at ten second intervals.

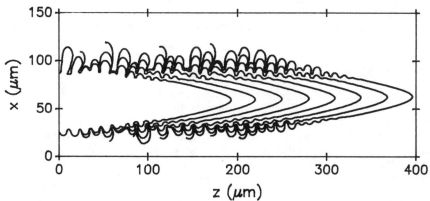

Fig. 1: Contours of an ammonium bromide dendrite growing into supersaturated solution. The contours are drawn at ten second intervals.

These dendritic patterns are macroscopic in size, even though the characteristic intrinsic length scale is microscopic, on the order of angstroms. This intrinsic scale, known as the capillary length, reflects the interfacial surface tension, and is an equilibrium property of the material. The macroscopic scale of the pattern results from a competition between this microscopic equilibrium scale and the macroscopic diffusion scale set by the dynamics of the growth process. An important question to address is whether the complex shapes seen in physical systems can be understood in terms of simple models of the microscopic dynamics, or whether the full complexity of the microscopic processes must be included in order to understand the macroscopic patterns.

Recent theoretical models have predicted that the steady state properties of growing dendritic tips depend on the small *anisotropy* in the interfacial surface tension.[2,3] This anisotropy reflects the symmetry of the underlying crystal lattice. The cubic structure of NH_4Br is clearly evident in Fig. 1, where the *sidebranches* grow at right angles to the main dendrite stem. (An additional set of branches grows perpendicular to the plane of the figure, but is not shown.) This *anisotropy* is ordinarily quite small, however, so it is rather remarkable that the macroscopic tip shape and speed depend on such a small microscopic effect.

Similarly, the sidebranches appear to result from the selective amplification of microscopic fluctuations in the solution near the tip, although the origin of these fluctuations is not known.[4] The resulting macroscopic sidebranch structure is thus also sensitive to microscopic effects.

In this paper I will focus on a quantitative test of recent predictions for the steady state dendritic properties. It is important to test whether the macroscopic continuum model provides an accurate description of realistic physical systems, or whether a more complete microscopic approach is necessary. I find that the predictions of the model are consistent with experiments on the dendritic growth of NH_4Br from solution,[5] so that fairly simple models for the microscopic dynamics appear to capture many of the essential features of steady state dendritic growth. This implies that the complexity of fully developed patterns need not arise from complex microscopic dynamics, but instead could arise from the macroscopic evolution of instabilities in a non-linear, non-equilibrium system.

Although considerable progress has been made towards an understanding of the steady state dendritic properties, the complex time-dependent properties and the long-time evolution of dendritic structures still pose many interesting questions for future study.

2. Steady State Model

In the standard continuum model for dendritic crystal growth from supersaturated solution, the growth of the crystal is assumed to be limited by the diffusion of the depositing substance towards the interface. The interface is assumed to be microscopically rough and approximately at thermodynamic equilibrium. It is assumed that kinetic effects at the crystal surface can be ignored, and that the solution is not stirred during the growth. Since thermal diffusion is generally much more rapid than chemical diffusion, the latent heat of crystallization is also ignored.[1]

The concentration of the depositing substance in the solution, $C(\vec{r})$, is expressed in terms of a dimensionless concentration field $u(\vec{r}) = [C(\vec{r}) - C_{eq}]/\Delta C$, where C_{eq} is the equilibrium saturation concentration (mass per unit volume) at the operating temperature, and $\Delta C = \rho_s - C_{eq}$, where ρ_s is the density of the solid. In the solution, $u(\vec{r})$ satisfies the diffusion equation

$$D\nabla^2 u = \frac{\partial u}{\partial t}, \tag{1}$$

where D is the diffusion constant for the depositing substance in solution.

If the surface tension is isotropic, then the concentration at the interface is given by the *Gibbs Thompson* relation $u_i = d_0\kappa$, where κ is the curvature, and d_0 is known as the capillary length.[1] If the surface tension is *anisotropic*, however, the GibbsThompson relation must be modified, and the concentration at the interface is then[2]

$$u_i = d_0(\theta_1, \theta_2)(\kappa_1 + \kappa_2) + \frac{\partial^2 d_0(\theta_1, \theta_2)}{\partial^2 \theta_1}\kappa_1 + \frac{\partial^2 d_0(\theta_1, \theta_2)}{\partial^2 \theta_2}\kappa_2, \qquad (2)$$

where θ_1 and θ_2 are angular displacements along the principal axes of the surface, and κ_1 and κ_2 are the corresponding principal curvatures. The capillary length $d_0(theta_1, \theta_2)$ is given in terms of the surface tension $\gamma(\theta_1, \theta_2)$ and the chemical potential μ of the depositing substance by[1]

$$d_0(\theta_1, \theta_2) = \gamma(\theta_1, \theta_2)/(\Delta C)^2(\partial\mu/\partial C).$$

For a cubic crystal with the [001] direction along the z axis, the capillary length in the (100) plane is assumed to be of the form[2]

$$d_0(\theta) = \bar{d}_0(1 + \epsilon_4 \cos 4\theta + \cdots), \qquad (3)$$

where θ is the angle from the [001] direction, $\bar{d}_0$ is the average of $d_0(\theta)$ over all angles, and the dots indicate higher harmonics.

Far from the interface, the concentration tends to the asymptotic value $u_\infty = \Delta$, where Δ is the supersaturation. Finally, from conservation of mass, the normal velocity of the interface is $v_n = D\nabla u \cdot \hat{n}$, where $\hat{n}$ is a unit vector normal to the interface.

2.1. THE MULLINS-SEKERKA INSTABILITY[6,7]

The above equations can be solved exactly for the growth of a planar interface. A steady state solution exists only for the case of $\Delta = 1$. In that case, the interface advances with uniform velocity v in the z direction. The concentration field in the solution is given by $u(z) = 1 - \exp(-2z/\ell)$, where z is the distance from the interface, and $\ell = 2D/v$ is known as the diffusion length. However, the velocity can not be determined in this analysis.

The flat interface is unstable to perturbations over a wide range of wavelengths. In a linear stability analysis in the quasi-static approximation (where the derivative $\partial u/\partial t$ is set to 0 in Eq. (3), the wavelength of the most rapidly growing mode is[1]

$$\lambda_m = 2\pi\sqrt{3/2}\sqrt{d_0\ell}. \qquad (4)$$

At large wavelengths, the perturbations grow due to the enhanced diffusion towards the protrusions into the fluid. At small wavelengths, the interface is stabilized by surface tension. The characteristic wavelength of the instability scales as the geometric mean of the two competing length scales in the problem: the capillary length and the diffusion length. Similar scaling will also be seen in steady state dendritic growth.

2.2. DENDRITIC GROWTH

In the limit of vanishing surface tension, one steady-state shape of a growing crystal is a smooth parabolic needle crystal propagating at constant speed. In general, the solution is a paraboloid of elliptical cross-section[8]; the special case of a paraboloid of revolution is known as the Ivantsov solution.[9] In the Ivantsov solution, the dendrite tip speed vel and radius of curvature ρ are given in terms of the supersaturation by $\Delta = pe^p E_1(p)$, where $p = \rho v/(2D)$ is the Peclet number, and $E_1(p)$ is the exponential integral.

In this zero surface tension limit, there is a continuous family of solutions, so that for a given Δ, only the product ρv can be predicted. Experimentally, however, a unique tip radius and speed are found at a given Δ. The experiments of Glicksman and co-workers on the solidification of pure succinonitrile[10] established that the relationship between p and Δ given by the Ivantsov formula is essentially correct. Subsequent experiments[11] further established that the combination $v\rho^2$ is approximately constant over a wide range of Δ. This can be expressed in terms of a dimensionless stability constant[1]

$$\sigma^* = \frac{2\bar{d}_0 D}{v\rho^2}. \tag{5}$$

For succinonitrile,[11] σ^* is found to be 0.0195.

This result could alternatively be formulated as

$$\rho = \left(\frac{\ell \bar{d}_0}{\sigma^*}\right)^{1/2}, \tag{6}$$

where $\ell = 2D/v$, is the diffusion length corresponding to the tip velocity v. This empirical relation has the same form as that obtained for the linear instability of a flat interface, Eq. (4), namely that the characteristic length scale is given by the geometric mean of the capillary and diffusion lengths. The theoretical challenge is then to predict the value of σ^*, and its dependence (if any) on the materials parameters.

Recent theoretical work including surface tension has shown that if the surface tension is anisotropic, there exists a single linearly stable needle crystal solution.[12-19] This solution has an approximately parabolic tip that propagates at constant velocity. Near the tip, the shape is well approximated by the Ivantsov solution for the corresponding Peclet number. Moreover, σ^* is approximately constant for a range of Δ. A key prediction of the model is that σ^* is a strong function of the anisotropy ϵ_4 in the surface tension. For materials of experimental interest, where $\epsilon_4 \sim 0.01$, the value for σ^* obtained from numerical calculations is found to vary approximately linearly[20] with ϵ_4.

In this model, the anisotropy in the surface tension provides a mechanism to define a preferred direction of growth. In general, it appears that the precise mechanism may not be important, only that there be a preferred direction. For example, it has been shown that if the surface tension is isotropic, but a curvature dependence is introduced into the miscibility gap, ΔC, then again a discrete family of solutions is found, and the one with the highest speed is linearly stable.[21] Similarly, in the presence of moderate amounts of impurities, where both thermal and chemical diffusion are important, a boundary layer of impurities can develop around the needle crystal. This boundary layer can introduce anisotropy into the system that provides a preferred direction of growth.[22] In many experimental situations, however, it appears that the dominant effect is the surface tension anisotropy.

It is worth emphasizing the the curvature correction in Eq. 2 is ordinarily quite small. For the NH_4Br dendrites studied in this work, for example, $\bar{d}_0\kappa$ is typically on the order of 0.01Δ. Furthermore, the anisotropy ϵ_4 represents about a 1.6% variation of $d_0(\theta)$ with orientation. In this light, it is perhaps remarkable that the overall morphology is predicted to be so strongly dependent on the surface tension anisotropy.

2.3. SIDEBRANCHES

The theory described above is for steady state needle crystals, that is, for dendrites without sidebranches. However, in nearly all experiments on free dendritic crystal growth, sidebranches are observed. Thus a complete theory of dendritic growth should predict the sidebranch behavior and whether the tip remains stable even in the presence of sidebranching.

One hypothesis is that the tip is linearly stable, but unstable to finite amplitude perturbations, which might be provided by finite amplitude microscopic noise. The noise (e.g., concentration fluctuations in the solution) gives rise to small interfacial perturbations in the vicinity of the tip. In the frame of reference moving with the tip, these perturbations propagate away from the tip and are amplified with distance. The amplification rate will in general depend upon the perturbation wavelength, and thus a characteristic spacing of the sidebranches near the tip, λ, may develop.[23] Depending on the details of the process, this amplification may not select a unique wavelength, so the sidebranching structure could be nonperiodic and have a range of amplitudes and spacings. Such irregularity is, in fact, observed[4] in the sidebranch structure of NH_4Br.

If the linearly stable needle crystal obtained in the model is subjected to a finite amplitude perturbation, it is found that the perturbation is strongly amplified for a broad range of wavelengths. If λ is estimated to be the wavelength of the most rapidly growing perturbation, then the ratio λ/ρ depends only on the crystalline anisotropy. For $\epsilon_4 \sim 0.01$, this ratio is approximately proportional to $sqrt\epsilon_4$. However, the exact values for λ/ρ and the amplification rate depend somewhat on the details of the perturbation, so the theoretical situation is not entirely settled.[24-27]

3. Experimental Results

The key difficulty in making a quantitative test of the theory is the need for measurements of the capillary length and anisotropy. The capillary length is typically on the order of angstroms, and the anisotropy is typically on the order of a few percent. Nevertheless, both quantities can be determined from careful measurements of the properties of a small, nearly spherical crystal approximately in equilibrium in the solution. Digital image processing techniques were used to make precise measurements of the interfacial contours.[4,5]

3.1. DETERMINATION OF $\bar{d}_0$.

The capillary length can be determined from the rate at which the crystal dissolves when Δ is small. In the quasistatic approximation, the radial velocity of a slowly

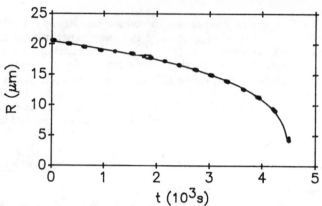

Fig. 2: Time dependence of the measured radius of a slowly dissolving crystal ($\Delta \sim 10^{-5}$) along with a fit to Eq. (7).

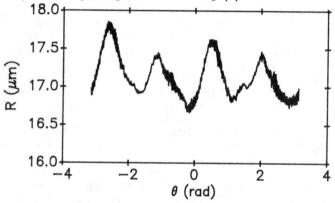

Fig. 3: Angular dependence of the radius of a slowly dissolving crystal. The four-fold component of the shape anisotropy averaged over several crystals is $\epsilon_4 = 0.016 \pm 0.004$.

dissolving spherical crystal of radius R is given by

$$v_R = \frac{D}{R} \cdot \left(\frac{\Delta - 2\bar{d}_0}{R} \right). \tag{7}$$

The measured radius as a function of time for a crystal of NH_4Br is shown in Fig. 2, along with a fit to Eq. (7). The capillary length is found to be $\bar{d}_0 = 2.8 \pm 0.4 \times 10^{-4} \mu m$.

3.2. DETERMINATION OF ϵ_4

The anisotropy can be determined from the shape of a crystal in equilibrium with the solution. If the capillary length in the (100) plane is given by Eq. (3), then the radius of the crystal in the (100) plane (to first order in ϵ_4) is

$$R(\theta) = R_0(1 + \epsilon_4 \cos 4\theta + \cdots), \tag{8}$$

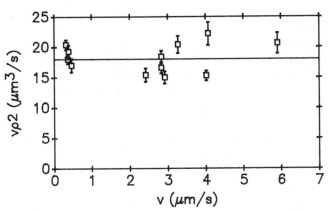

Fig. 4: Value of $v\rho^2$ for a range of tip velocities. The solid line indicates the average value of $18 \pm 3\mu\mathrm{m}^3/s$.

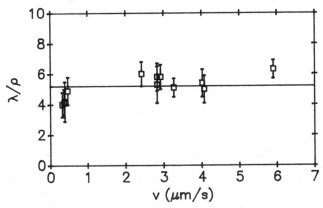

Fig. 5: Value of the ratio λ/ρ for the same dendrites as in Fig. 4. The solid line indicates the average value of 5.2 ± 0.8.

where R_0 is the value of R averaged over all θ. Because the equilibrium size $R_0 = 2\bar{d}_0/\Delta$ is actually unstable, it is in practice necessary to measure the anisotropy of a slowly growing or slowly dissolving crystal.[5] The angular dependence of the radius of a slowly dissolving crystal is shown in Fig. 3. The average value of the anisotropy determined from several crystals is found to be $\epsilon_4 = 0.016 \pm 0.004$.

3.3. TIP RADIUS AND SPEED

The tip radius ρ is determined by fitting to the tip a parabola of the form $(z - z_{\mathrm{tip}}) = (x - x_{\mathrm{tip}})^2/(2\rho)$, where $(x_{\mathrm{tip}}, z_{\mathrm{tip}})$ is the location of the tip, and $z - z_{\mathrm{tip}}$ is the distance on the dendrite axis behind the tip. The tip speed v is determined from successive measurements of the tip location at known time intervals. For a range of tip velocities, the value of $v\rho^2$ is shown in Fig. 4. This combination is approximately constant, with an average value of $18 \pm 3\mu\mathrm{m}^3/s$. The corresponding value for the stability constant is $\sigma^* = 0.081 \pm 0.02$.

3.4. SIDEBRANCHES

The mean initial sidebranch spacing, λ, is determined from the variations of the width of the dendrite measured at a fixed distance z behind the tip.[4] This method gives a statistically stationary measurement of the average sidebranch spacing near the tip. For the same dendrites as in Fig. 4, the value of the ratio λ/ρ is shown in Fig. 5.

This ratio is approximately constant, with an average value of 5.2 ± 0.8. The sidebranches are found to contain a broad range of wavelengths.[4] This irregularity is consistent with the hypothesis that the sidebranches result from the selective amplification of microscopic noise near the tip.

4. Comparison with Theory

In the steady state models discussed above, the value for σ^* is a function of the anisotropy ϵ_4. For NH_4Br, with $\epsilon_4 = 0.016 \pm 0.004$, Kessler and Levine predict that stability constant is[28] $\sigma^* = 0.065 \pm 0.02$, where the uncertainty is primarily due to that in ϵ_4. This is in agreement with the experimental value of $\sigma^* = 0.081 \pm 0.02$.

Using the same anisotropy, Kessler and Levine estimate the mean initial sidebranch spacing to be[28] $\lambda/\rho \approx 3.7$. This estimate depends on the particular details of the sidebranch model, but it is slightly less than the experimental value of $\lambda/\rho = 5.2 \pm 0.8$.

5. Conclusions

The complex macroscopic shapes seen in dendritic crystal growth arise from a delicate combination of microscopic and macroscopic effects. The tip radius rho of a growing dendrite is found to be proportional to the geometric mean of the capillary length $\bar{d}_0$, which is an equilibrium microscopic length, and the diffusion length ℓ, which is a macroscopic dynamically determined length. The value of the proportionality constant, σ^*, is predicted to depend on the anisotropy ϵ_4 in the interfacial surface tension. This prediction is in good agreement with the steady state measurements of the dendritic growth of NH_4Br at moderate supersaturation.

These results illustrate that the steady state properties of growing dendrites can be determined from a continuum model of diffusion-limited growth that neglects most of the complex microscopic dynamics. Thus general features of pattern formation in dendritic growth, such as sensitivity to microscopic noise, might be expected to be seen in other systems, such as viscous fingering, even though the microscopic processes are quite different.

The possible limitations of the continuum model can be seen in the development of the sidebranch structure. In order to produce sidebranches similar to those seen in experiments, fluctuations are ordinarily introduced into the model, although the origin of the fluctuations in the physical system is not known.[27] A complete understanding of the complex sidebranch structure may require a microscopic understanding of the nature and effect of these fluctuations on the microscopic dynamics, including the role of impurities and crystal defects. The role of such microscopic effects in the macroscopic pattern formation remains an important area for future research.

Acknowledgements: This research was supported by National Science Foundation Low Temperature Physics Grant No. DMR 8503543. I thank H. Levine for providing the theoretical predictions used in this work, and J. P. Gollub for advice on every stage of this project.

1. For a general review, see J. S. Langer, Rev. Mod. Phys. 52, 1 (1980).
2. For recent reviews, see D. A. Kessler, J. Koplik, and Herbert Levine, Adv. Phys., to appear, and the following reference.
3. J. S. Langer, "Lectures in the theory of pattern formation," in *Chance and Matter*, ed. J. Souleti, J.¥Anninemus, and R. Stora, Les Houches XLVI 1986 (Elsevier Science Pub. B. V., 1987).
4. A. Dougherty, P. D. Kaplan, and J. P. Gollub, Phys. Rev. Lett. 58, 1652 (1987).
5. A. Dougherty and J. P. Gollub, Phys. Rev. A (To appear).
6. W. W. Mullins and R. F. Sekerka, J. Appl. Phys. 34, 323 (1963).
7. W. W. Mullins and R. F. Sekerka, J. Appl. Phys. 35, 444 (1964).
8. G. Horvay and J. W. Cahn, Acta Metall. 9, 695 (1961).
9. G. P. Ivantsov, Dokl. Akad. Nauk USSR 58, 567 (1947).
10. M. E. Glicksman, R. J. Schaefer, and J. D. Ayers, Metall. Trans. A 7A, 1747 (1976).
11. M. E. Glicksman and S.-C. Huang, Acta Metall. 29, 701 (1981).
12. D. A. Kessler and H. Levine, Phys. Rev. Lett. 57, 3096 (1986).
13. D. Meiron, Phys. Rev. A 33, 2704 (1986).
14. P. Pelce and Y. Pomeau, Stud. Appl. Math. 74, 245 (1986).
15. M. Ben-Amar and Y. Pomeau, Europhys. Lett. 2, 307 (1986).
16. D. C. Hong and J. S. Langer, private communication.
17. David A. Kessler and Herbert Levine, Phys. Rev. A 36, 4123 (1987).
18. B. Caroli, C. Caroli, B. Roulet, and J. S. Langer, Phys. Rev. A. 33, 442 (1986).
19. B. Caroli, C. Caroli, C. Misbah, and B. Roulet, J. Phys. 48, 547 (1987).
20. David A. Kessler and Herbert Levine, private communication.
21. M. A. Lemieux and G. Kotliar, Phys. Rev. A 36, 4975 (1987).
22. M. Ben-Amar and P. Pelce, private communication.
23. Roger Pieters and J. S. Langer, Phys. Rev. Lett. 56, 1948 (1986).
24. D. A. Kessler and H. Levine, Europhys. Lett. 4, 215 (1987).
25. Michael N. Barber, Angelo Barbieri, and J. S. Langer, Phys. Rev. A 36, 3340 (1987).
26. B. Caroli, C. Caroli, and B. Roulet, J. Phys. 48, 1423 (1987).
27. J. S. Langer, Phys. Rev. A. 36, 3350 (1987).
28. David A. Kessler and Herbert Levine, private communication.

GUY DEUTSCHER AND ROLAND LENORMAND

RELAXATION OF EXCITATIONS IN POROUS SOLIDS

J. M. DRAKE,* J. KLAFTER,† and P. LEVITZ‡

*Exxon Research and Engineering Company
Clinton Township
Route 22 East
Annandale, NJ 08801 USA

†School of Chemistry
Tel Aviv University, 69978
Tel Aviv, Israel

‡C.N.R.S. – C.R.S.O.C.I.
45071 – Orleans
Cedex, France

ABSTRACT. The relaxation of an initially excited donor molecule due to the presence of acceptor molecules embedded in porous solids has been studied. The time evolution of the relaxation process has been related to the geometrical restrictions imposed on the participating molecules by the porous structures.

1. Introduction

A large effort has been recently devoted to the understanding of relaxation dynamics of excited molecules embedded in restricted spaces.[1-7] The main emphasis has been on studying molecular probes in porous solids with pore dimensions on length scales small enough to influence molecular relaxation. Due to the complexity of most porous solids, which are both geometrically and chemically disordered, there is a need for simple models in order to interpret experimental data. Such models have to capture the basic geometrical and chemical characteristics of the studied systems. Here we concentrate on the geometrical aspects of these materials. One way that has been used to characterize the structure of porous solids is the classical approach which approximates the complex structures through regular geometrical shapes.[5,8] In this approch one usually starts from a primary picture of local pores and a spatial distribution of these primary elements is then added. The important parameters that are necessary to account for structure and dynamics in the framework of the classical approach are the pore size distribution, mean pore size and tortuosity.[9] Much of the recent interest in these materials has evolved around the idea that it is possible to describe them as having self-similar, fractal interfaces or pore structures. This idea of self similarity is the basis for a different way to model the geometry, a way which has been extensively applied to analyze properties of porous solids. The fractal concept appeals to one's intuition of rough surfaces and its possible applicability to such cases has caught the attention of several groups.[6,10]

In this contribution we focus on the geometrical nature of porous silicas as reflected through the relaxation dynamcis of probe molecules placed in these systems. What we seek to understand are the following: (1) To what extent can morphology of a porous system be deduced from the dynamical properties of probe molecules? (2) Which of the approaches (classical vs. fractal) is consistent with the

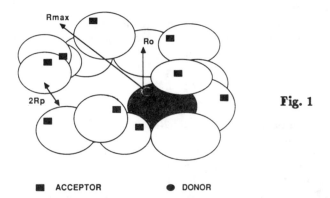

Fig. 1

■ ACCEPTOR   ● DONOR

results obtained for porous silica gels? and (3) What are the relevant morphological parameters necessary to interpret both structural and dynamical observations?
The relaxation of initially excited molecules has been widely applied in order to
gain some insight into the complexity of a broad number of materials such as a
polymer,[11] microemulsions,[12] porous oxides,[5,6] and various molecular assemblies.[13]
In all these examples, the probe molecules serve as sensors of microenvironments.

We will present two different scenarios for relaxation in donor-acceptor systems embedded in porous silica gels. We believe that both cases are related to the
morphology of porous silicas and that the dynamical process in each case directly
senses the local structure. In the first case the donor and acceptors are adsorbed on
the surfaces of the porous solids. Thus, the locations of the participating molecules
are dictated by the underlying geometries. In the second case, the donor is still
adsorbed on the interface, but the acceptors diffuse within the pore network with a
diffusion constant which is controlled by scattering from the intereface, introducing
again a geometrical contribution. All studies have been conducted on well characterized families of porous silicas, each family composed of a series of silicas having
different mean pore sizes but sharing the same morphological structure within the
same family. Structural observables such as surface areas of the porous solids and
small angle scattering intensities have been shown in the framework of the classical
approach with the smaller pore size R_p as scaling parameters.[14,18]

2. Direct Energy Transfer (DET)

The basic idea behind the DET measurement, as shown schematically in Fig. 1, is
to tag a silica surface with a random distribution of donor (rhodamine 6G, R6G)
and acceptor (malachite green, MG) molecules at concentrations low enough to
allow only for a one step transfer of the initially excited donors to the acceptors.
Excitation transfer among the donors or among the acceptors themselves is excluded
in the concentration range and time scale of the experiment. One then follows the
fluorescence of the donors. Within a generalized Forster model[15] for DET the
survival probability of the donor has been calculated and has the form[16,17]:

$$\Phi(t) = \exp\left[-\frac{t}{\tau_D} - p \int dr \rho(r)\{1 - \exp[-tw(r)]\}\right], \tag{1}$$

where p is the density of acceptors ($p \ll 1$), τ_D is the fluorescence lifetime of the
isolated donor and $\rho(r)$ is the density of sites on the structure. Equation (1) relates

the time evolution of the excited donor to the spatial arrangement of the molecules involved in the energy transfer. We assume dipolar donor-acceptor interaction,

$$w(r) = \frac{3}{2}\chi^2 \left(\frac{R_0}{r}\right)^6 \cdot \frac{1}{\tau_D},$$ (2)

where χ is the anisotropy factor and R_0 is a critical radius which is determined from the spectral overlap of donor fluorescence and the absorption of the acceptor on the silica surface. R_0 provides and estimate for the lengthscale which can be probed by DET and it does not usually exceed 100Å. R_0 is a length, which, together with R_p, determines the characteristics of the relaxation in each silica.

Equation (1) applies in cases of DET in regular geometrical shapes[32] (spheres, cylinders, etc.) which mimic simple pores. The overall decay usually fits an expression of the form

$$\Phi(t) = \exp\left[-\frac{t}{\tau_D} - Ac\Gamma\left(1 - \frac{D_s}{6}\right)\left(\frac{t}{\tau_D}\right)\frac{D_s}{6}\right],$$ (3)

with D_s being an effective dimension.[33] Ac is related to the surface concentration of the acceptors[5] and should scale with both R_p and R_0.

DET, we believe, provides a spectroscopic method to elucidate the special organization of adsorbed molecules in relationship to local pore geometries. It is capable of sensing the location of acceptors relative to donors up to a time, t_{max} ($\simeq$ 20 nsec in our studies) which is determined by the experimental limitations of the detections system. Using Eq. (2) a length scale, R_{max}, related to R_0 is determined,[5,18]

$$R_{max} = R_0 \left(\frac{t_{max}}{\tau_D} \cdot \frac{2}{3\frac{2}{\chi}}\right)^{1/6},$$ (4)

which provides a more realistic estimate to lengths probed by DET. The value of R_{max} establishes the upper scale sensed. The relationship between R_p (from pore size distributions and scattering) and R_{max} (from DET) determines what features of the morphology can be probed by the DET process. A series of silicas, Si-40, Si-60 and Si-100, has been studied experimentally (for charactaerization details see Refs. 14, 18):

(a) When $2Rp/R_{max} \gg 1$ (Si-100), the DET probes a length scale less than R_p. The DET process therefore senses only a portion of the local spherical surface with a radius curvature similar to R_p. Here D_s is nearly 2, which means an almost two-dimensional DET.[13,18]

(b) For $2Rp/R_{max} \ll 1$ (Si-40), the probing length scale is well above Rp. We are then sensitive to the pore network. In other words, the local surface appears as a space filling interface with an effect D_s near 3. A transition to $D_s = 2$ is possible at higher acceptor concentrations and short time.

(c) In the limit of $Rp/R_{max} = 1$ (Si-60), the problem is less clear. The DET process probes on average the local vicinity of the primary building blocks.

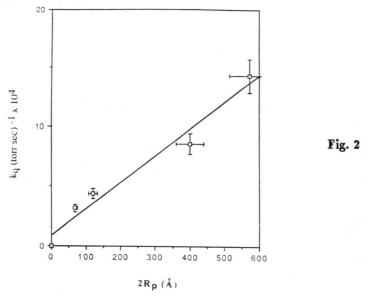

Fig. 2

Here, we expect a more pronounced crossover effect. The curious reader is referred to Ref. 18 for more details.

3. Donor Quenching by Diffusing Acceptors

We now concentrate on the situation where the acceptor molecules *move* with the spatial confinement of the pore network and quench the excited adsorbed donor upon encounter. In order for the geometry to play a major role one has to make sure it is not molecule-molecule collisions among the quenchers that dominate the diffusion but rather the scattering of the individual quenchers off the pore boundaries. Such a behavior is expected for the quenchers in the gas phase at relatively low pressures in the Knudsen regime.[9] A criterion for Knudsen diffusion is that the mean free pather due to gas phase molecule-molecule scattering, λ, is larger than the mean pore size, Rp;

$$\lambda = \frac{k_B T}{\sqrt{2\pi\sigma_p^2}}.$$

Here σ is the collision diameter of the molecules and p the gas pressure. The problem of donor quenching due to Knudsen diffusion is complementary to the direct energy transferproblem, discussed in the previous section. In the energy transfer case the boundaries restricted the location of the adsorbed acceptors. Here the same boundaries restrict the motion of the acceptors. In order to calculate the survival probability of the donor due to acceptors in the Knudsen regime, we assume that the pore space in porous silicas can be viewed as a three dimensional homogeneous, yet tortous, network. The quenching of a low concentration, of randomly adsorbed donors by gas phase quenchers can be modeled through a three-dimensional target picture where (a) the acceptors diffusion is determined by collisions with the solid walls of the material; (b) the porosity and tortuosity are folded into the diffusion as a structure factor; (c) possible contributions from surface diffusion of adsorbed acceptor is negligible. It has been shown that under these conditions the rate of donor quenching is given by[19]

$$K_q \sim gnvR_p r_{AB} = \frac{gvr_{AB}}{k_B} \cdot R_p \cdot P. \tag{5}$$

r_{AB} is a reaction radius and g a structure factor that depends on a porosity and tortuosity. n is the quencher gas density and v its velocity. The donor quenching rate scales with the mean pore size of the silicas and with the gas pressure. Equation (5) can be also derived by calculating the frequency of collisions of an ideal gas of density n with a section of length r_{AB} on the wall of a cylinder of radius R_p.[20]

Experimentally, the decay of initially excited triplet benzophenone by oxygen molecules has been studied.[19] The survival probability has been determined by measuring the relaxation of the benzophenone triplet population using time-resolved diffuse reflectance transient adsorption. The silica series, Si-100, Si-300, Si-500 and Si-1000, which has been discussed and characterized in Ref. 14 has been applied for the Knudsen case. At the range of pressures and temperatures of the experiment (room temperature) the mean free path λ is much larger than any R_p, i.e., for $\sigma = 3.45$Å and $P = 250$ torr Eq. (10) yields $\lambda \simeq 2200$Å. We are then safely in the Knudsen limit. Figure 2 presents the quenching rate as a function of Rp. The predicted scaling is observed.

In conclusion, we have presented structural and dynamical results on homologous series of silica gels which strongly support the classical description of these silicas. The classical approach accounts well for the relaxation of electronlically excited molecules embedded in the silicas. No self similarity has been observed.

This work was supported by the U.S.-Israel Binational Science Foundation.

1. W. D. Dosier, J. M. Drake and J. Klafter, Phys. Rev. Lett. 56, 197 (1986).
2. R. Kopelman, in *Transport and Relaxation in Random Materials*, Eds. J. Klafter, R. J. Rubin and M. F. Shlesinger (World Scientific, Singapore, 1986).
3. J. Warnock, D. D. Awschalom and M. W. Shafer, Phys. Rev. 34, 475 (1986).
4. R. Kopelman, S. Parus and J. Prasad, Phys. Rev. Lett. 56, 1742 (1986).
5. P. Levits and J. M. Drake, Phys. Rev. Lett. 58, 686 (1987).
6. D. Rojanaki, D. Huppert, H. D. Bale, Xie Dacai, P. Schmidt, D. Farin, A. Seri-Levy, and D. Avnir, Phys. Rev. Lett. 56, 2505 (1986).
7. J. Klafter, J. M. Drake and A. Blumen, in *Excited State Spectroscopy in Solids*, Eds. Grassano and Tersi (North Holland, Amsterdam, 1986).
8. R. Sh. Mikhail and E. Robens, *Microstructure and Thermal Analysis of Solid Surfaces* (John Wiley, Singapore, 1982).
9. E. L. Cussler, *Diffusion: Mass Transfer in Fluid Systems* (Cambridge University Press, 1984).
10. D. Avnir and P. Pfeiffer, J. Chem. Phys. 79, 3588 (1983).
11. M. D. Ediger and M. D. Fayer, Macromolecules 16, 1839 (1983).
12. P. E. Zinsli, J. Phys. Chem. 91, 618 (1987).
13. A. Tekami and N. Mataga, J. Phys. Chem. 91, 618 (1987).
14. J. M. Drake, P. Levits and S. Sinha, in S. Mat. Res. Soc. Symp. Proc. 73 (1986).
15. T. Forster, Natureforsh. Teil A 4, 321 (1949).
16. J. Klafter and A. Blumen, J. Chem. Phys. 80, 874 (1984).
17. J. Klafter and A. Blumen, J. Lumin. 34, 77 (1985).
18. P. Levits, J. M. Drake and J. Klafter, J. Chem. Phys. (in press).
19. J. M. Drake, P. Levits, N. J. Turro, K. S. Nitsche and F. Cassidy, J. Phys. Chem. (in press).
20. W. G. Pollard and R. D. Present, Phys. Rev. 73, 762 (1948).

SOAP BUBBLES - A SIMPLE MODEL SYSTEM FOR SOLIDS

G. HELGESEN[1] and ARNE T. SKJELTORP[2]

[1] *Department of Physics, University of Oslo*
Blindern, N-03416 Oslo 3, Norway

[2] *Institute for Energy Technology*
N-2007 Kjeller, Norway

ABSTRACT. The purpose of this demonstration is to show how rafts of monodisperse soap bubbles can be used to study static and dynamic effects in solids.

Within the last years a lot of work have been done in studying the physical properties of colloidal systems, i.e. systems with particle size in the range 0.01 μm - 10 μm (silica particles, polystyrene spheres)[1] and using these as model systems for various effects in real atomic systems.[2] To study these systems one usually needs a microscope, video camera, light scattering equipment etc. to detect and visualize what is going on because of the small size of the particles. But there is indeed a classical experiment which visualizes a lot of the same effects using monodisperse soap bubbles.[3]

A simple set-up for producing the equal sized soap bubbles is shown in Fig. 1. It consists mainly of a container for the soap solution with a screw for adjusting the water flow speed and two capillary tubes with a small opening in between to allow inlet of air. By careful adjustments of the flow speed and the positions of the two capillary tubes it is possible to vary the diameters of the soap bubbles from about 0.5 mm to 2 mm. The tubes can either be in direct contact with one shifted about half the tube diameter sideways, or the lower one can be placed at a distance 2-3 mm directly below the upper one. The bubbles are collected in a petri dish which can be placed directly on an overhead projector for visualization to a larger audience. The soap solution is made of about 10% ordinary liquid kitchen soap and 5% glycerol in water.

The spheres will crystallize in a triangular lattice which usually contains a variety of defects well-known in solid state physics: vacancies, interstitials (a different sized bubble in between), disclinations (one bubble surrounded by 5 or 7 other bubbles), dislocations (bound pairs of 5- and 7-disclinations) and grain boundaries. During the formation of the crystal it is easy to observe motion of dislocations (soliton-like sliding of layers), interactions between dislocations and reformation of the grain boundaries.

It is also possible to observe frustration effects by introducing a larger bubble in the middle of a crystalline area, or at the edge due to the cylindrical form of the petri dish. Frustrations in the packing on a curved surface is seen by collecting the bubbles on a part of a watch glass and carefully removing the excess water by a micro-pipette. Recrystallization and annealing is easily demonstrated by stirring the bubble raft. At the beginning the raft is broken into small crystallites but because of the high strain in this state, teh system will recrystallize during about 1/2 hour to much larger grains.

Introduction of a small number of bubbles with a different size will destroy the crystalline order, and the system goes into a state with short range translational

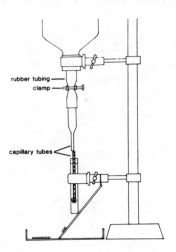

Figure 1

but long range orientational order or a glassy state depending on the concentration ratio.

Acknowledgements: We would like to thank Nils Skogen for practical advices in how to construct the bubble-machine.

1. Pa. Pieranski, Contemp. Phys. **24**, 25 (1983).
2. A. T. Skjeltorp, in Time dependent effects in disordered materials edited by R. Pynn and T. Riste (Plenum, New York, 1987), p. 1.
3. L. Bragg and J. F. Nye, Proc. Roy. Soc. (London) A190, 474 (1947).

INTRODUCTION TO MODERN IDEAS ON FRACTURE PATTERNS

HANS J. HERRMANN
Service de Physique Theorique
CEN Saclay, 91191 Gif-sur-Yvette, Cedex, France

ABSTRACT. The fracture of real materials, i.e., materials that have some disorder, is a very common phenomenon in daily life. Its technological importance justifies the existence of a whole branch of material science with an over one hundred year long tradition.[1] But the complexities of the mechanisms and the enormous variety of material dependent effects has not yet allowed for the formulation of a universally valid theory which would explain patterns and force-displacement characteristics encountered in fracture. Promising progress has, however, been made recently due to the concepts of fractals and scaling laws that have been introduced into the field largely by statistical physicists.[2]

1. Experimental Reality

Real fracture can occur in materials as diverse as concrete, textiles, glass, paper, or soil and depends very much on the way an external stress is applied to the sample. In Fig. 1 we see the fracture patterns of three different situations. In Fig. 1a we see what happens when wet clay is subjected to an external shear imposed on the bottom and the top of the sample.[3] Fig. 1b shows what happens when the stress appears everywhere in the system due, in this case, to the evaporation of water from drying concrete.[4] In Fig. 1c the stress is concentrated on one point, in this case chemically.[5] Clearly the three patterns look very different simply because the distribution of external stress is different. Not only the shape of a crack but also the way the cracks grow depends on the external stress. Three classical types of cracks are distinguished in three dimensions as shown in Fig. 2. But besides the external stress many other factors determine the growth of the cracks. Type and amount of disorder in the sample are crucial. So it depends if the material is porous, fibrous, or amorphous; if it has grain boundaries, dislocations, or microcracks; if it is a composite, an eutecticum, or a polymer, etc. There are also other effects which can occur simultaneously with fracture and render the situation less transparent like plasticity, adhesion, chemical reactions, or fatigue. In the following we want to disregard these effects.

2. Modelization

In order to understand the fundamental laws of fracture, it is necessary to construct models where all input parameters are well controlable, having of course the drawback that they simplify the real world. First of all, all our models will be defined on a finite, regular lattice. Each bond of the lattice should model the reality on a mesoscopic level and we are going to study how the macroscopic system, i.e. the lattice breaks, if one defines for these bonds an irreversible breaking rule. So, what we want to study is a collective phenomenon because the overall breaking is, of course, a consequence of the intricate interplay between the individual breaking of bonds.

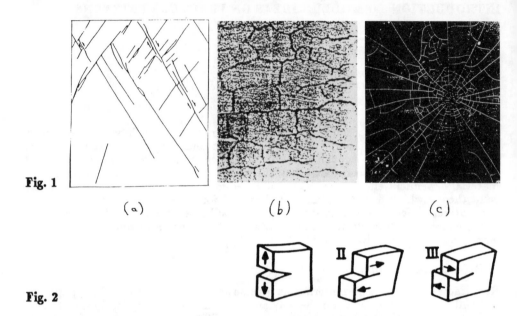

Fig. 1

(a)   (b)   (c)

Fig. 2

Essential to get interesting results is the introduction of disorder into the models. One can introduce it either via a probabilistic rule or by having a quenched disorder and a deterministic breaking rule. Two types of probabilistic rules have been used: a bond is broken with a probability either proportional to the force applied on it (same rule as used in DLA)[6,7] or proportional to a Boltzmann factor $e^{-\beta E}$ where E is the elastic energy in the bond.[8,9] The last rule corresponds to thermally activated breaking which is found for instance when wet sand dries in the desert. In deterministic rules each bond is supposed to be ideally fragile, i.e. to have a linear elastic dependence between force and displacement up to a certain threshold displacement λ_c. The quenched disorder can now be introduced by randomly choosing for each bond either the elastic constant[10,11] or the thresholds λ_c,[11-17] or one can set for a random fraction of bonds the elastic constants to unity and for the rest to zero[18-20] (dilution). In addition to all these different types of disorder one can also choose the random variables according to different probability distributions, the most common being the power law distribution, $P(x) = (1-r)x^{-r}$ with $x \in (0,1)$ which contains the uniform distribution ($r = 0$) and the Weibull distribution $P(x) = (m/x_o^m)x^{m-1}e^{-(x/x_o)^m}$ with $x \in (0,1)$. The Weibull distribution is largely known among material scientists as a distribution that well fits experimental data for $2 \leq m \leq 10$; x_o is an irrelevant normalization constant.

3. The Dual of Dielectric Breakdown

The simplest breaking model one can imagine and which can serve as a pedagogical prototype for all models is, in fact, dual to the well-known dielectric breakdown model[21] (DBM) which is, besides some details of boundary conditions, the same as DLA.[22] Let us consider the DBM between two plates; one on top grounded and the other on bottom with an imposed voltage V_o (Fig. 3) on a square lattice with

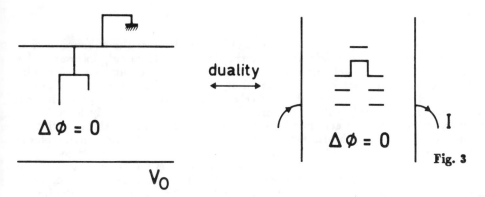

Fig. 3

periodic boundary conditions in the horizontal direction. On each side of the electric potential ϕ that fulfills the Laplace equation $\Delta\phi = 0$. One can see the dielectricum as made out of resistors, ie, each non-broken bond having the same resistance $r = 1$. Now, when a bond is broken, with a probability proportional to $\nabla\phi$, it is actually replaced by a superconducting bond $(r = 0)$. For resistor networks a dual transformation can be defined[23] by placing the sites of the dual lattice into the areas of the original lattice which for the square lattice gives again a square lattice. On each bond of the dual lattice a resistor of resistance $\tilde{r} = r^{-1}$ is placed where r is the resistance of the bond that is crossed in the original lattice. Replacing voltage drops by currents and vice versa and inverting properly the boundary conditions (Fig. 3) one can show that in two dimensions the dual network is equivalent to the original one[23]. If now one breaks first one dual bond as a seed, setting $\tilde{r} = \infty$, i.e., making him an insulator, one can proceed as in the DBM by breaking one of its six neighbors with a probability proportional to the current flowing through it. Note only that a neighbor is now one of the two parallel bonds which is not connected in the usual sense or one of the four bonds that orthogonally enter its endpoints. But for convenience let us redefine the concept of connectivity for our purpose by calling bonds connected that are neighbors in the above sense. By iterating this dual breaking process in the same way as for the DBM we obtain therefore a connected structure which would be exactly the dual of the usual breakdown pattern had we used in both cases the same sequence of random numbers.

Although we have simply transformed via duality the usual DBM and got in two dimensions even identical structures, i.e. also the same fractal dimensions etc., our dual model has an interesting new physical interpretation: it is a network of fuses which burn if the current gets too high and so one forms a crack which once it wraps around the system (don't forget that we have periodic boundary conditions in the vertical direction of the dual lattice) it will break the system in two pieces interrupting the externally imposed current. This model is thus a fracture model and it is this type of model that is going to interest us in the following. We see also that the fracture models being so closely related to DLA will have the same kind of non-local effects as DLA and will be of the same degree of difficulty. In general one can define to each fracture model the corresponding DBM model even if for instance due to boundary conditions the duality relation cannot be fully implemented anymore. Since in general the two models will not give identical results we will distinguish them by calling "f" the fracture model and "d" its dielectric breakdown version.

4. Elastic Effects

Rarely one is tempted to burn a network of electrical fuses; the common case is the breaking of an elastic medium. While the electric problem only has one scalar variable, namely ϕ, on each site elastic problems are tensorial and more variables must be assigned to each site. The simplest elastic model one can imagine is a network of Hookean springs which can freely rotate around the sites of the lattice. This model is commonly called the central force model[24] since only force in the direction of the bond can act. In two dimensions where mostly a triangular lattice is used there are on each site two variables namely the two coordinates of the displacement of the site. The central force model seems to be a good model for systems made out of fibers.[9] Unfortunately this model is not independent of the lattice. On a square lattice for instance it does not have any resistance to an external shear. Actually this model is not a pure vectorial model[25]. On the triangular lattice one can assign to each site a scalar ψ which is ψ on the six radial bonds and $-\psi$ on the six outer bonds of the hexagon surrounding the site. One can then see that the force f in a bond is given by $f = \frac{1}{4}(3\partial_\perp - \partial_\parallel)\psi$ where $\partial_\perp(\partial_\parallel)$ are derivatives orthogonal (parallel) to the bond and finds that the condition that the vectorial sum of the forces entering a site is zero yields a discretized version of the equation $\Delta^2\psi = 0$. In this sense the central force model is a scalar model which fulfills the bi-Laplacian equation instead of the Laplacian equation that we know from the DBM.

If one wants to take into account elastic effects correctly one must consider Lamé's equation[26] $(\lambda + \mu)\vec{\nabla}(\vec{\nabla}\vec{u}) + \mu\Delta\vec{u} = 0$ where $\vec{u}$ is the displacement field and λ and μ the Lamé's cofficients. This vectorial equation is actually identical to Navier's equation which describes low Reynold number hydrodynamics and which in a Hele-Shaw cell geometry reduces to Darcy's law. This equation can be correctly implemented on a lattice through the beam model[27]. The difference between this approach and the central force model from the preceeding paragraph is that here bond-bending effects[28] are taken into account: a bond cannot anymore rotate freely around a site without being energetically

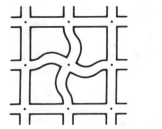 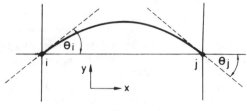

Fig. 4

In the beam model[27], which can be implemented on any lattice, one assigns in two dimensions to each site i in addition to the two coordinates x_i and y_i of the displacement also one angle θ_i. One imagines a rigid cross placed on the site which is rotated just by this angle θ_i and on the bonds elastic beams that tangentially enter the sites in the direction of the crosses. In Fig. 4a we see how if only the central site has an angle $\theta_i \neq 0$ the bonds will have to bend. In Fig. 4b the shape of a beam is shown between sites i and j. The bending of the beam costs elastic energy which can be expressed as a function of the variables that sit on the sites i

and j. One finds the linear relations[15]

$$F = a(x_i - x_j) \tag{1a}$$

for the force needed to stretch a beam along his axis,

$$S = b\left(y_i - y_j + \frac{1}{2}(z_i + z_j)\right) \tag{1b}$$

for the shear exerted on the beam and

$$M_i = c\left(z_i - z_j\right) + \frac{b}{2}\left(y_i - y_j + \frac{2}{3}z_i + \frac{z_j}{3}\right) \tag{1c}$$

for the moment that acts on site i where a, b and c are material dependent constants that involve the Young- and shear moduli and the cross-section and moment of inertia of the beam and where $z_i = \theta_i \ell$, $\ell = 1$ being the length of a beam in equilibrium. This beam model automatically fulfills Lamé's equation since each beam and each site where the beam joins is thought of as made out of an elastic solid that obeys these equations. This model is really vectorial having two independent elastic constants and cannot be reduced to a scalar model if no additional constraint is imposed. The beam model can be used to describe solids like concrete.

Summarizing we have defined three different possible natures of bonds a scalar one ("sc") that fulfills the Laplace equation, the central force case ("cf") which is described by the bi-Laplacian and the beam model ("b") which obeys the vectorial Laplacian equation, namely the Lamé's equation. In all three cases we assume that the bonds be ideal in the sense that for any force applied on them they have a linear response, i.e. no non-linear effects, until of course they break irreversibly and their response is set to zero (or to infinity if a model of type dielectric breakdown is considered). This linear law is given by Ohm's law for the "sc" model, by Hooke's law for the "cf" model and by Eq. (1) for the "b" model. In the first two cases their is only one variable on the bond and therefore only one breaking mode possible but a beam can actually be broken in various ways (in lowest order in two ways) typically by stretching or by bending. To define a unique breaking criterion for a beam one can use from material science the classical von Mises yielding criterion and one obtains that a beam between sites i and j breaks if[29]

$$\left(\frac{F}{t_F}\right)^2 + \frac{\max(|M_i|, |M_j|)}{t_M} \geq 1 \tag{2}$$

where t_F and t_M are two independent breaking thresholds characteristic of the material. So in the case of quenched randomness in the thresholds one has two independent random variables for each beam.

5. Classification of Models

We have already defined two types of models: fracture ("f") and dielectric break-down ("d"); three natures of bonds: scalar ("sc"), central force ("cf"), and beam ("b"); five different ways to implement disorder: a probabilistic one proportional to the gradient of the field ("gr"), a thermal one proportional to a Boltzmann factor ("bo"), and the three quenched disorders in the thresholds ("th"), in the elastic constants ("el"), and dilution ("di"). But we know that patterns also depend on the way the external stress is applied (Fig. 1) and so let us distinguish between the most common boundary conditions: Uni-axial elongation or compression ("u"), shear applied on two plates as seen in Fig. 1a ("s"), homogenous dilation as the one of Fig. 1b ("h") and a geometry with a center as Fig. 1c ("c"). We note that the scalar models can only be in situations "u" or "c". Finally, we have mentioned that connectivity is a matter of definition. In dielectric breakdown nature imposes a connectivity that allows the transport of charge but in fracture connectivity seems a rather artificial condition which is usually even removed altogether. Let us characterize connectivity by a number which is 1 in the usual DBM and 0 if no connectivity is imposed and then cracks can form everywhere in the system. Various connectivity conditions have been analyzed in Ref. 6 and 7 and are discussed in this volume by Meakin.

The above classification allows for 180 different combinations many of them being however of no physical interest or of quite predictable outcome. What one would like to know is how the various variations modify the result and if some of them are irrelevant leading to some kind of universality classes. Unfortunately already our starting model, namely DBM, can presently only be treated numerically and only the fractal dimension[30] d_f of the pattern has been systematically calculated. In addition, the numerical simulation of these Laplacian type growth models is very heavy due to the fact that after each breaking the equations must be solved again and these solutions can only be found via rather time consuming algorithms (eg. relaxation techniques or matrix inversions). For this reason only two dimensional results are presently available. In Table 1 we summarize the results that to my knowledge have been obtained at this time.

As already mentioned the numerical data correspond to two dimensions and we are not very precise since corrections to scaling, anisotropy, etc. are not taken care of. In all cases we show data for $\eta = 1$ (η is the exponent to which the gradient of the field can be raised[21] and which tunes d_f). In the case the crack is not connected one can also ask how the mass of all the cracks together scales with the size of the system and this defines the exponent d_t, while d_f is always the fractal dimension of one crack.

Unfortunately, the present data only allow for very rough conclusions. In the case of dilution-disorder cracks are quite clearly one-dimensional and if disorder is in the thresholds also the dimension of the cracks is close to unity but might be logarithmical since the asymptotic regime is reached slower. None of the other variations of the models considered seem to alter these conclusions for these two types of disorder. For the other three types of disorder the relevant parameters among the ones considered are the geometry and the nature of the bonds. A scalar model in a central geometry seems to give the largest fractal dimension ($\simeq 1.7$); if one takes instead a central force model or a uni-axial geometry the dimension of a crack slightly decreases to about 1.6; and if both things are changed: central force model in a uni-axial geometry, then the dimension jumps down to about 1.2.[7] There are no indications that going from a fracture model to a dielectric breakdown model is going to change anything.

6. Fracture with Random Thresholds

Let us in the following consider the models discussed in Refs. 12-16 namely fracture

type	nature	disorder	b.c.	connect.	d_f	d_t	Ref.
d	sc	gr	c	1	1.7		[21] (DBM)
d	sc	$\left\{\begin{smallmatrix} th \\ gr+th \end{smallmatrix}\right\}$	c	1	$\left\{\begin{smallmatrix} >1.1 \\ 1.7 \end{smallmatrix}\right\}$		[17]
d, f	sc	el	u	0	1.6		[10]
d	sc	di	u	0	≈ 1		[19]
f	sc	gr	u	1	1.65		[7]
f	cf	gr	u, s	0.8-1	1.2		[7]
f	cf?	gr	c, s	1	1.6		[6]
f	cf	gr	c, s	0.6-1	1.4-1.6		[6]
f	cf	bo	u	0	1.3		[9]
f	cf	bo	h	0	fragmentation		[8]
f	sc	di	u	0	≈ 1	$L^2/\sqrt{\ln L}$	[18]
f	cf	di	u	0	≈ 1	$L^2/\sqrt{\ln L}$	[20]
f	sc	th	u	0	> 1.1	1.7	[12],[16]
f	cf	th	u, s	0	> 1.1	1.7	[13],[14]
f	b	th	u, s	0	> 1.1	1.7	[15]

models with disorder in the thresholds without constraints on the connectivity of the cracks to which uni-axial elongation or shear is applied. In these models usually first one must steadily increase the external stress to break the next bond and these first broken bonds are quite independently randomly distributed in space and so this first regime might be called *disordered regime*. After n_c bonds have been broken the system will continue breaking without any increase in the external stress, so that if experimentally one imposes the external stress the system will rupture in this second regime all by itself and very fast until it falls apart and for this reason we call it the *catastrophic regime*. Experimentally one can of course also impose the external strain[31] and in this case it is usually possible to watch also this regime and measure how the force needed to sustain the process indeed decreases. Just when one goes from the disordered to the catastrophic regime, i.e., after n_c bonds have been cut, one is applying the largest external force f_c, i.e., the force needed to break the system apart, which technologically is actually the most interesting quantity.

Since it is usually our belief that pattern are self-similar and fractal dimensions can be defined it is reasonable to think that there will also be scaling laws in system size L for other quantities like external stress or strain. Actually it seems plausible that in order to better understand the scaling behavior of the spatial crack pattern it is useful to look for scaling laws in the whole process of fracture. This will be the aim of the rest of this article and as a preliminary step to this goal we will next study in more detail n_c and f_c.

7. Unstable Crack and Dilute Crack Approximations

Let us consider for simplicity the scalar model and a uniform distribution $P(i_c)$ of the threshold currents i_c between i_- and i_+, i.e. having a width $W = i_+ - i_-$. If the first bond is broken the value of the electrical potential on the sites changes and more explicitly it will be the sum of the value of the potential without any broken bond plus the potential due to a dipole sitting on the position of the broken bond in a system that has the same boundaries but no externally imposed current. Using this fact one can calculate the factor α_1 by which the current in the parallel neighboring bond to the broken bond is increased. On the square lattice in the limit of an infinitely large system one has for instance $\alpha_1 = 4/\pi$. If one has one linear crack of length k one can calculate the asymptotic expression $\alpha_k \sim 1 + ck^{1/2(d-1)}$ for large k, d being the dimension of the system. Consequently any crack of length k that fulfills $k > k_0$ with $i_-\alpha_{k_0} = i_+$ will be unstable, i.e. its growth cannot be stopped without increasing the external current and it can therefore only appear in the catastrophic regime. (Note that k_0 does not depend on the size L of the system.)

Let us derive criteria[12] for crack growth. Suppose we are at the beginning of the breaking process and have cut n single bonds; the current i_n to do so is simply $i_n = i_- + nW/2L^2$. In order to have a neighboring bond to break (a crack of size two) a current $i_{nn} = i_- + W/(2n+1)$ is needed. So when $i_n\alpha_1 = i_{nn}$ a crack of size two will form. This gives the equation

$$(\alpha_1 - 1)i_- = \left[\frac{1}{2n_1 + 1} - \frac{n_1}{2L^2}\right]W \tag{3}$$

for the number n_c of bonds that one must break to grow a crack of size two. If $i_c \neq 0$, i.e. if the distribution of thresholds has a finite cutoff n_1 will be a finite number for any system size L. Since this argument can be applied to cracks of any length and since there is a maximum length k_0 for a crack to be stable we have shown that $n_c \sim L^0$ if the distribution of thresholds has a lower cutoff.

If the distribution has no lower cutoff, i.e., $i_- = 0$ then $n_1 \simeq L/\sqrt{\alpha}$ follows from Eq. (3). So, if one makes the *unstable crack approximation*,[12] that a crack once it starts growing cannot be stopped again one would in this case have $n_c \propto L$. But since this approximation underestimates the number of cuts before the catastrophic regime sets in one really has a lower bound for the growth of n_c with L.

Let us make another type of argument[12,18]. Suppose one has already cut a fraction p of all the bonds in an essentially random way since one is in the disordered regime. Then $N(k) = p^k L^2$ is the average number of cracks of length k and therefore $k = \ln L$ is the typical largest crack one will find since then $N(k)$ is of order unity ("extreme value statistics"). So, the system becomes unstable when $k_0 = \ln L$, i.e. for a current $i_c = i_+/(1 + \sqrt{\ln L})$ which leads us to a number $n_c \propto L^2/\sqrt{\ln L}$ of bonds cut. This "dilute crack approximation"[12] neglects interactions between cracks which tend to organize them in space such that the system gets catastrophic earlier and is therefore really an upper bound for the growth of n_c with L.

Using the above approximation we could bound n_c between L and $L^2/\sqrt{\ln L}$ which is actually a very rough bound. Our subsequent numerical work will show

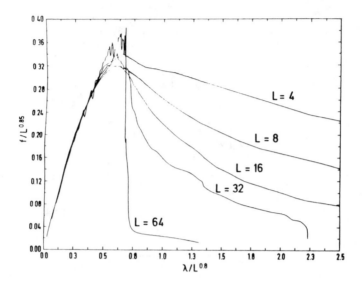

Fig. 5

that both approximations are quite bad and that we must be very careful using hand-waving arguments when we deal with the tricky problems of fracture.

8. Scaling of the Breaking Characteristic

To see how n_c really grows with L and to find the scaling laws for stress and strain we simulated[13] in two dimensions using the conjugate gradient method[32], the scalar model[16], the central force model[14], and the beam model[15]. Typically we averaged over 50000 samples for $L = 4$ and less than 10 samples for $L = 64$. We used a power law distribution for the thresholds $P(\lambda_c) \propto \lambda_c^{-r}$ and for the scalar case also the Weibull distribution with $m = 2, 5$ and 10. Each time a bond is broken we monitor the external stress f and the external strain λ averaging them for constant number of broken bonds which when plotted gives the so-called breaking characteristics. In Fig. 5 we see this characteristic for the beam model in the case of uni-axial elongation and a power law distribution with exponent $r = -1$. We plot the result of various system sizes L and rescale both axes with a power of L such as to find a collapse of the data from all sizes on a single curve in the disordered regime, i.e. in the regime before the maximum is reached. After the maximum, i.e. in the catastrophic regime it is not possible to scale the data probably because the curve seems to become vertical when $L \to \infty$. Fig. 5 is numerical evidence for a scaling law

$$f = L^\alpha \varphi(\lambda L^{-\beta})$$ (4)

where φ is a scaling function and α and β are two exponents that we found to be both of order 0.8. In the same way one can also find a scaling relation with the

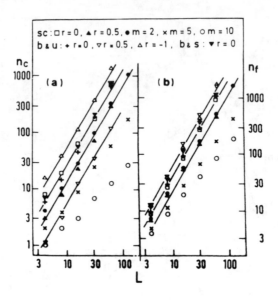

Fig. 6

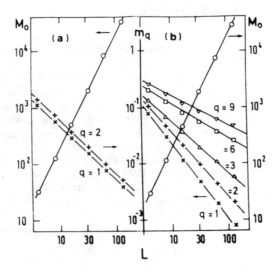

Fig. 7

number n of bonds cut

$$n = L^\gamma \tilde{\varphi}(\lambda L^{-\beta}) \qquad (5)$$

and a value of γ of about 1.7. What is particularly striking is that the exponents α, β, and γ are independent of the nature of the bonds (sc, cf, or b) if one has elongation or shear and independent on the distribution used for the thresholds within the numerical error bars. So we have numerical evidence for a universality class.

Next we show in Fig. 6a how n_c scales with size L for several different models and distributions of thresholds and see that $n_c \sim L^{1.7}$ except if the disorder gets very small ($m = 5$ and 10) where one feels the expected crossover to $n_c \sim L^0$. Similarly one can also analyze n_f the total number of bonds that are broken to fully break the system apart (Fig. 6b) and one finds the same exponent 1.7. Only the crossover for small disorder goes in this case to $n_f \sim L$ as it should. So we verified that the number of broken bonds scales during the whole process with a dimension 1.7 which by accident is close to the fractal dimension of DBM but note that in our case the bonds do not form a connected cluster. We also analyzed the fractal dimension of one single connected crack and found that it is very close to unity.

9. Multifractal Distribution of the Local Strain

Mechanical engineers are often interested how the strain or stress is locally distributed in a sample that is about to break in order to better understand the mechanisms and to try to build less fragile materials. Since numerically we have to calculate these local properties anyway to implement our breaking rule we analyzed the distribution $n(v)$ of the local voltage drops in the scalar case[16] and similarly of local forces and shears in the vectorial models.[14,15] Precisely $n(v)$ is defined as the number of bonds that carry a voltage drop between v and $v + \delta v$. We calculate then the moments $m_q = (M_q/M_0)^{1/q}$ with $M_q = \sum_v v^q n(v)$ of the distribution and plot them in Fig. 7 log-log against L for the case of the scalar model with a Weibull distribution of the thresholds with $m = 2$. In Fig. 7a we see what happens after n_c bonds have been cut, i.e. at the maximum of the breaking characteristics and in Fig. 7b we see the situation just before the last bond breaks before the system fully falls apart. In the first case all moments for $q > 0$ scale in L with the same exponent while just before the system breaks apart each moment has a different exponent x_q: defined by $m_q \sim L^{x_q}$ and it seems that $x_\infty = 0$. The situation encountered in the second case where one finds an infinity of different exponents x_q is called multifracality.[33] Physically it means that the regions of very large strain lie on fractal subsets and the fractal dimensions of these subsets depends on how these strains increase with increasing system sizes. Since they are fractal the regions of high strain become also less frequent if larger systems are considered. All models considered have multifractality just before the system breaks apart and the exponents x_q do not seem to depend on the distribution of thresholds but seem to be different if the model is scalar or vectorial. We do not know if the multifractality is related to the scaling laws of Eqns. 4 or 5.

10. Outlook

We have seen that since nature is so diverse, various models must be used to describe fracture and this even without taking into account many effects that often accompany the fracture of real materials. All these models are however variants of Laplacian growth models. Besides the fractal dimension we also analyzed scaling laws in the breaking characteristics and found even multiscaling behavior in the distribution of local strains just before the system breaks apart.

1. See e.g. H. Liebowitz (ed.) *Fracture*, Vols. I-VII, (Academic Press, New York, 1984).

2. See e.g. R. Englman and Z. Jaeger (eds.) *Fragmentation, Form, and Flow in Fractured Media*, Ann. Israel Phys. Society 8 (Adam Hilger, Bristol, 1986).

3. Z. Recher, p. 42 in Ref. 2.

4. P. Acker and S. Roux, preliminary results from experiments at ENPC, Paris.

5. G. Ananthakrishna, experiments at the Atomic Research Centre in Kalpakkam, India.

6. E. Louis, F. Guinea and F. Flores in *Fractals in Physics*, eds. L. Pietronero and E. Tossatti (Elsevier, Amsterdam, 1986), p. 177 and P. Meakin, G. Li, L.M. Sander, E. Louis and F. Guinea, J. Phys. A (submitted); see also Meakin's contribution in this volume.

7. E. L. Hinrichsen, A. Hansen, and S. Roux, Europhs. Lett. (submitted).

8. P. Meakin, *Thin Solid Films* 151, 165 (1987).

9. Y. Termonia, P. Meakin, and P. Smith, *Macromolecules* 18, 2246 (1985).

10. H. Takayasu in *Fractals in Physics*, eds. L. Pietronero and E. Tossatti (Elsevier, Amsterdam, 1986), p. 181; H. Takayasu, Phys. Rev. Lett 54, 1099 (1985).

11. M. Sahimi and J. D. Goddard, Phys. Rev. B 33, 7848 (1986).

12. B. Kahng, G. G. Batrouni, S. Redner, L. de Arcangelis and H. J. Herrmann, Phys. Rev. B 37, 7625 (1988) and for the case of infinite disorder: S. Roux, A. Hansen, H. J. Herrmann and E. Guyon, J. Stat. Phys. 52 251 (1988).

13. L. de Arcangelis, A. Hansen, H. J. Herrmann and S. Roux, Phys. Rev. Lett. (submitted); H. J. Herrmann, in Proceedings to Les Houches.

14. A. Hansen, S. Roux and H. J. Herrmann, preprint.

15. H. J. Herrmann, A. Hansen and S. Roux, preprint.

16. L. de Arcangelis and H. J. Herrmann, preprint.

17. F. Family, Y. C. Zhang and T. Vicsek, J. Phys. A 19, L733 (1986).

18. L. de Arcangelis, S. Redner and H. J. Herrmann, J. Physique Lett. 46, L585 (1985); P. M. Duxbury, P. D. Beale and P. L. Leath, Phys. Rev. Lett. 57, 1052 (1986); P. M. Duxbury, P. L. Leath and P. D. Beale, Phys. Rev. B 36, 367 (1987); A. Gilabert, C. Vanneste, D. Sornette and E. Guyon, J. Physique 48, 763 (1987); P. M. Duxbury and P. Leath, J. Phys. A 20, L411 (1987).

19. P. D. Beale and P. M. Duxbury, Phys. Rev. B 37, 2785 (1988).

20. P. D. Beale and D. J. Srolovitz, Phys. Rev. B 37, 5500 (1988).

21. L. Niemeyer, L. Pietronero and H. J. Wiesmann, Phys. Rev. Lett. 52, 1033 (1984).

22. T. A. Witten and L. M. Sander, Phys. Rev. Lett. 47 1400 (1981).

23. J. P. Straley, Phys. Rev. B 15, 5733 (1977).

24. S. Feng and P. N. Seu, Phys. Rev. Lett. 52, 216 (1984).

25. S. Roux, private communication.

26. L. D. Landau and E. M. Lifshitz, *Theory of Elasticity* (Pergamon, Oxford, 1975).

27. S. Roux and E. Guyon, J. Physique Lett. 46, L999 (1985).

28. Y. Kantor and I. Webman, Phys. Rev. Lett. 52, 1891 (1984).

29. see e.g. J. Salençon, *Calcul à la rupture et analyse limite* (Presses de l'ENPC, Paris, 1983), p. 110.

30. B. B. Mandelbrot, *The Fractal Geometry of Nature*, (Freeman, New York, 1982).

31. L. Benguigui, P. Ron and D. J. Bergmann, J. Physique 48, 1547 (1987).

32. G. G. Batrouni, A. Hansen and M. Nelkin, Phys. Rev. Lett. 57, 1336 (1986).

33. G. Paladin and A. Vulpiani, Phys. Rep. 156, 147 (1987).

RUPTURE IN RANDOM MEDIA

D. SORNETTE, A. GILABERT and C. VANNESTE
Laboratoire de Physique de la Matière Condensée
CNRS UA190, Faculté des Sciences
Parc Valrose, 06034 NICE Cedex FRANCE

ABSTRACT. Rupture in random media is studied by focusing on two problems: (i) the complexity of structure and (ii) the statistics of extremes: rupture often occurs on the weakest "link" and is therefore extremely sensitive to fluctuations. We will illustrate these questions on discrete and continuous ("Swiss cheese" and "blue cheese") percolation models.

Introduction

Understanding the physical properties of heterogeneous or random media constitutes a major challenge of present research activities in the physical and natural sciences. Investigations on structural characterization (density, correlations, geometrical features, connectivity ...), on transport (electric, fluid ...), on wave propagation (acoustic, electromagnetic) in a vast class of inhomogeneous media have unravelled a wealth of new qualitative behavior demonstrating that disorder cannot be seen only as a perturbation with respect to order but must be fully addressed.[1] Furthermore, non-linear regimes in transport and deformations are beginning to be studied systematically.

Among these problems, rupture phenomena are extreme cases which occur when some elements of an heterogeneous system exhibit an irreversible evolution: breakdown of a random electrical network, tearing of tissues or cracks nucleation and propagation in mechanics. These phenomena have important applications since they condition the security, the wearing out and lifetime of many industrial systems. They also raise fundamental questions: determination of the fracture topology as a function of the nature and disorder of the elements, the distribution of the rupture thresholds as a function of imposed stresses, the amplification and the screening of stresses, and non-linear phenomena A rupture is indeed a true irreversible growth process which can be related to coagulation phenomena, fluid imbibition in porous media, electrostatic breakdown or dendritic growth. These phenomena are presently the object of an intense investigation in order to determining the different universality classes, namely, the analogies and the differences between these problems.

In this paper, we illustrate these ideas on percolation models. We compare the critical behavior of the failure threshold in discrete and continuous percolation models (§1). The sensitivity of rupture to fluctuations is emphasized. In particular, we introduce a crack percolation problem developed as the strong disorder limit of crack deteriorated systems (§2). Many new properties of this models are discussed as well as experimental results. In conclusion (§3), we discuss open problems.

1. Non-universality[2] and Weibull distribution[3] of failure threshold in continuous percolation

As stated above, percolation models correspond to the strong disorder limit of random systems. Such systems are appealing as they are described by the powerful percolation theories. They describe the scaling properties of highly heterogeneous systems and stress the influence of the complexity of the system on its rupture properties.

1.1. DIFFERENCE BETWEEN LATTICE AND CONTINUUM FAILURE THRESHOLD[2]

In breakdown phenomena, the effect of disorder is markedly more pronounced than in transport phenomena. The failure threshold is dominated by extreme fluctuations, in contrast to transport and the elastic coefficients which are related typically to the second moment of the distributions.

This crucial sensitivity is particularly obvious in the difference between lattice and continuum failure thresholds in percolation. We illustrate this point by determining failure thresholds near the percolation transition of a class of "Swiss-cheese" continuum models, where spherical empty holes of radius "a" are randomly distributed in an otherwise uniform electric or elastic medium. The exponents governing the electrical and mechanical rupture properties of these percolation structures are shown to be quite different from the corresponding ones in the conventional discrete-lattice percolation models. This results from the existence of very weak bonds which control the rupture in contrast to the discrete case for which the weakest link is a lattice bond. A similar statement has recently been presented for transport exponents.[4] The difference between continuum and lattice exponents is all the more pronounced for rupture. In particular, in 2D, even when the conductivity exponent is the same for both cases, the failure exponent is different and greater in the continuous lattice case (see below).

The analysis adapts the discussion of Ref. 4 to the case of failure. It relies on the well-known node-link-blob picture of the percolation backbone[5] which consists of a network of quasi-one-dimensional string segments (or macro-links) tying together a set of nodes whose typical separation is the percolation length $\xi \approx (p - p_c)^{-\nu}$ (Fig. 1).

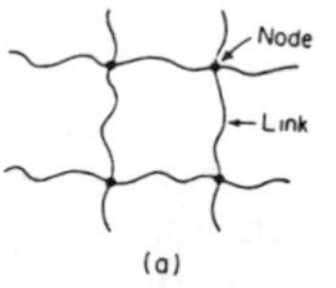

Fig. 1a 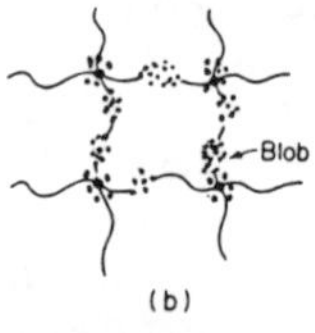   Fig. 1b 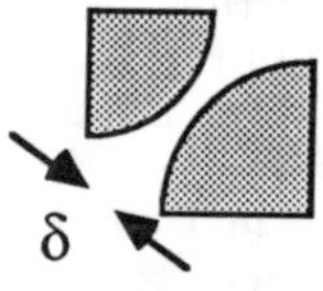   Fig. 2

Each string consists of several sequences of singly connected bonds (or micro-bonds) of total number $L_1 \approx (p - p_c)^{-d}$ with $d = 1$,[6] in series with thicker regions or blobs. Using the mapping of the Swiss-cheese model onto a discrete random network as discussed in Ref.4, one can identify the singly connected bonds as the channels of width δ_i (Fig. 2).

Let us examine the electrical failure and treat as an illustration the case in which the system is submitted to a *given current density* j. Suppose that rupture will occur in any bond of the system if the Ohmic losses in that bond become larger than a threshold value P_c, leading to the burning out of the bond. From the node-link-blob structure of percolation, each macro-link can be viewed as a thread supporting an equal fraction of the total applied stress. Since we are interested in scaling laws, all macro-links can be considered identical and the global failure is equivalent to finding the failure criterion for each macro-link. We therefore neglect the influence of fluctuations in macro-links strength.[7] As the macro-links are made of micro-bonds in series (blobs are neglected), we have to find the strength of the weakest micro-bond along a macro-link. Therefore, the distribution of the bonds as a function of their strength must be determined. This distribution depends on the nature of the system. For the example of spherical holes distributed in an uniform medium, the channel width δ_i controls the strength of the bond i. Hence, we must determine the channel distribution as a function of their width, especially in the limit of small δ_i which correspond to the weakest bonds. The essential ingredient is to recognize that in the case where the holes are randomly distributed, the δ are distributed according to a continuous probability distribution $p(\delta)$ which approaches a finite *non-zero* limit $p(0)$ for $\delta \to 0^+$. As shown below, this allows to estimate the typical minimum value δ_{min} of the width δ of a micro-bond along a string of L_1 singly connected bonds.

Consider a random variable X taking values in the interval $[0, +\infty]$ and governed by a normalized density probability distribution $p(X)$. Let us note $P(X_n > x) = 1 - P(X_n < x)$ the probability that, in the set of n trials X_n of the random variable X, no values of X_n are found smaller than x. $P(X_n > x)$ obeys the following recursion relation:

$$P(X_{n+1} > x) = P(X_n > x)\left\{1 - \int_0^x p(X)dX\right\},\tag{1}$$

since $\left\{1 - \int_0^x p(X)dX\right\}$ is the probability that X be larger than x at the $(n+1)^{th}$ trial. For large n, the solution of Eq. (1) is[3,8]

$$P(X_n > x) \propto \left\{1 - \int_0^x p(x)dX\right\}^n \sim \exp\left\{-n\int_0^x p(x)dX\right\}.\tag{2}$$

Using this type of argument in our case, we obtain the probability

$$P(\delta_{min} > \epsilon) \propto \left\{1 - \int_0^\epsilon p(\delta)d\delta\right\}^{L_1} \approx e^{-p(0)\epsilon L_1}.\tag{3}$$

Hence, ϵ scales like L_1^{-1} and yields

$$\delta_{min} \approx \epsilon \propto (p - p_c)^d.\tag{4}$$

This argument can be cast slightly differently: $\int_0^\epsilon p(\delta)d\delta$ is the probability that $\delta < \epsilon$. The condition $L_1 \int_0^\epsilon p(\delta)d\delta \leq 1$ writes that less than one event $\delta < \epsilon$ has occured over L_1 trials. This allows to identify δ_{min} as L_1^{-1}.

Using the node-link-blob picture of percolation, one can relate the current I flowing into a macro-bond of length ξ to the current density j imposed on the boundary by[9]

$$I \propto j\xi^{(d-1)},\tag{5}$$

where d is the space dimension. The Ohmic loss in a single bond of conductance Σ is

$$P = \Sigma^{-1}I^2, \tag{6}$$

where Σ is controlled by the "strength" of the bond. A bond corresponding to a narrow neck of width δ can be approximated as a thin parallelogram of width δ and length $\ell \approx (a\delta)^{1/2}$ where a is the hole radius (Fig. 2). The corresponding conductance is[4]

$$\Sigma \propto \left(\frac{\delta}{a}\right)^{d-3/2}. \tag{7}$$

Inserting (5) and (7) in (6) yields the Ohmic losses for the weakest singly connected bond:

$$P_{sc} \sim \delta_{\min}^{3/2-d} j^2 \xi^{2(d-1)}. \tag{8}$$

According to the failure criterion, the rupture will develop macroscopically when $P_{sc} = P_c$ yielding the failure threshold for the current density

$$j_c \approx (p - p_c)^f, \tag{9}$$

with

$$f = \left(d - \frac{3}{2}\right)\frac{d}{2} + (d-1)\nu. \tag{10}$$

For $d = 2$, $f = 19/12$ with $\nu = 4/3$. For $d = 3$, $f \approx 2.51$ with $\nu \approx 0.88$.[10] Note that the discrete case $f = (d-1)\nu$ is recovered when $d = 0$.[9] The difference between the continuum and the discrete case is $f_{\text{cont}} - f_{\text{disc}} = (d - 3/2)d/2$ and is relevant for $d > 3/2$.

1.2. WEIBULL FAILURE DISTRIBUTION INDUCED BY FLUCTUATIONS[3]

Not only are the failure threshold critical exponents changed from discrete to continuous percolation[2] as discussed in the previous section, but the corresponding ensemble distribution is also modified. Near the percolation transition of a discrete finite lattice, the distribution of failure thresholds is created by the "intrinsic" fluctuations in the mesh sizes L around the typical value equal to the percolation correlation length ξ. It can be shown that such fluctuations lead to a very steep failure distribution of the form (exponential of an exponential)[3]

$$P(j_c \leq x) = 1 - \exp\{-ae^{bx}\} \qquad \text{for} \quad -\infty < x < +\infty, \tag{11}$$

where a and b are two constants.

On the other hand, in continuum percolation like the Swiss cheese percolation model, the continuous distribution of holes creates another source of fluctuations: the width, and therefore the strength of the micro-bonds in the percolating structure. In contrast to the discrete-lattice case, the weakest bond fluctuations dominate over the "intrinsic" mesh size fluctuations and control the failure thresholds distribution. It follows that the distribution of failure thresholds transforms from an exponential of an exponential to an exponential of a power law (Weibull-like distribution) as we now discuss. These ideas also apply to other continuous percolation model such as the "blue-cheese" model which will be addressed in §2.

As discussed in §1.1, one can argue that in the limit of large mesh sizes ξ, macroscopic failure occurs when each macro-link has suffered failure.[7] The corresponding failure probability distribution is therefore obtained from the distribution of failure threshold on macro-links which itself is controlled by the distribution of micro-bond strength. It is given by Eq. (3) which was obtained from Eq. (2) with $n = L_1$ and $p(\delta) \sim p(0)$ for δ small:

$$P(\delta > \delta_m) \sim \exp\{-L_1 p(0)\delta_m\}. \tag{12}$$

In order to find the failure distribution as a function of the bias current density j, we use Eq. (8) where the Ohmic power P_{sc} is replaced by the threshold value P_c. This leads to the failure current density $j_c \propto \xi^{-(d-1)}\delta^\alpha$ with $\alpha = (d-3/2)/2$. By inversion,

$$\delta \propto \xi^{(d-1)/\alpha} j_c^{1/\alpha}. \tag{13}$$

Inserting (13) in (12) yields the macroscopic failure threshold distribution

$$P(j) \propto \exp\{-c\xi^{[(d-1)/\alpha+1/\nu]} j^{1/\alpha}\}, \tag{14}$$

which is of the Weibull type $[F(x) = 1 - \exp(-x^\gamma)]$! If one takes into account of the fluctuations in mesh sizes of the node-link-blob macro-lattice on percolation, ξ must be replaced by $L_{typ} \approx \xi(\log \Im)^y$ where $\Im$ is the total size of the system and $y = d/D(d-1)$ with $D = d - \beta/\nu$. Then, (14) is replaced by[3]

$$P(j) \approx \exp\{-c[\xi(\log \Im)^y]^{[(d-1)/\alpha+1/\nu]} j^{1/\alpha}\}. \tag{15}$$

The typical and ensemble average macroscopic failure thresholds scales as

$$\langle j \rangle \approx (p - p_c)^{(d-1)\nu+\alpha}(\log \Im)^{-y\{(d-1)+\alpha/\nu\}}. \tag{16}$$

We obtain a system size dependence characteristic of brittle materials. The logarithmic dependence $(\log \Im)^{-y\{(d-1)+\alpha/\nu\}}$ is rather slow and will be difficult to verify numerically or experimentally.

2. The "Blue Cheese" Continuum Crack Percolation Model[11]

We now introduce a new class of continuum percolation systems ("blue cheese" model), where the transport medium is the space between randomly distributed clefts. The critical behavior of the electrical conductivity, fluid permeability, elastic constants and failure thresholds have been discussed in Ref.11 and are, for most of them, distinct from their counterparts both in the discrete-lattice and "Swiss-cheese" continuum percolation models. Furthermore, it has been argued[11] that the asymmetric elastic response of a crack submitted to compression or extension leads, for the macroscopic mechanical properties, to a new percolation threshold at a crack concentration higher than that of usual geometric percolation. This is due to the presence of "hooks" or overhang crack configurations which transform

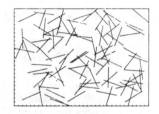

Fig. 3: A typical random crack structure before geometric disconnection in two dimensions obtained by a computer with the algorithm described in the text.

locally the macroscopic applied extensional stress into a compressional stress. In two dimensions, an analogy with directed percolation has been suggested. When the cracks have a non-vanishing width, one recovers the usual geometric percolation threshold and the analysis of the transport properties is similar to that of the Swiss cheese model developed by Halperin et al.[4]

The blue-cheese model is defined as follows (Fig. 3). Empty rectangle holes of length ℓ_c in two dimensions or disk-like holes of diameter ℓ_c in three dimensions with thickness $a \ll \ell_c$ are randomly distributed in an uniform electric or elastic medium. As the number N of cracks per unit surface (in 2d) or per unit volume (in 3d) increases, a percolation threshold N_c is reached corresponding to the geometrical disconnection of the sample in multiple fragments. In three dimensions, the cracks form a percolating structure at a first threshold N_1 strictly smaller than N_c, before disconnecting the sample in pieces. We will be interested only in the regime $N > N_1$. The relevance of the blue-cheese model is suggested from the crack structure of many mechanical[12] (damaged crystals, solids, ceramics, rocks ...) and natural systems[13] (arrays of faults in geology in relation to oil recovery, geothermics and earthquakes[14] ...) which often consists in random arrays of micro-cracks of vanishing thickness.

In the following, we discuss the different modes of rupture which may occur in a system deteriorated by cracks in the vicinity of the percolation threshold and which are submitted to an applied stress.

2.1. BRITTLE GRIFFITH FAILURE

In a system containing cracks, brittle rupture following the Griffith mechanism (see Ref. 15, below) will occur under tensile stress (mode I[15]). This rupture scenario is expected to be prominent for cracks with vanishing thickness. In this failure mechanism, the concentration of the stresses at the apex of a crack tends to open the two edges and lengthen the crack, thus leading to failure. This mechanism must be considered when the tips of the cracks are sharp and no screening due for example to the presence of rounding or of a cavity occurs at the tips of the cracks.

The Griffith criterion states that crack growth is governed by a balance between the mechanical energy released and the fracture surface energy spent as the crack propagates. In the blue cheese model, one can argue that the typical generic configuration for the weakest part of the system involves one crack in close proximity (to within a distance d) to a border created by other cracks (Fig. 4).

This leads to a local stress amplification proportional to $\delta^{-1/2}$ and thus to a failure force $\mathbf{F}_c \approx \delta^{1/2}$.[11] If σ is the stress applied at the boundary, the force $\mathbf{F}$ results from the torques $M_\xi \sim F_\xi \cdot \xi$ with $F_\xi \sim \sigma \cdot \xi$ exerted on a scale ξ. Hence,

Fig. 4: (a) Illustration of a generalized Voronoi tessalation for a micro-bond configuration in the blue cheese model: the dashed line schematizes the equivalent discrete-lattice micro-bond; (b) Typical topology of *weak* micro-bonds in three dimensions. The width a of the cracks has been taken zero.

$\mathbf{F} \cdot \ell_c \approx M_\xi \approx \sigma \xi^2$ or $\mathbf{F} \approx \sigma \xi^2 / \ell_c$. Therefore, the stress failure threshold is, with δ given by Eq. (4)

$$\sigma_\rho \approx (p - p_c)^{\mathbf{F}m}, \tag{17}$$

with $\mathbf{F}_m = 2\nu + 1/2$ larger than the discrete case by $1/2$.

In 3d, one has to evaluate the concentration of the elastic energy due the presence of the two cracks. In Ref. 11, we obtain $\mathbf{F}_m = d\nu + (d-1)/2$ with a correction $(d-1)/2$ to the failure exponent of the discrete-lattice case.

2.2. OUT OF PLANE TWO-DIMENSIONAL RUPTURE

2.2.1. *Crack opening.* In two dimensions, in-plane deformations can only occur via the opening or overlapping of cracks or bending of the micro-bonds. In the case where the width a of the cracks is very small (smaller than the smallest bond width δ), the bending of a micro-bond involves a crack opening or overlapping by a given angle proportional to the applied stress. In this case, the failure problem can be analysed with the tools of Ref. 2. 2d-mechanical failure is dominated by the weakest bond in a macro-link. Supposing that rupture will occur if the bending elastic energy $E \approx \gamma^{-1} M_\xi^2$ in that bond become larger than a threshold value E_c where[11] $\gamma \propto \delta^2$ and $M_\xi \approx \sigma \xi^d$, one obtains the failure threshold

$$\sigma \approx (p - p_c)^{\mathbf{F}m}, \tag{18}$$

with $\mathbf{F}_m = 2\nu + 1$ larger than the discrete result $F_m = 2\nu$ by one. If rupture occurs when the angle θ by which the neck is bent is larger that a threshold value θ_c (second criterion of Ref. 2), one has $\mathbf{F}_m = 2\nu + 2$.

2.2.2. *Buckling failure.* If the macroscopic 2d-plate is not rigidly clamped, it can become preferable for the elastic plate not only to have the crack edges to overlap slightly but even to buckle out of the plane in order to relax the in-plane elastic stresses. In this case, the screening of the strain by buckling is so efficient that one recovers the discrete-lattice exponents $f_m = d\nu$.[11]

We have performed experiments in our laboratory[16] on the rupture of aluminium sheets of thickness $\sim 100\mu$m and typical size 30×30cm. Each sheet is deteriorated by N identical cracks of length in the range 2-4cm. As we apply an

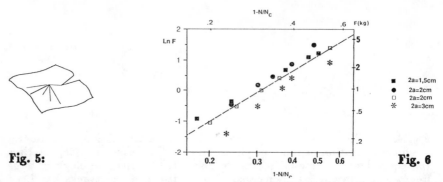

Fig. 5:   **Fig. 6**

increasing tensile stress at the two end of the sheet, we observe a complex buckling structure which develops and ends eventually in the complete tearing of the sheet. Figure 5 represents the failure threshold stress as a function of the number of cracks in the vicinity of the percolation threshold N_c. We measure a rupture exponent $F_m = 2.5 \pm 0.1$ in good agreement with the expected result $2\nu = 2.66$. It is remarkable that we are able to reach the critical regime in such small systems ($N < 1000$). This can be explained by the very efficient screening of the stress concentration by the numerous random cracks surrounding each crack. In a similar experiment where the cracks are put in two orientations and on a regular lattice, the apparent failure exponent is around 1 which signals that we are outside the critical region. In this case, stress concentration screening is much less efficient and long-range interactions between cracks are present. These problems and other related ones are under active study and will be reported in.[16]

3. Conclusion

We have given a few illustrations of the questions encountered in failure. Many problems are still unsolved:

(1) Is there universal behavior in the breaking characteristics of a disordered system which are independent of the details of the models? What are the corresponding universality classes? Can we define a renormalization group governing the evolution with scale of the distribution of rupture strength?

(2) What is the relative importance of the initial quenched disorder and the "growth" aspect of the failure? In the infinite disordered limit,[17] one expects that failure should be controlled by the disorder. In this limit, failure becomes a purely geometrical problem whereas for finite disorder, there is a transition to a regime where local stress fluctuations control largely the failure.

(3) The problem of dynamics of rupture has been tackled in very simple situations. Can we develop analogies with dendritic growth or diffusion limited aggregation or dielectric breakdown? How can we understand the non-local propagation of a crack in presence of large quenched disorder?

These opened questions are likely to receive interesting and profound answers in the coming years.

1. See for example *Chance and Matter*, Les Houches, Session XLVI, eds. J. Souletie, J. Vannimenus and R. Stora (North Holland, Amsterdam, 1987) and references therein.

2. D. Sornette, J. Phys. Paris 48, 1843 (1987).

3. D. Sornette, J. Phys. Paris 49, 889 (1988).

4. B. I. Halperin, S. Feng and P. N. Sen, Phys. Rev. Lett. 54, 2391 (1985).

5. A. Skal and B. Shklovskii, Sov. Phys. Semicond. 8, 1029 (1976); P. G. de Gennes, J. Phys. Lett. 37, L1 (1976).

6. A. Coniglio, Phys. Rev. Lett. 46, 250 (1981); R. Pike and H. E. Stanley, J. Phys. A 14, L169 (1981).

7. After the breaking of a macro-link, the stress is transfered to the other macro-links. In reality, the macro-links are not exactly identical. It is therefore possible that after a single macro-link has failed, the other macro-links will resist under the redistributed total applied stress. This problem involves the competition between (i) interaction between links (screening and enhancement effects as well as the detailed stress transfer mechanism) and (ii) the failure distribution strength of the macro-links. A similar problem has been addressed in L. De Arcangelis, A. Hansen, H. J. Herrmann and S. Roux, "Scaling Rules in Fracture," preprint.

8. E. J. Gumpel, *Statistics of Extremes* (Columbia University Press, New York, 1958); J. Galambos, *The Asymptotic Theory of Extreme Order Statistics* (J. Wiley, New York, 1978).

9. P. M. Duxbury, P. D. Beale and P. L. Leath, Phys. Rev. Lett. 57, 1052 (1986); A. Gilabert, C. Vanneste, D. Sornette and E. Guyon, J. Phys. Paris 48, 763 (1987); E. Guyon, S. Roux and D. J. Bergman, J. Phys. Paris 48, 903 (1987).

10. D. Stauffer, in *On Growth and Form*, eds. H. E. Stanley and N. Ostrowsky (Martinus Nijhoff, Amsterdam, 1986).

11. D. Sornette, J. Phys. Paris 49, 1365 (1988).

12. E. Guyon, see Ref. 1.

13. See for example, S. Crampin, Nature 328, 491 (1987).

14. C. J. Allegre, J. L. Le Mouel and A. Provost, Nature 197, 47 (1982).

15. H. D. Bui, *Mécanique de la rupture fragile* (Masson Ed., 1978); G. C. Sih (ed.), *Mechanics of Fracture 1: Methods of Analysis and Solutions of Crack Problems* (Noordhoff International Publ., Leyden, 1973).

16. M. Benayad, "Etude expérimentale des propriétés mécaniques (rupture) et électrique de systèmes de percolation," Thèse troisieme cycle, Nice (1988); A. Gilabert, M. Benayad, D. Sornette and C. Vanneste, to be published.

17. S. Roux, A. Hansen, H. J. Herrmann and E. Guyon, "Rupture of Heterogeneous Media in the Limit of Infinite Disorder," J. Stat. Phys. (in press).

NICOLE OSTROWSKY AND NAEEM JAN

FRACTURE EXPERIMENTS ON MONOLAYERS OF MICROSPHERES

ARNE T. SKJELTORP

Institute for Energy Technology
N-2007 Kjeller, Norway

ABSTRACT. The development of fracture patterns in thin films is explored by using a model system of uniformly sized microspheres confined to regular monolayers between two glass plates. Fracture is introduced due to a shrinking of the spheres during drying. The results show a succession of slow and fast crack propagation. As an increased number of cracks is formed, they become more and more irregular and finally percolating the whole sample. The final fracture patterns are shown to have an effective fractal dimensionality $D = 1.68 \pm 0.06$.

1. Introduction

A wide variety of irregular and sometimes fractal[1] objects are found in nature as a result of processes far from equilibrium. A special class of phenomena of this type is fracture. This is of considerable interest in material science[2] and it has been suggested that fracture surfaces of metals may have fractal character.[3] Various model simulations of fracture in regular two-dimensional (2D) systems subjected to expansion or shear have shown fractal patterns.[4,5] Other model work has stressed the importance of irregular grain boundaries for producing fractal fracture.[6]

The purpose of this presentation is to report the results of cracking in a monolayer of uniformly sized microspheres. This is an interesting and simple model system to gain insight into fracture processes and it also resembles real systems like paint films and metals coated with ceramics. Meakin has recently introduced a model for elastic fracture of thin films[7,8] believed to be close to the present physical realization.

2. Experimental

The production of planar lattices for the fracture experiments was realized using uniformly sized sulfonated polystyrene spheres[9] of effective diameter $d_1 = 3.4\mu m(\pm1\%)$ dispersed in water. By confining the spheres between planar glass plates it was possible to form a polycrystal with relatively large crystalline grains containing typically $10^5 - 10^6$ spheres. An optical microscope with video-camera attachment allowed direct long-term observations and digital analysis of the patterns with use of a frame grabber with 512×512 pixels resolution.

The fracturing was realized as the result of particle shrinking during a slow drying process reducing the sphere diameter to $d_2 = 2.7\mu m$. This produced a strained film with strong bonding between the spheres and relatively weak bonding to the glass surfaces.

3. Results and Discussion

The evolution of fracture in the whole sample after the monolayer has dried, shows many characteristic features: The first cracks are formed along the grain boundaries where fewer bonds have to be broken than inside the grains. The cracks also have a tendency to pass lattice defects like vacancies and impurity inclusions. However, for the results to be discussed in the following, observations are made inside regular grains with defects limited to dislocations. Initially, the whole film is strained leading to a slow formation of local defects which grow. This produces high stresses at the tips of some microcracks resulting in rapid propagation of more or less linear cracks, Figs. 1(a) - (c). The number of cracks increases with an intricate succession of branching into uncracked regions, Fig. 1(d). As no forces are transmitted across the existing cracks, no cracks penetrate other cracks. Gradually, the average stress field in the system is reduced and the cracking slows down. A typical final crack pattern is shown in Fig. 2 for two different magnifications. The patterns look intricately random with successively smaller and more irregular cracks reflecting the evolution of an increasingly irregular global stress field with time.

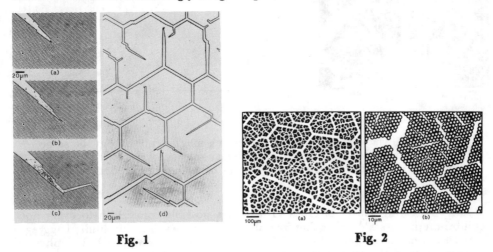

Fig. 1

Fig. 2

Fig. 1: Optical micrographs of the initial stages of the crack formation inside a single grain of a much larger system: (a) - (c) propagation of one crack; (d) later stage in the crack propagation process.

Fig. 2: Optical micrographs of a typical final fracture pattern for (a) low and (b) high magnification.

The similar appearance of the cracks for different magnifications, signifies a possible random fractal scaling with dimensionality D. To investigate this, D was calculated from the final fracture patterns for four different magnifications, Figs. 3(a) - (d), using the box counting technique.[10] The number of $L \times L$ square boxes, N, needed to cover the cracks was counted and averaged over different center points. For a fractal structure it is expected that

$$N \propto L^{-D}. \tag{1}$$

Figure 3(e) shows a log-log plot of N versus L changing by a factor of about 300. As may be seen, the data appear to fall on a straight line with slope $D = 1.68 \pm 0.06$ in accordance with Eq. (1).

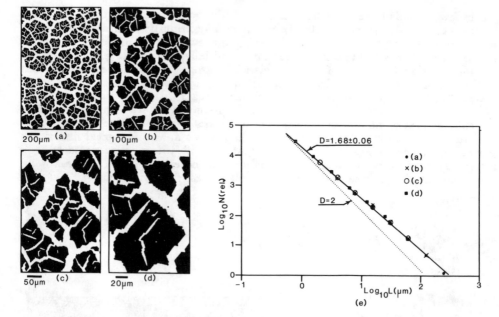

Fig. 3: Final fracture patterns for increasing magnification from the same area of the sample [(a) through (d)]. Determination of the fractal dimension D of the fracture patterns (a)-(d), as discussed in the text, is shown in (e).

An interesting aspect of crack growth is a possible connection to other non-equilibrium processes[11] producing fractal objects such as diffusion-limited aggregation (DLA), dielectric breakdown (DB), and viscous fingering (VF). The physical reasoning behind the conjectured universality among the growth patterns in these processes is that the Laplace equation applies for generalized scalar fields equal to the probability density of random walkers in DLA, electric potential for DB and pressure for VF. Recently, a two-dimensional computer model[4] has also been used to relate crack growth to Laplacian fractal growth[11] introducing a vectorial displacement field. The resulting fracture patterns appear to have fractal character with $D \simeq 1.6 - 1.7$ and thus quite close to the value found from two-dimensional DLA simulations with $D = 1.71$.[11]

Although the crack patterns in the present experiments seem to have an effective fractal dimensionality quite close to the DLA-value, there appears to be no direct formal connection to the DLA-related pattern growth which is a non-local process. In contrast, the weak bonding between the microsphere monolayer and substrate for the present system localizes the effects of defects. Indeed, without this bonding the cracking would stop after a few cracks had relieved the stress. A more detailed discussion of these aspects is given by Meakin in these proceedings.

4. Conclusions

In conclusion, we have introduced an experimental model system of microsphere monolayers to study the development of fracture. The observations support the concept of fractal scaling for the fracture patterns with an effective fractal dimensionality $D = 1.68 \pm 0.06$. The results also support recent model simulations for elastic fracture in films.

Acknowledgements. This research has been supported in part by Dyno Industrier A/S and NAVF. The supply of samples from John Ugelstad and collaborators at SINTEF and valuable comments from Paul Meakin are gratefully acknowledged.

1. B. B. Mandelbrot, *The Fractal Geometry of Nature* (Freeman, San Francisco, 1982); J. Feder, *Fractals* (Plenum, New York, 1988).

2. R. M. Latanison and R. H. Jones (eds.), *Chemistry and Physics of Fracture* (Nijhoff, 1987).

3. B. B. Mandelbrot et al., Nature 308, 721 (1984).

4. E. Louis and F. Guinea, Europhys. Lett. 3, 871 (1987).

5. H. Takayasu, in *Fractals in Physics*, eds. L. Pietronero and E. Tosatti (Elsevier, Amsterdam, 1986), p. 181.

6. C. W. Lung, in *Fractals in Physics*, eds. L. Pietronero and E. Tosatti (Elsevier, Amsterdam, 1986), p. 189.

7. P. Meakin, Thin Solid Films 151, 165 (1987).

8. A. T. Skjeltorp and P. Meakin Nature (in press).

9. J. Ugelstad et al., Adv. Colloid Interface Sci. 13, 101 (1980); produced by Dyno Particles, N-2001 Lillestrøm, Norway.

10. R. F. Voss, in *Scaling Phenomena in Disordered Systems*, eds. R. Pynn and A. Skjeltorp (Plenum, New York, 1985), p. 1.

11. P. Meakin, Phase Transitions 12, 335 (1988).

BENOIT MANDELBROT, GENE STANLEY AND PREBEN ALSTRØM

SIMPLE MODELS FOR COLLOIDAL AGGREGATION, DIELECTRIC BREAKDOWN AND MECHANICAL BREAKDOWN PATTERNS

PAUL MEAKIN

Central Research and Development Department
E. I. du Pont de Nemours and Company
Wilmington, DE 19898 U.S.A.

ABSTRACT. Some simple models for colloidal aggregation, dielectric breakdown and mechanical breakdown patterns are described. These non-equilibrium growth and aggregation models frequently lead to the formation of complex structures which have a random fractal geometry. They provide a basis for understanding a wide range of phenomena of both practical and scientific importance. In some instances experimental realizations of these simple models have been found. The models discussed in these lectures also provide a sound basis for the development of more elaborate, but more realistic, models which can be applied to a broader range of processes.

In recent years considerable interest has developed in the formation of random patterns under non-equilibrium conditions. A rich phenomenology is found in most pattern formation processed, but in a variety of important cases quite simple models provide a basis for understanding these processes.[1] The purpose of these lectures is to show how simple models can be used to explore the generation of complex (often fractal[2]) patterns by both aggregation and material failure processes.

1. Colloidal Aggregation

The aggregation of small particles to form large structures is important in many systems of both scientific and practical importance. More than 20 years ago simple models for colloidal aggregation were developed by Vold[3] and Sutherland[4-7]. In recent years a strong resurgence of interest in this approach towards development of a better understanding of non-equilibrium growth and aggregation processes has occurred. This development was stimulated by the discovery of the diffusion-limited aggregation (DLA) model by Witten and Sander[8] which demonstrated that very simple processes could lead to the generation of complex fractal patterns which closely resemble those generated by natural processes. The diffusion-limited aggregation model led to the development of a variety of models for colloidal aggregation including the diffusion-limited cluster-cluster aggregation model[9,10] and the reaction-limited aggregation model.[11,12] Despite the fact that these models generate a wide variety of fractal and compact patterns associated with a broad range of physical processes, they are all closely related and can be described in terms of a single general model. In this model we start with a large number (N_o) of particles (single particle clusters) which form a list of clusters. As the simulation proceeds, pairs of clusters are selected from the list and combined irreversibly to form a larger cluster which is returned to the list (which now contains one less cluster). As the simulation proceeds, the clusters get larger and larger and fewer and fewer. The models differ in the way the clusters are selected and the way in which they are combined. In particle-cluster aggregation models such as the Vold-Sutherland (ballistic aggregation)[3-5], DLA[8] and Eden[13] models a single particle is always added to the largest cluster in the system. The most important characteristic of the way in which the clusters are brought together is the fractal dimension D_t of the trajectories which they follow. The dimension of the trajectory is 2 (diffusion), 1 (ballistic) and 0 for the DLA, Vold-Sutherland and Eden models respectively. Simulations have also been carried out using fractal (Levy flight[14] and Levy walk[15]) trajectories with dimensions in the range $1 < D_t < 2$. In almost all cases both off-lattice and lattice models have been explored. In most cases they generate clusters with the same fractal dimension (within the accuracy of the simulation). However, for the DLA model lattice anisotropy has an important effect on the cluster shape and fractal dimension[16−18].

Figure 1 shows clusters generated using two dimensional off-lattice versions of the Eden, Vold-Sutherland and DLA models. The effective fractal dimensions for $d = 2$, 3 and 4 are summarized in Table I.[19]

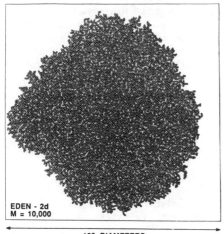

EDEN - 2d
M = 10,000

150 DIAMETERS

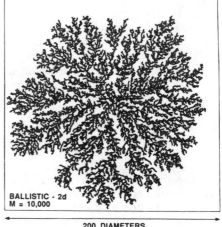

BALLISTIC - 2d
M = 10,000

200 DIAMETERS

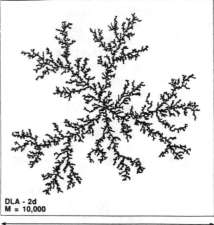

DLA - 2d
M = 10,000

350 DIAMETERS

Fig. 1: Typical clusters generated using two dimensional off-lattice particle-cluster aggregation models. Except for the dimension of the particle trajectories (D_t), these models are very similar. Figure 1a shows a cluster generated using the Eden $(D_t = 0)$ model. Figures 1b and 1c show clusters generated using the ballistic aggregation $(D_t = 1)$ and diffusion-limited aggregation $(D_t = 2)$ models respectively.

In cluster-cluster aggregation models clusters of more or less the same size are irreversibly combined. In hierarchical models[20] which were first investigated by Sutherland[6,7] the simulation is started with $N_o = 2^m$ particles which are used to form 2^{m-1} binary clusters, then 2^{m-2} clusters of four particles each and in the n^{rmth} stage 2^{m-n} clusters each containing 2^n particles. In more realistic cluster-cluster aggregation models pairs of clusters are selected with probabilities which are determined by appropriate cluster mass dependent reaction kernels or by their collision cross-sections.[21] Figure 2 shows clusters generated using two dimensional (polydisperse) models of this type with trajectory dimensions of $D_t = 2$ (diffusion-limited cluster-cluster aggregation), $D_t = 1$ (ballistic cluster aggregation) and $D_t = 0$ (reaction-limited cluster aggregation). The effective fractal dimensions associated with these models are given in Table I. Recent simulation results[22,23] and theoretical considerations[24-26] indicate that the fractal dimension of cluster-cluster aggregates is smaller for hierarchical (monodisperse) models than for the more realistic (polydisperse) models. This means that the range of universality for these models is at best quite small (The fractal dimension, D, depends on model details). Nevertheless, essentially the same fractal dimension is found for quite different three-dimensional systems undergoing colloidal aggregation under fast aggregation conditions (diffusion-limited aggregation, $D \simeq 1.80$) and slow aggregation conditions (reaction-limited aggregation, $D \simeq 2.10$).[27-30]

The simple reaction-limited and diffusion-limited cluster aggregation models have proven to be quite successful in describing the results of aggregation experiments under well controlled conditions. However, most aggregation processes lead to structures which do not closely resemble those predicted by these simple models. However, these models do provide a basis for the development of more realistic models which may enable us to obtain a better understanding of more complex aggregation processes. For example, in real aggregation processes clusters may be held together by weak Van der Waals forces before they become irreversibly and rigidly bound to each other via metallic bonding, sintering, silicate bridge formation, covalent bonding, etc. Under these conditions pairs of clusters may be able to reorganize after they come into contact with each other. Figure 3 shows the results of a two dimensional simulation[31] in which pairs of rigid clusters are allowed to reorganize after contact. In this model cluster 1 is rotated about the center of the contacting particle in cluster 2 until a second contact is formed between the two clusters and (if possible) cluster 2 is then rotated about the center of the contacting particle in cluster 1 until a third contact is formed. This type of restructuring process has been observed directly[32] in two dimensional aggregation experiments and leads to structures which quite closely resemble that shown in Figure 3.

For two dimensional cluster-cluster aggregation models the effects of the restructuring mechanisms discussed in the previous paragraph are quite small (not much larger than the statistical uncertainties). In three dimensions rotational restructuring takes place in three stages (bending, folding and twisting). In this case the effects on the fractal dimensions of the clusters are somewhat larger.[23,33] Results obtained for the effects of restructuring on three dimensional polydisperse models for reaction-limited, ballistic and diffusion- limited aggregation are given in Table II.

It has been recognized for a long time that attractive and repulsive interactions can have an important effect on the structure of colloidal aggregates. For example, Hurd and Schaefer[34] have investigated the aggregation of silica particles confined to an air/water interface and attributed the low fractal dimension ($D = 1.20 \pm$

Reaction Limited $D_t = 0$

dimension of space of lattice	Particle-cluster	Polydisperse cluster-cluster	Hierarchical cluster-cluster
2	$2.0^{0,\ell,t}$	1.61^0	1.54^0
3	$3.0^{0,\ell,t}$	2.10^0	2.00^0
4	4.0^t	2.49^0	2.37^0

Ballistic $(D_t) = 1$

dimension of space of lattice	Particle-cluster	Polydisperse cluster-cluster	Hierarchical cluster-cluster
2	$2.0^{0,\ell,t}$	1.55^0	1.51^0
3	$3.0^{0,\ell,t}$	1.95^0	1.89^0
4	$4.0^{0,\ell,t}$	2.24^0	2.22^0

Diffusion-Limited $(D_t) = 2$

dimension of space of lattice	Particle-cluster	Polydisperse cluster-cluster	Hierarchical cluster-cluster
2	1.71^0	1.45^0	1.44^ℓ
3	2.50^0	1.80^0	1.78^0
4	3.40^0	2.10^0	2.02^ℓ

ℓ — lattice model
o — office lattice model
t — theoretical

Table I: Fractal dimensions obtained from some simple aggregation models. Results from the most extensive large scale simulations availabble in June 1988 are shown here for each model.

0.15) of the aggregates to anisotropic repulsive interactions. Similarly, Mors, Botet and Jullien[35,36] have used computer simulations to explore the effects of dipolar interactions on three-dimensional cluster-cluster aggregation processes and have applied their results to the experiments of Kim et al.[37,38] on the aggregation of iron and cobalt particles. They found that D varies continuously from a value of about 1.35 for large dipole moments to about 1.78 for small dipole moments. These results seem to be consistent with those obtained by Kim et al. ($D \simeq 1.34$ for iron (large dipole moment) and $D \simeq 1.72$ for cobalt (small dipole moment)). Niklasson et al.[39] have investigated the structure of cobalt particle aggregates formed from Co vapor in argon at pressures in the range of 0.25 to 10 Torr. Under these conditions ballistic cluster-cluster aggregation leading to aggregates with a fractal dimension of about 1.95 would be expected in the absence of long range interactions. Instead, aggregates with effective fractal dimensions in the range $1.35 \leq D \leq 1.60$ were formed for large particles. For small particles (with radii ≤ 8 nm) clusters with fractal dimensions in the range $1.95 \leq D \leq 2.05$ were formed. The small fractal

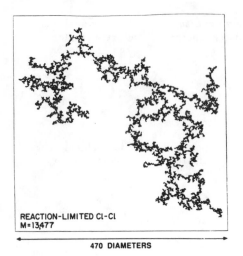

REACTION-LIMITED Cl-Cl
M=13,477

470 DIAMETERS

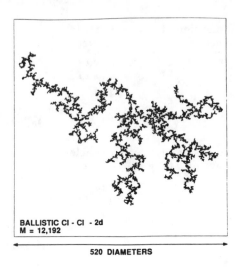

BALLISTIC Cl - Cl - 2d
M = 12,192

520 DIAMETERS

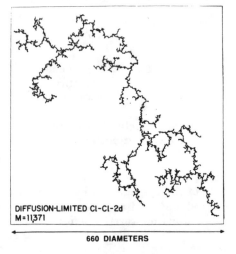

DIFFUSION-LIMITED Cl-Cl-2d
M=11,371

660 DIAMETERS

Fig. 2: Representative two dimensional clusters obtained from polydisperse cluster-cluster aggregation models. Figures 2a-2c show clusters obtained using cluster trajectories with fractal dimensions (D_t) of 0 (reaction-limited aggregation), 1 (ballistic aggregation) and 2 (diffusion-limited aggregation) respectively.

dimensions for the aggregates of large particles was attributed to the effects of magnetic dipole interactions when the critical size for ferromagnetism is exceeded.

Figure 4 shows some results from simulations carried out using a two dimensional off-lattice model for reaction-limited cluster-cluster aggregation[40] with pairwise particle-particle interactions. In the reaction-limited cluster-cluster aggregation models pairs of clusters are selected randomly from all possible bonding configurations (contacting but non-overlapping configurations) and joined irreversibly. In the models used to obtain the clusters shown in Fig. 4 the bonding configurations do not all have equal probabilities. Instead the interaction between pairs of clusters

Fig. 3: A cluster generated using a two dimensional diffusion-limited cluster-cluster aggregation of rigid clusters with restructuring immediately after contact. In this model the particles in one cluster are rotated about the center of the contacting particle in the other cluster until an additional contact between the two clusters has been formed. The smaller of the two possible angles of rotation was selected and two stages of restructuring (in which first one cluster and then the other is rotated) were attempted.

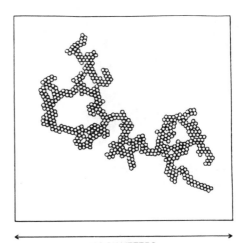

60 DIAMETERS

Model	Number of Restructuring Stages			
	0 (None)	1 (Bending)	2 (Bending +Folding)	3 (Bending+Folding +Twisting)
Reaction-limited	2.09	2.18	2.24	2.25
Ballistic	1.95	2.13	2.18	2.19
Diffusion-limited	1.80	2.09	2.17	2.18

Table II: Effective fractal dimensions obtained from three dimensional polydisperse. Cluster-cluster aggregation models with smallest angle restructuring.

of size i and j is assumed to be given by

$$E_{ij} = A \sum_{k=1}^{i} \sum_{\ell=1}^{j} \frac{1}{r_{k\ell^a}}.$$ (1)

In Eq. (1) the interaction energy is in units of kT and $r_{k\ell}$ is the distance between the center of the kth particle in cluster 1 (of size i) and the ℓth particle in cluster 2 (of size j). The probability that this bonding configuration will be selected is then proportional to the corresponding Boltzmann factor. Figure 4 shows that attractive interactions (negative A) lead to more compact clusters and that repulsive interactions (positive A) lead to more "stringy", less compact, clusters. The results

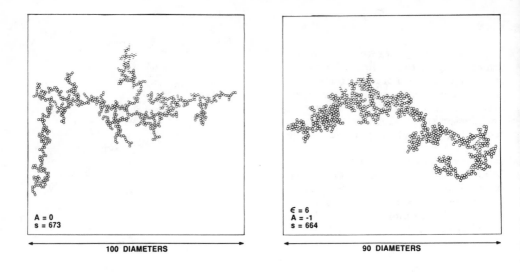

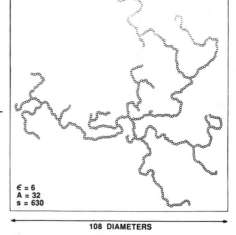

Fig. 4: Clusters obtained from two dimensional off-lattice simulations of reaction-limited cluster-cluster aggregation with no interactions (Fig. 4a), attractive interactions (Fig. 4b) and repulsive interactions (Fig. 4c).

shown in Figure 4 were obtained using a value of 6 for the exponent ϵ in the interaction potential [Eq. (1)]. However, similar results were obtained for other values of ϵ (2 and 4) and from three dimensional simulations (for $\epsilon = 2$, 4 and 6).

Clusters like those shown in Fig. 4 can be more quantitatively characterized via the particles, s). Figure 5 shows the dependence of $\ln(R_g/s^{0.7})$ on $\ln(s)$ obtained from simulations carried out using a value of 6 for the exponent ϵ and several different values for the interaction strength parameter A. The results displayed in Fig. 5 indicate that for the values of A investigated in these simulations the effect on the effective fractal dimension is quite small despite the quite dramatic changes in the cluster geometries shown in Fig. 4. Similarly, for $\epsilon = 4$ the effect of attractive and repulsive interactions on the effective fractal dimension is small,

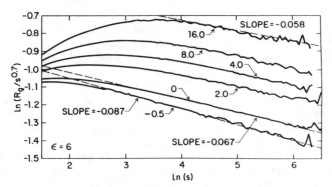

Fig. 5: Dependence of $\ln(R_g/s^{0.7})$ on $\ln(s)$ obtained from two dimensional simulations carried out using a polydisperse off-lattice model for reaction-limited aggregation with both attractive (negative A) and repulsive (positive A) interactions. In these simulations a value of 6 was used for the exponent ϵ. The value of the interaction parameter A is shown by each curve.

but for $\epsilon = 2$ the fractal dimension is changed substantially. An effective fractal dimension of about 1.25 is obtained for the largest values of A for which simulations are practical.

This lecture has been confined to a discussion of some aspects of the use of computer simulations to explore the structure of colloidal aggregates. The kinetics of cluster aggregation and the physical properties of fractal aggregates are subjects which are of at least equal interest and importance. They also have been studied exclusively using computer simulations. Recent reviews concerned with these and other aspects of non-equilibrium aggregation processes can be found in Ref. 1, 18, 19, 21, 41-43.

2. Dielectric and Mechanical Breakdown Models

In the diffusion-limited aggregation (DLA) model of Witten and Sander[8] particles are added to a growing cluster, one at a time, via random walk trajectories. In the original square lattice model random walkers are launched from outside of the region occupied by the growing cluster and follow random walks on the lattice until they either move far from the cluster or enter an unoccupied perimeter site (an unoccupied site with one or more occupied nearest neighbors). In the latter event the unoccupied perimeter site is filled and the cluster grows by one site. Witten and Sander[44] discussed the analogy between DLA and electrostatics. This analogy was explicitly exploited by Niemeyer et al.[45] to develop a model for dielectric breakdown. In this model the Laplace equation ($\nabla^2\phi = 0$) is solved numerically on a lattice with the boundary condition $\phi = 1$ on some distant boundary enclosing the cluster and $\phi = 0$ for sites occupied by the cluster. The growth probabilities (P_i) for each of the unoccupied perimeter sites are then given by

$$P_i \alpha z \phi_i^\eta, \tag{2}$$

where ϕ_i is the potential at the ith size and z is its coordination number (number of occupied nearest neighbors). For the case $\eta = 1$ clusters generated using this model seem to have the same fractal dimension as clusters grown using the corresponding DLA model[45,46]. Dielectric breakdown model simulations (for $\eta = 1$) can also be carried out using random walkers.[47] In this model growth occurs in the last unoccupied site entered by a random walker before it moves onto an occupied site and the walk is terminated. Although the dielectric breakdown and DLA models seem to have the same asymptotic scaling behavior on Euclidean lattices (but not necessarily on fractal lattices[48]) the structures are quite different on short length scales. This is particularly apparent in noise reduced growth models[47,49] in which perimeter sites must be selected m times before they are finally occupied. In these models the random walkers are terminated when they reach the surface of a growing cluster and the "score" associated with the corresponding unoccupied perimeter site is incremented by one. When this score finally reaches a value of m the perimeter site is filled and any new perimeter sites created by this process start with a score of zero. Figure 6 shows clusters grown on a square lattice with noise reduction parameters of 10,000 using both the DLA (Fig. 6a) and dielectric breakdown (Fig. 6b) boundary conditions (both simulations were carried out using random walkers). The cluster shown in Fig. 6a consists of a solid cross-shaped core[50] with side branches beginning to develop. The cluster shown in Fig. 6b has a quite different shape. At this stage there is no side branching but each arm of the cluster has undergone tip splitting at about the same stage in its growth.

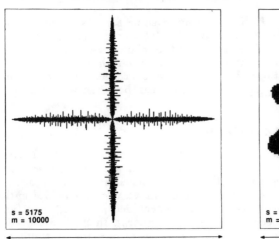
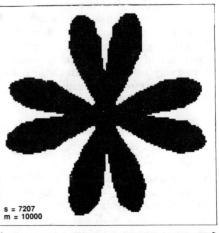

s = 5175
m = 10000

s = 7207
m = 10000

360 LATTICE UNITS   140 LATTICE UNITS

Fig. 6: Figure 6a shows a cluster of 5175 sites grown on a square lattice using the DLA model with a noise reduction parameter (m) of 10,000. In Fig. 6b a cluster of 7207 sites (also generated with $m = 10,000$) which was grown using dielectric breakdown boundary conditions is shown. Both clusters were grown using random walkers to simulate a field obeying the Laplace equation.

The development of the dielectric breakdown version of the DLA model suggests that it might be possible to represent other non-equilibrium breakdown processes in which random growth processes are controlled by a non-local field by similar models. In particular, it is reasonable to expect that a similar approach might be of value in developing a better understanding of mechanical failure processes controlled by the vector stress and strain fields in a linear elastic medium. In this case the stress and strain fields are related by the Navier equation[51]

$$(\lambda + \mu)\partial_i \left[\sum_j \partial_j U_j \right] + \mu \left[\sum_j \partial_j^2 \right] U_i = 0, \tag{3}$$

where λ and μ are the Lame coefficients and U_i is the ith component of the displacement field.[52]

Louis and Guinea[53] have developed a model for crack propagation which is quite closely related to the DLA and dielectric breakdown models. Here results from a closely related model[54] are discussed. In this model the elastic medium in which crack propagation is occurring is represented by a triangular network of nodes connected by Hookean springs. At the start of a simulation, each of the nodes (except for those at the edge of the system) is connected to six nearest neighbors. For this system the elastic energy (E) is given by

$$E = \frac{1}{2} \sum_{ij} k_{ij}(\ell_{ij} - \ell_0)^2, \tag{4}$$

where ℓ_{ij} is the length of the bond joining the ith and j^{th} nodes and k_{ij} is the force constant associated with the bond joining these nodes. Here $k_{ij} = k$ if the nodes are joined and $k_{ij} = 0$ otherwise. At the start of each simulation the array of bonds and nodes is either isotropically dilated (typically by 0.1%) or sheared in either the X or Y directions by the transformation $X_i \rightarrow X_i + \delta X_i$ or $Y \rightarrow Y_i + \delta Y_i$. In most of the simulations a value of 0.01 was chosen for δ. For both dilation and shear, small strains were used to ensure that the simulation results were not influenced by non-linear effects. Simulations with larger strains were also carried out to explore the effects of these non-linearities. The results showed that for the values of the dilation and shear strains indicated above the simulations are well within the linear regime.

At the start of a simulation, a bond near to the center of the network is broken and the system is relaxed to mechanical equilibrium (the elastic energy in (4) is minimized) using standard relaxation methods[55] including block relaxation and overrelaxation. In addition, extra relaxation cycles were used for those bonds in the vicinity of the (last) broken bond. The relaxation procedure was stopped when the largest displacement of any node from its local equilibrium position is smaller than 0.01 times the initial bond strain. This would not ordinarily be sufficient to guarantee an accurate convergence, but relaxation of the whole system continues as other bonds are broken and the additional relaxation near to the last broken bond ensures almost complete relaxation in this region. Tests with different values for the largest local displacement indicated that the criterion for stopping the relaxation was adequate. After the network has been relaxed, the strains (δ_i) associated with each of the bonds at the surface of the "crack" formed by the broken bond (s) are obtained and one of these bonds is randomly selected with probabilities given by

$$P_i = \frac{(\delta_i)\eta}{\sum_{i=1}^N (\delta_i)\eta}, \tag{5}$$

where δ_i is the strain associated with the ith bond at the crack surface and N is the number of surface bonds. The crack propagation process is then stimulated by a sequence of random bond breaking and relaxation stages. The positions of the nodes at the edges of the network are maintained at fixed positions throughout the simulations.

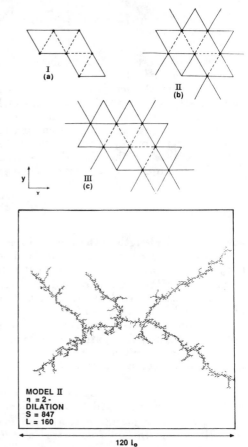

Fig. 7: Definition of the surface bonds used in models I, II and III respectively (Figs. 7a, 7b and 7c). The broken bonds are indicated by dashed lines and the damaged nodes by large dots. The solid lines indicate those bonds (the crack surface bonds) in the triangular network which may be broken in the next stage of the crack growth simulations.

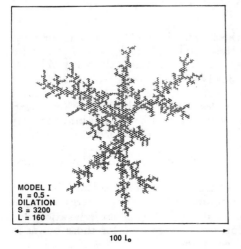

Fig. 8: Typical cracking patterns obtained using the model illustrated in Fig. 7 and described in the text. Here η is the growth probability exponent [Eq. (5)], s is the crack size (number of broken bonds) and L is the size of the network.

Several versions of the model were investigated in which different definitions of the crack surface bonds (bonds which may be broken) were used. In model I (Fig. 7a) only those bonds which are at the edge of the crack may be broken. In model II (Fig. 7b) those bonds associated with "damaged" nodes (nodes for which one or more of the associated bonds is already broken) may be broken and in model III (Fig. 7c) all of the bonds associated with all of the nodes at the crack surface

Growth Exponent (η)	Stress Field	Model	D_{eff}
1.0	Dilation	I	1.35
1.0	Dilation	II	1.51
1.0	Dilation	III	1.66
1.0	Shear	I	1.42
1.0	Shear	II	1.62
1.0	Shear	III	1.65
2.0	Dilation	I	1.12
2.0	Dilation	II	1.16
2.0	Dilation	III	1.45
2.0	Shear	I	1.17
2.0	Shear	II	1.49
2.0	Shear	III	1.40

Table III: Effective fractal dimensions obtained from the crack growth models illustrated in Fig. 7 and described in the text.

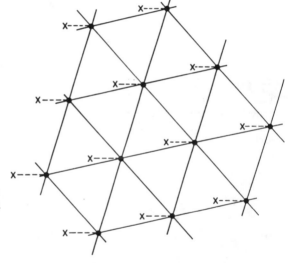

Fig. 9: A schematic representation of the model for elastic fracture in thin films. The nodes (large dots) are connected by strong bonds to form a triangular lattice. Each node is joined to the underlying substrate by a weak bond (---) at its original position at the start of the simulation. Throughout the simulation the distance from the nodes to the underlying substrate is constant (only horizontal motion is allowed). This figure shows the original configuration in which each node is associated with six bonds.

may be broken. In most of the simulations, the triangular network of bonds and nodes consisted of 160 rows of 160 nodes (the equilibrium size of the system is 160ℓ in the X direction and $160(\sqrt{3/2})\ell$ in the Y direction). The crack propagation process was stopped before the crack tips approached closely to the edges of the network. Typical cracking patterns generated by this model are shown in Fig. 8. In both cases dilational strain was used. In Fig. 8a a crack generated using model I (Fig. 8a) with a value of 0.5 for the growth probability exponent η [Eq. (5)] is shown. The pattern shown in Fig. 8b was obtained using model II (Fig. 8b) and a value of 2.0 for η. These figures show the location of the broken bonds in the undistorted network. Simulations have been carried out for all three models with

Fig. 10: The results of a simulation of elastic fracture in a surface film attached to a rigid substrate. This figure shows the locations of the broken bonds (in the unstrained network) after 5000 bonds have been broken. The long linear cracks were the first to grow and these cracks were connected at a later stage by the slower growth of the more irregular cracks. The model parameters used in the simulation are indicated on the figure.

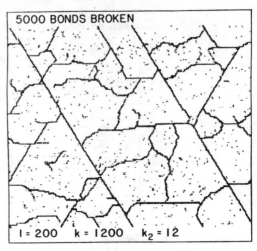

Fig. 11: The results of a simulation of elastic fracture in a surface film attached to a rigid substrate using a square network instead of a triangular network.

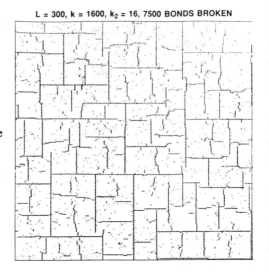

both dilation and shear stress for $\eta = 0.5$, 1.0 and 2.0. Effective fractal dimensions (D_β) can be obtained from the dependence of the radius of gyration (R_g) of the cracking patterns on the number of broken bonds (s). In all cases the dependence of R_g on s can be described quite well by

$$R_g \sim s^\beta, \tag{6}$$

and the effective fractal dimension D_β is then given by $D_\beta = 1/\beta$. For all the simulations carried out with $\eta = 0.5$ effective fractal dimensions in the range $1.90 \leq$

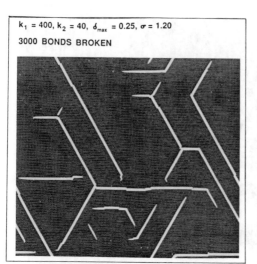

k$_1$ = 400, k$_2$ = 40, δ_{max} = 0.25, σ = 1.20

3000 BONDS BROKEN

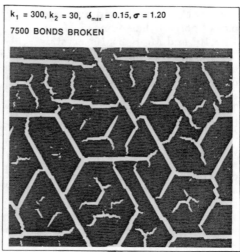

k$_1$ = 300, k$_2$ = 30, δ_{max} = 0.15, σ = 1.20

7500 BONDS BROKEN

Fig. 12: Cracking patterns generated by a model used to simulate the cracking of a monolayer of sulfonated polystyrene spheres. Here δ_{max} is the maximum dislacement of a node from its point of attachment to the surface and σ is the initial strain ($\sigma = \ell/\ell_o$). Figures 12a and 12b show cracking patterns generated using two different sets of model parameters.

Fig. 13: Typical patterns obtained from a two dimensional model for tensile deformation of polymer films. For Fig. 13a, 13b, 13c and 13d the entanglement spacing parameters (ϕ) were 0.004, 0.02, 0.1 and 1.0 respectively, the widths of the samples were 3.2, 1.6, 0.7 and 0.2μm respectively. For cases b-d the draw ratio was 2.7 and for case a the sample failed before reaching this draw ratio. This figure was provided by Y. Termonia and is taken from Ref. 63.

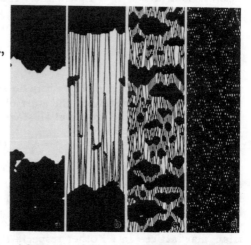

$D_\beta \leq 2.0$ were obtained. The results obtained from simulations carried out with $\eta = 1.0$ and $\eta = 2.0$ are shown in Table III. The results shown in this table indicate that the effective values for D_η depend on both η and on the boundary conditions. The dependence on η is expected by analogy with the dielectric breakdown model.[45,46] The dependence on boundary conditions may be a result of different sensitivities to lattice anisotropy (for dilational strain) and the anisotropy of the shear field (for shear strain). The apparent sensitivity to boundary conditions may result from a crossover from fluctuation dominated to lattice dominated patterns.

Fig. 14: Micrographs of drawn samples of polyethylene films of $M_w = 1.5 \times 10^6$ and $M_n \sim 2 \times 10^5$ crystallized from solutions in decalin and from the melt these samples were drawn to a macroscopic draw ratio of approximately 3 at 100°C. The initial polymer volume fractions were respectively (a) $\phi = 0.005$; (b) $\phi = 0.02$; (c) $\phi = 0.1$; (d) $\phi = 1$. Prints (a), (b) and (d) are optical micrographs taken under crossed polarizers (except b) and (c) is a scanning electron micrograph. The width of all strips shown is 0.1 mm. This figure was provided by Y. Termonia and is taken from Ref. 63.

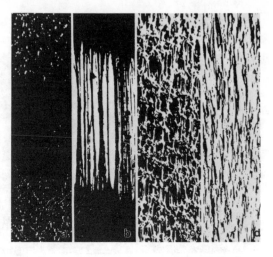

A simple model for elastic fracture in thin films[56] is illustrated in Fig. 9. As in the model described above the two dimensional elastic medium is represented by a triangular network of nodes and springs. However, in this model the elastic network is attached to a rigid underlying substrate by bonds which have a small force constant. Each node is attached to the underlying substrate by a weak bond and the force exerted by the weak bond on the ith node is given by

$$f = k_2 |\mathbf{S}_i - \mathbf{S}_i^0|. \tag{7}$$

Here $\mathbf{S}_i$ is the position of the ith node and $\mathbf{S}_i^0$ the position of its attachment to the substrate (its position at the start of the simulation). In this model all of the bonds are eligible to be broken at all stages in the simulation and the bond breaking probabilities are given by

$$P_i \sim e^{(k \delta_i^2 / 2)}. \tag{8}$$

In addition, the initial bond extension is typically about 10% of ℓ_o instead of 0.1% ℓ_o where ℓ_o is the equilibrium bond length. This model is intended to represent the mechanical failure of a high modulus film attached to a low modulus substrate or a high modulus film weakly bonded to a rigid substrate. At the start of a simulation each bond has a length of 1.0 and the network of bonds and nodes forms a perfect triangular lattice. In all other respects the model is very similar to that of Louis and Guinea[53] and Meakin et al.[54] described above. The simulation proceeds by means of a sequence of bond breaking and relaxation steps.

Figure 10 shows the results of a simulation carried out using a network of 200×200 nodes and 120,000 bonds. In this simulation the equilibrium bond length was assumed to be 0.90909—corresponding to an initial extension ratio of 1.1 (the initial value for δ is 0.090909—for each of the bonds). In this simulation the value for the force constant k is quite large ($k = 1200$) and the bonding to the substrate is quite weak ($k_2 = 12$). Under these conditions a relatively slow crack initiation period is followed by the rapid growth of a few very linear cracks which reduce the strain in the surface layer. The cracks which grow at a later stage are less linear and propagate more slowly.

Figure 11 shows the results of a similar simulation carried out using a network 300×300 of nodes joined by 180,000 bonds to form a square network. In most cases square lattices cannot be used in cracking models which employ a network of Hookean springs because such a system has a zero shear modulus. In this case the attachment to the substrate allows realistic results to be obtained. The cracking patterns shown in Figs. 10 and 11 resemble quite closely those which occur in real systems such as ceramic coated metals, paint films, etc.

This model for elastic fracture in thin films has been modified to represent the experiments of Skjeltorp[57-59] in which a monolayer of uniformly sized spheres of sulfonated polystyrene, initially dispersed in water, is confined between parallel planar glass sheets. A strained film was produced by allowing the monolayer to dry slowly. During the drying process the diameter of the spheres is reduced from $3.4\mu m$ ($\pm 1\%$) to $2.7\mu m$. To model this process the point of attachment of the nodes to the underlying substrate is moved if the distance $|S_i - S_i^o|$ exceeds a value of δ_{max}. In this event the point of attachment is moved towards the current position of the node until $|S_i - S_i^o|$ becomes equal to δ_{max}. To avoid "overshoots" in the movement of the position of attachment to the substrate overrelaxation was not used in these simulations. Figure 12 shows two cracking patterns obtained from this model. These patterns closely resemble those found in the experimental system.[57-59]

This model is based on the idea that the polymer microspheres are relatively strongly bonded to each other and weakly bonded to the glass surfaces which confine the microspheres to a monolayer. In the model we consider attachment to only one surface, but this is equivalent to attachment to both surfaces with bonds which have a force constant of $1/2k_2$. We also assume that the bond breaking probabilities (crack propagation rates) are very sensitive to the local stress field. However, the qualitative aspects of our results are not sensitive to the precise form of the relationship between P_i and $(\ell_i - \ell_o)$ provided that P_i increases sufficiently rapidly with $(\ell_i - \ell_o)$.[56]

The models described in this lecture can all be quite easily made time dependent and used to simulate the kinetics of crack initiation and propagation.[56,58] Unfortunately, length (time) restrictions preclude the discussion of this important aspect of these models.

3. Discussion

Both colloidal aggregation and mechanical failure are complex processes which are of considerable practical and scientific importance. In recent years considerable progress has been made towards developing a working understanding of these processes based on simple computer models and concepts such as fractal geometry and scaling. However, these simple models themselves present theoretical challenges which have not yet been fully met. In the case of colloidal aggregation a variety of efficient computer models has been developed which provides a sound basis for the development of more elaborate but more realistic models. In the case of mechanical failure processes simple models have also been developed. However, our present algorithms and computer resources do not allow us to reach reliable conclusions concerning a quantitative description of the patterns which the models generate.

In this case these models also provide a basis for the development of more realistic models and important steps have been made in this direction. For example, Termonia et al.[59-63] have developed a series of models for the failure and deformation of polymers. Like the model for elastic fracture in thin films described above, these are Monte Carlo models based on the Eyring theory for thermally activated processes. These models include effects such as entanglements, molecular weight distributions, breaking and reformation of Van der Waals bonds, etc. They are fully time dependent and can be used to explore phenomena such as the dependence of mechanical properties on strain rates. Figure 13 shows typical patterns obtained from two dimensional simulations of the tensile deformation of entangled polymers. Here the parameter ϕ is a measure of the mean spacing between entanglements

$$\phi = \frac{1900}{M_e}, \tag{9}$$

where M_e is the mean molecular weight between entanglements and the numerical factor of 1900 is the molecular weight between entanglements in a linear polyethylene melt. For comparison, Fig. 14 shows micrographs of drawn polyethylene films which were prepared from solutions in decalin to control the degree of entanglement.

Acknowledgements: I am indebted to a large number of collaborators and colleagues who have helped me to obtain a better understanding of the topics discussed in these lectures. In particular, the work on colloidal aggregation was carried out in collaboration with R. Jullien and the work on mechanical failure models was carried out in collaboration with F. Louis, G. Li, E. Louis, L. M. Sander, A. T. Skjeltorp and Y. Termonia. Figures 13 and 14 were also provided by Y. Termonia.

1. *On Growth and Form: Fractal and Non-Fractal Patterns in Physics* H. E. Stanley and N. Ostrowsky, eds. (Martinus Nijhof, Dordrecht, 1986).
2. B. B. Mandelbrot, *The Fractal Geometry of Nature* (W. H. Freeman & Company, NY, 1982).
3. M. J. Vold, J. Colloid Sci. 18, 684 (1963).
4. D. N. Sutherland, J. Colloid Interface Sci. 22, 300 (1966).
5. D. N. Sutherland, J. Colloid Interface Sci. 25, 373 (1967).
6. D. N. Sutherland, Nature 226, 1241 (1970).
7. D. N. Sutherland and I. Goodarz-Nia, Chem. Eng. Sci. 26, 2071 (1971).
8. T. A. Witten and L. M. Sander, Phys. Rev. Lett. 47, 1400 (1981).
9. P. Meakin, Phys. Rev. Lett. 51, 1119 (1983).
10. M. Kolb, R. Botet and R. Jullien, Phys. Rev. Lett. 51, 1123 (1983).
11. M. Kolb and R. Jullien, J. Physique Lett. 45, L977 (1984).
12. R. Jullien and M. Kolb, J. Phys. A 17, L639 (1984).
13. M. Eden, in *Proc. 4th Berkeley Symp. Math, Statistics and Probability, Vol. IV*, F. Neyman, ed. (University of California Press, Berkeley 91960), p. 133.
14. P. Meakin, unpublished.
15. P. Meakin, Phys. Rev. B 29, 3722 (1984).
16. L. Turkevich and H. Scher, Phys. Rev. Lett. 55, 1026 (1985).
17. R. C. Ball, Physica 140A, 62 (1986).
18. P. Meakin, in *Phase Transitions and Critical Phenomena*, C. Domb and J. L. Lebowitz, eds., Vol. 12 (1988), p. 335.
19. P. Meakin, *Ann. Rev. Phys. Chem.*, H. L. Strauss, ed., Vol. 39 (1988).
20. R. Botet, R. Jullien and M. Kolb, J. Phys. A 17, L75 (1984).

21. P. Meakin, in *Time-Dependent Effects in Disordered Materials*, NATO ASI Series B 167, R. Pynn and T. Riste, eds. (Plenum Press, New York, 1987), p. 45.

22. P. Meakin and R. Jullien, J. Chem. Phys. 89, 246 (1988).

23. W. D. Brown and R. C. Ball, J. Phys. A 18, L517 (1985).

24. R. C. Ball and T. A. Witten, J. Stat. Phys. 36, 873 (1984).

25. S. P. Obukhov, Sov. Phys. JETP 60, 1167 (1984).

26. R. Botet, J. Phys. A 18, 847 (1985).

27. D. A. Weitz, private communication.

28. D. A. Weitz, J. S. Huang, M. Y. Lin and J. Sung, Phys. Rev. Lett. 54, 1416 (1985).

29. C. Aubert and D. S. Cannell, Phys. Rev. Lett. 56, 738 (1986).

30. H. M. Lindsay, M. Y. Lin, D. A. Weitz, R. C. Ball, R. Klein and P. Meakin, preprint.

31. P. Meakin and R. Jullien, J. de Physique 46, 1543 (1985).

32. A. T. Skjeltorp, Phys. Rev. Lett. 58, 1444 (1987).

33. R. Jullien and P. Meakin, J. Colloid Interface Sci. xx, xxxx (1988).

34. A. J. Hurd and D. W. Schaefer, Phys., Rev. Lett. 54, 1043 (1985).

35. P. M. Mors, R. Botet and R. Jullien, J. Phys. A 20, L975 (1987).

36. R. Jullien, R. Botet and P. Mors, Faraday Discuss. Chem. Soc. 83, 125 (1987).

37. S. Kim and J. R. Brock, J. Applied Phys. 60, 509 (1986).

38. S. Kim and J. R. Brock, J. Colloid Interface Sci. 116, 431 (1987).

39. G. A. Niklasson, A. Torebring, C. Larsson, C. G. Grangvist and T. Farestam, Phys. Rev. Lett. 60, 1735 (1988).

40. P. Meakin and M. Muthukumar, unpublished.

41. *Kinetics of Aggregation and Gelation*, F. Family and D. P. Landau, eds. (North Holland, Amsterdam, 1984).

42. R. Jullien and R. Botet, *Aggregation and Fractal Aggregates* World Scientific, Singapore, 1987

43. P. Meakin, in *Adv. Colloid and Interface Science*, Vol. 28, A. C. Zettlemoyer, ed. (1988), p. 249.

44. T. A. Witten and L. M. Sander, Phys. Rev. B27, 5686 (1983).

45. L. Niemeyer, L. Pietronero and H. J. Wiesmann, Phys. Rev. Lett. 52, 1033 (1984).

46. P. Meakin, J. Theor. Bio. 118, 101 (1986).

47. L. Pietronero, private communication.

48. A. Aharony, private communication.

49. R. C. Ball, private communication.

50. J. P. Eckmann, P. Meakin, I. Procaccia and R. Zeitak, preprint.

51. A. H. England, *Complex Variable Methods in Elasticity* (William Cloves and Sons, London, 1971).

52. L. D. Landau and E. M. Lifshitz, *Theory of Elasticity* (Pergamon, Oxford, 1975).

53. E. Louis and F. Guinea, Europhys. Lett. 3, 871 (1987).

54. P. Meakin, G. Li, L. M. Sander, E. Louis and F. Guinea, preprint.

55. de G. D. M. Allen, *Relaxation Methods* (McGraw Hill, New York, 1954).

56. P. Meakin, Thin Solid Films 151, 165 (1987).

57. A. T. Skjeltorp, preprint.

58. A. T. Skjeltorp and P. Meakin, preprint.

59. A. T. Skjeltorp, these proceedings.

60. Y. Termonia, P. Meakin and P. Smith, Macromolecules 18, 2246 (1985).

61. Y. Termonia, P. Meakin and P. Smith, Macromolecules 19, 154 (1986).

62. Y. Termonia and P. Smith, Macromolecules 20, 835 (1987).

63. Y. Termonia and P. Smith, Macromolecules xx, xx (1988).

DIELECTRIC BREAKDOWN PATTERNS WITH A GROWTH PROBABILITY THRESHOLD

E. ARIAN,[1,2] P. ALSTRØM,[2] A. AHARONY,[1,2] and H. E. STANLEY[2]

[1] *School of Physics and Astronomy*
Beverly and Raymond Sackler Faculty of Exact Sciences
Tel Aviv University, Tel Aviv 69978, ISRAEL

[2] *Center for Polymer Studies and Department of Physics*
Boston University, Boston, MA 02215 USA

The η model of dielectric breakdown divides the Laplace patterns into universality groups. It is possible that an aggregate will grow as an η_1 aggregate for $r < R_c$, and as an η_2 aggregate for $r > R_c$. We define R_c as the crossover radius. The simplest example for an algorithm leading to such crossover is the probability cutoff, i.e., allowing growth only for probabilities larger than a cutoff probability P_0. We find that at a radius R_c the aggregate crosses over from $\eta = 1$ (normal DLA) into $\eta = \infty$ spiky-type aggregate. We find $R_c \sim P_0^{-1}$. Cutting off the gradients lower than ∇_0 leads to a different relation, $R_c \sim \nabla_0^{-1/(2-D)}$ where D is the fractal dimension.

In order to understand these exponents, one must learn the multifractal structure of the growth probability distribution (or gradient distribution). The peak of the function $f(p) = n(p)p$, where $n(p)$ is the growth probability density distribution, is at a probability P_1. When the size of the system goes to infinity all growth takes place on the sites having this probability P_1, which can be shown to be

$$P_1 = \exp\left(\sum_i p_i \ln p_i\right),$$

where the sum is over all the growth probabilities p_i.

We expect that during the early stages (starting with ~ 100 particle clusters, the growth will take place primarily near P_1 (in the probability space). In this case we predict that crossover will occur when P_1 crosses the cutoff probability P_0, i.e., when $P_1 = P_0$ (P_1 is a monotonically decreasing function). P_1 scales with an exponent α_1, i.e. $P_1 \sim L^{-\alpha_1}$, which equals the information dimension D_1 ($D_1 = 1$ for connected sets in 2 dimensions); therefore $R_c \sim P_0^{-1}$. In the gradient cutoff case, the equivalent of P_1 is $\nabla_1 \equiv \nabla_{\max} P_1/P_{\max}$, where $\nabla_{\max}$ and $P_{\max}$ are the maximal gradient and growth probability. ∇_1 is shown to scale according to $\nabla_1 \sim L^{D-2} = L^{-0.3}$. After the crossover, most of the growth takes place in the region close to P_0 (or ∇_0), giving a spiky-type ($\eta = \infty$) aggregate.

THE STATISTICAL MECHANICS OF CRUMPLED MEMBRANES

DAVID NELSON
Department of Physics
Harvard University
Cambridge, MA 02138 USA

ABSTRACT. An enterprise of considerable current interest in theoretical physics is the study of interfaces and membranes. In condensed matter physics, an "interface" usually means a boundary between two phases, whose fluctuations can be studied by methods adapted from equilibrium critical phenomena. The statistical mechanics is typically controlled by a surface tension, which insures that such surfaces are relatively flat. Recently, however, there has been increasing interest in membrane-like surfaces. "Membranes" are composed of molecules different from the medium in which they are imbedded, and they need not separate two distinct phases. Because their microscopic surface tension is small or vanishes altogether, membranes exhibit wild fluctuations. New ideas and new mathematical tools are required to understand them.

We first sketch the physics of "flat" interfaces, and then discuss crumpled tethered membranes which are natural generalizations of linear polymers. More generally, the large distance behaviors of membranes fall into a variety of universality classes, depending, for example, on whether the local order is liquid or crystalline. We show that membranes with a nonzero shear modulus differ from their liquid counterparts in that they exhibit a flat phase with long-range order in the normals at sufficiently low temperatures. Because entropy favors crumpled surfaces with decorrelated normals, there must be a transition to a crumpled phase at sufficiently high temperatures. We also discuss the energies of disclinations and dislocations in flexible membranes with local crystalline order. Unlike crystalline films forced to be flat by a surface tension, it is energetically favorable for membranes to screen out elastic stresses by buckling into the third dimension. Dislocations, in particular, are predicted to have a finite energy. We conclude that a finite density of dislocations must exist at all nonzero temperatures in nominally crystalline but unpolymerized membranes. The result macroscopically is a hexatic membrane, with zero shear modulus, but extended bond orientational order. The elastic energy which controls undulations in hexatic membranes is discussed briefly.

Related problems arise in field theory models of elementary particles.[1] In contrast to these models of quantum mechanical strings, however, most of the models discussed here have explicit experimental realizations in condensed matter physics. Much of the vitality of this subject arises because of a delicate interplay between theory and experiment: theoretical predictions can, in principle, be checked by inexpensive but revealing laboratory experiments in a matter of months.

1. Flat Surfaces

1.1. THE ROUGHENING TRANSITION

Interesting problems in statistical mechanics arise even for surfaces constrained by surface tension to be fairly flat. A particularly well-studied example is the roughening transition of crystalline interfaces.[2] As shown in Figure 1, we imagine a crystal in equilibrium, with, say, its own vapor. The position of the interface is described by a height function $h(x^1, x^2)$. Such a description implicitly ignores "overhangs" (which cannot be described by a single-valued $h(x^1, x^2)$), islands of crystal in the vapor phase, and islands of vapor in the crystal. These complications are believed to be irrelevant variables in the long wavelength limit.[2] Microscopically, the interface height is quantized in units of the spacing between the Bragg planes normal to the h-axis.

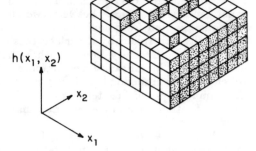

Fig. 1: Height function $h(x^1, x^2)$ used to describe the configuration of a crystal-vapor interface.

At high temperatures, this discreteness is washed out by thermal fluctuations, and we can describe the free energy of the interface by a surface tension σ. It is a useful pedagogic exercise to describe this free energy using differential geometry, which, although inessential here, is often the language of choice for crumpled membranes. For an arbitrary parameterization of the surface $\vec{r}(\zeta^1, \zeta^2)$, the free energy is the surface tension times the surface area,

$$F = \sigma \int \sqrt{g} \, d^2\zeta \tag{1.1}$$

where g is the determinant of the metric tensor, $g = \det g_{ij}$,

$$g_{ij} = \frac{\partial \vec{r}}{\partial \zeta^i} \cdot \frac{\partial \vec{r}}{\partial \zeta^i} \tag{1.2}$$

The formula for the surface area in terms of the metric tensor is derived in many textbooks.[3] For the particular parameterization embodied in Fig. 1, $i.e.$,

$$\vec{r}(x^1, x^2) = (x^1, x^2, h(x^1, x^2)) \ , \tag{1.3}$$

we have

$$g_{ij} = \begin{pmatrix} 1 + (\frac{\partial h}{\partial x^1})^2 & , & (\frac{\partial h}{\partial x^1})(\frac{\partial h}{\partial x^2}) \\ (\frac{\partial h}{\partial x^1})(\frac{\partial h}{\partial x^2}) & , & 1 + (\frac{\partial h}{\partial x^2})^2 \end{pmatrix} \ . \tag{1.4}$$

With this coordinate system (called the Monge representation in differential geometry), Eq. (1.1) assumes the familiar form

$$F = \sigma \int d^2x \sqrt{1 + |\vec{\nabla}h|^2} \ . \tag{1.5}$$

At temperatures sufficiently high so that (1.1) is an appropriate description, we can expand the square root in (1.5),

$$F \approx \text{const.} + \frac{1}{2}\,\sigma \int d^2x\, |\vec{\nabla}h|^2 \tag{1.6}$$

and calculate, for example the height-height correlation function

$$\langle (h(\vec{y}) - h(\vec{0}))^2 \rangle = \frac{\int \mathcal{D}h(\vec{x})|h(\vec{y}) - h(\vec{0})|^2 e^{-F/k_B T}}{\int \mathcal{D}h(\vec{x}) e^{-F/k_B T}} \ . \tag{1.7}$$

The effects of higher order gradients in Eq. (1.6) can be absorbed into a renormalized surface tension. The Gaussian functional integral is easily carried out in Fourier space, with the result

$$\langle (h(\vec{y}) - h(\vec{0}))\rangle^2 = \frac{2k_B T}{\sigma} \int \frac{d^2q}{(2\pi)^2} \frac{1}{q^2} (1 - e^{i\vec{q}\cdot\vec{y}})$$

$$\approx \frac{k_B T}{\pi\sigma} \ell n(y/a), \qquad \text{as} \quad y \to \infty\,, \tag{1.8}$$

where a is a microscopic length. The large y behavior is the signature of a high temperature rough phase.

At low temperatures, on the other hand, one might expect a "smooth" interface, i.e., one that has become localized at an integral multiple of a, the spacing between Bragg planes. To see how quantization of the interface height affects the prediction (1.8), we add a periodic perturbation to Eq. (1.6) which tends to localize the interface at $h = 0, \pm a, \pm 2a, \ldots$, and consider the free energy

$$F = \text{const.} + \frac{1}{2} \int d^2x \left[\sigma|\vec{\nabla}h|^2 + 2y(1 - \cos(2\pi h/a)) \right] \ . \tag{1.9}$$

This sine-Gordon model can be solved directly by renormalization group methods, or by first mapping the problem via a duality transformation onto an XY-model or the two-dimensional Coulomb gas.[2,4] There is a finite temperature roughening transition which is in the universality class of the Kosterlitz-Thouless vortex unbinding transitions. At sufficiently high temperatures ($T > T_R \simeq \pi\sigma a^2/k_B$), the periodicity is irrelevant and the interface behaves according to Eq. (1.8). For $T < T_R$, however, the interface localizes in one of the minima of the periodic potential and the effective free energy at long wavelengths can be approximated by expanding the cosine

$$F \approx \text{const.} + \frac{1}{2} \int d^2x \left[\sigma|\vec{\nabla}h|^2 + \left(\frac{4\pi^2 y}{a^2}\right) h^2 \right] \ . \tag{1.10}$$

It is easily shown from Eq. (1.10) that height-height correlation function (1.7) now tends to constant,

$$\langle (h(\vec{y}) - h(\vec{0}))^2 \rangle \approx \text{const.} , \qquad \text{as} \quad y \to \infty , \qquad (1.11)$$

in contrast to Eq. (1.8).

The analogy with vortex unbinding transitions leads to many detailed predictions about the roughening transition.[2] This analogy is only approximate, however, so it is important to have rigorous proofs of phase transitions in this and related models.[5] Although Eq. (1.9) is a plausible model of roughening, a more faithful representation of the microscopic physics is the solid-on-solid model, where interface heights $\{h_i\}$ sit on a lattice of sites $\{i\}$ and are themselves quantized at all temperatures, $h_i = 0, \pm a, \pm 2a, \dots, \forall i$. The Hamiltonian is

$$H = J \sum_{\langle ij \rangle} |h_i - h_j| \quad , \qquad (1.12)$$

where the sum is over nearest neighbor lattice sites and $J > 0$ is a microscopic surface energy. Equation (1.12) measures directly the increase in interfacial area associated with discrete steps in the interface. We call (1.12) a " Hamiltonian" because it is a microscopic energy, in contrast to " free energies" like (1.9), which are supposed to be coarse-grained descriptions, embodying both energy and entropy.

1.2. WETTING TRANSITIONS

Interesting transitions in interfacial surfaces also occur in wetting layers.[6] Consider, in particular, the approach to a liquid-gas phase boundary in the presence of a wall which microscopically prefers to be wet by the liquid, as opposed to the gas. The interfacial profile is shown in Fig. 2a.

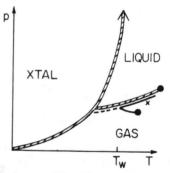

Fig. 2a: Density profile near a wall in the bulk gas phase close to liquid-gas coexistence. The density starts at a large value ρ_ℓ appropriate to the nearby liquid phase and drops to a smaller value ρ_g appropriate to the gas a distance ℓ from the wall.

Because the wall prefers the denser liquid, there is a thin layer of liquid present, even though the chemical potential of the bulk gas is slightly slower than the bulk liquid. This wetting layer extends a distance $\ell(T, p)$ into the gas phase, terminating at a liquid-gas interface whose width is comparable to the correlation length $\xi(T, p)$.

Two distinct behaviors are possible as the liquid-gas coexistence curve is approached from the gas phase (see Fig. 2b). Far from the critical point, $\ell(p, T)$ usually remains finite at the liquid-gas coexistence curve (i.e., along the dashed line in Fig. 2b). Closer to the critical point, however, $\ell(p, T)$ diverges (logarithmically, in simple model calculations) as the coexistence curve is approached (along the solid line in Fig. 2a). This divergence may be preceded by a first order " prewetting" transition in the bulk liquid signalled by an upward jump in the liquid density at the wall. The point at which $\ell(p, T)$ diverges to infinity along the coexistence curve, at $T = Tw$, locates a wetting transition, which has been the subject of considerable theoretical interest recently. This transition can be first order, or it occurs via a rather exotic second order transition.[7]

Fig. 2b: Pressure temperature phase diagram with regions of first order and continuous wetting transitions along the liquid-gas coexistence curve indicated by dashed and solid lines, respectively. A first order "prewetting" transition terminating in a critical point extends into the gas phase. The density profile in Fig. 2a corresponds to the situation near a wall at the point x.

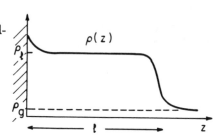

2. Crumpled Membranes

2.1. EXPERIMENTAL REALIZATIONS

Membranes can be regarded as two-dimensional generalizations of linear polymer chains, for which there is a vigorous theoretical and experimental literature.[8,9] Flexible membranes should exhibit even more richness and complexity, for two basic reasons. The first is that important geometric concepts like intrinsic curvature, orientability and genus, which have no direct analogue in linear polymers, appear naturally in discussions of membranes: Our understanding of the interplay between these concepts and the statistical mechanics of membranes is still in its infancy. The second reason is that surfaces can exist in a variety of different phases. The possibility of a two-dimensional shear modulus in membranes shows that we must distinguish between solids and liquids when these objects are allowed to crumple into three dimensions. We shall argue later that hexatic membranes, with extended six-fold bond orientational order, are another important possibility. There are no such sharp distinctions for linear polymer chains.

Figure 3 shows two examples of *liquid* membranes. Fig. 3a is a caricature of a eurethrocyte, or red blood cell. The cell wall is a membrane, composed of a bilayer of amphiphillic molecules, each with one or more hydrophobic hydrocarbon tails and a polar head group. The membrane has a spherical topology, as do artificial vesicles formed from bilayers. Although these membranes could, in principle, crystallize upon cooling, they exhibit an almost negligible shear modulus at biologically relevant temperatures. The small shear modulus that is observed for eurethrocytes may be due to an additional protein skeleton like spectrin.[10]

Figure 3b illustrates the topology of a microemulsion, which is a transparent solution in which oil (*e.g.*, dodecane) and water mix in essentially all proportions.[11] This remarkable mixing is only possible because of the addition of significant amounts of an amphiphile like SDS (sodium dodecyl sulfate), which sits at the interface between oil and water and reduces the surface tension almost to zero.

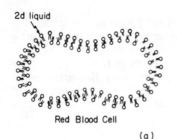

2d liquid

Red Blood Cell

Polar head group

Hydrocarbon tails

Lipid

(a)

Fig. 3: Examples of liquid-like membranes: **(a)** red blood cell and **(b)** microemulsion.

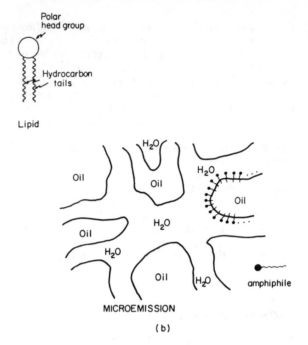

MICROEMISSION

(b)

The size of the oil-rich and water-rich regions, which are constantly shifting as the interface fluctuates, is of order 100 Angstroms. Usually, a cosurfactant like pentanol is necessary to stabilize the microemulsion.

For more about liquid membranes, see the lectures of W. Helfrich, L. Peliti, and D. Roux at this school, and the collections of papers in Refs. 12 and 13.

Although careful experimental investigations are only just beginning, there are also many examples of *solid* membranes. One can, for example, explore t he properties of flexible sheet polymers, the "tethered surfaces". Tethered surfaces can be synthesized by polymerizing Langmuir-Blodgett films or amphiphillic bilayers.[14] Although lipid monolayers polymerized at an air-water interface would be initially flat, they could be inserted into a neutral solvent like alcohol and their fluctuation s made visible by attaching a fluorescent dye. There are fascinating accounts of cross-linked methyl-methacrylate polymer assembled on and then extracted from the surface of sodium montmorillonite clays.[15]

Two less familiar examples of solid membranes are illustrated in Figure 4. Figure 4a shows a model of large sheet molecule believed to be an ingredient of glassy B_2O_3.[16] Similar structures, also in crumpled form, may exist i n chalgogenide glasses such as As_2S_3. Although it may be difficult to obtain dilute solutions in a good solvent, we might hope to produce a dense melt of such surfaces, in analogy with polymer melts or models of amorphous selenium.[17]

Figure 4b illustrates an idea for synthesizing a large number surfaces of two-di mensional polyacrylamide gel, which I have pursued in collaboration with R.B. Meyer at Brandeis University. We first form a lyotropic smectic liquid crystal of amphiphillic bilayers, simil ar to those discussed above. The bilayers are separated by water, and if necessary can be pushed further apart by the addition of oil or water.[18,19] If the lipids ha ve multiple double bonds, one could of course polymerize

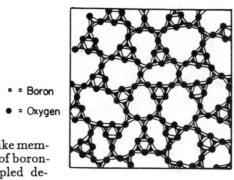

° = Boron

● = Oxygen

Fig. 4: Examples of solid-like membranes: **(a)** planar section of boron-oxide which, when crumpled describes a glass and **(b)** lyotropic smectic phase with polymerizable polyacrylamide lmonomer in the watery interstices.

(a)

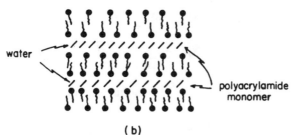

water

polyacrylamide monomer

(b)

the bilayers as discussed
above. An attractive alternative for producing flexible surfaces is to introduce polyacrylamide gel into the watery interstices between the bilayers. Meyer and I have succeeded in stabilizing a smectic phase in which each $\approx$ 20 Angstrom thick water-rich region contains about 15 weight percent acrylamide and bis-acrylamide monomers. By shining ultraviolet light on this mixture, it may be possible to produce many slabs of 2d cross-linked polyacrylamide gel. The lipid bilayers, which are used simply as spacers in this experiment, would then be washed away.

A third class of membrane surfaces is possible if we replace fixed covalent cross links like those in Fig. 4a by weaker van der Waals forces. Van der Waals interactions will tend to crystallize the lipid bilayers discussed above at sufficiently low temperatures. Although these surfaces will have a nonzero shear modulus when confined to a plane, they are unstable to the formation of free dislocations when allowed to buckle into the third dimension.[20] Dislocations necessitate broken bonds, and thus would require prohibitively large energies in covalently bonded systems. The presence of a finite concentration of unbound dislocations at any temperature means that unpolymerized lipid bilayers will in fact be hexatic liquids with residual bond-orientational order at low temperatures.[20,21] The properties of hexatic membranes are intermediate between liquid and solid surfaces, and will be discussed at the end of this chapter.

2.2. RESULTS FROM POLYMER PHYSICS

One route towards understanding crumpled membranes is to generalize various results from polymer physics.

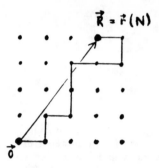

$$\vec{R} = \vec{r}(N)$$

Fig. 5: Polymer configuration extending from the origin to $\vec{R}$ on a square lattice.

As illustrated in Fig. 5, we can catalogue polymer configurations on a lattice by first counting the number of self-avoiding walks starting at the origin and terminating at position R. The function $\vec{r}(s)$ gives the position of the walk after the sth step. If $\mathcal{N}_N(\vec{R})$ is the number of walks of length N starting at the origin and terminating at $\vec{R}$, the *total* number of walks of length N is given by

$$\mathcal{N}_N^{\text{tot}} = \sum_{\vec{R}} \mathcal{N}_N(\vec{R}) \qquad . \tag{2.1}$$

A typical polymer size is given by the radius of gyration R_G,

$$R_G = \left[\frac{1}{N^2} \sum_{s=1}^{N} \sum_{s'=1}^{N} \langle |\vec{r}(s) - \vec{r}(s')|^2 \rangle \right]^{1/2} \qquad , \tag{2.2}$$

where the average is over all polymer configurations. Polymer critical exponents are defined by the asymptotic large N behavior of R_G and $\mathcal{N}_N^{\text{tot}}$,

$$R_G \sim N^\nu \tag{2.3}$$

$$\mathcal{N}_N^{\text{tot}} \sim (\bar{z})^N N^{\gamma-1} \qquad . \tag{2.4}$$

Here, $\bar{z}$ is a nonuniversal effective "coordination number", reduced from the actual coordination number by self-avoiding constraints. The radius of gyration exponent ν is increased by self-avoidance from the random walk result $\nu = 1/2$ to the universal result $\nu \approx 0.59 \approx 3/5$ in three dimensions. The exponent $\gamma \approx 1.18$ is also universal for polymers with free ends, although it changes for ring polymers.[9] The effect of self-avoidance on the exponents vanishes for $d > d_c = 4$, which is the upper critical dimension for linear polymers, and forms the basis for a $d = 4 - \epsilon$ expansion for the exponent ν and γ.[8,9]

A simple, but remarkably accurate estimate of the exponent ν can be obtained by working with a *continuum* model of self-avoiding polymers in a good solvent. The free energy F associated with a coarse-grained polymer configuration $\vec{r}(s)$ in d-dimensions is assumed to be[8]

$$F/k_B T = \frac{1}{2} K \int_0^N ds \left(\frac{d\vec{r}}{ds} \right)^2 + \frac{1}{2} v \int_0^N ds \int_0^N ds' \delta^{(d)}[\vec{r}(s) - \vec{r}(s')] \tag{2.5}$$

The first term represents a nearest-neighbor elastic energy, possibly entropic in origin, while the second counts the number of self-intersections and assigns them an excluded volume penalty v. Following a famous approximation scheme due to Flory,[8,9] we estimate the first term for a polymer of size R_G by dimensional analysis, and note that the second term should be proportional to the probability of self-intersection $(N/R_G^d)^2$, times the volume R_G^d over which a self-intersection is likely to occur. In this way, we find that

$$F/k_BT \approx \frac{1}{2} K R_G^2/N + \frac{1}{2} v \left(\frac{N}{R_G^d} \right)^2 R_G^d \ . \tag{2.6}$$

Upon minimizing with respect to R_G we obtain the classic Flory result, $R_G \sim N^\nu$, with

$$\nu = 3/(d+2) \ , \tag{2.7}$$

i.e., $\nu \approx 0.60$ is three dimensions.

C. Generalization to Tethered Surfaces

Membrane generalizations of the theory of polymer chains are conveniently presented in the language of differential geometry. The partition function of the resulting tethered surfaces without self-avoidance is a special case of a more general partition function which first arose in the study of bosonic strings, namely[22]

$$Z = \int \mathcal{D}g_{0,ab} \int \mathcal{D}\vec{r}(\zeta^1,\zeta^2) e^{-\frac{1}{2}K \int d^2\zeta \sqrt{g_0} g_0^{ab} \partial_a \vec{r} \cdot \partial_b \vec{r}} \ . \tag{2.8}$$

The "action" is composed of surface gradients $\partial_a \vec{r}$ contracted with a metric tensor g_0^{ab}. The integrations are over all possible metrics $g_{0,ab}$, as well as over all possible surface configurations $r(\zeta^1,\zeta^2)$. Although the underlying metric and the surface are independent variables, Polyakov[22] has shown that a relation analogous to Eq. (1.2), i.e.,

$$g_{0,ab} = \frac{\partial \vec{r}}{\partial \zeta^a} \cdot \frac{\partial \vec{r}}{\partial \zeta^b} \tag{2.9}$$

is recovered in the low temperature, strong coupling $(K \to \infty)$ limit.

A microscopic physical interpretation of Eq. (2.8) is illustrated in Fig. 6. For a fixed metric $g_{0,ab}$, the surface is represented by a fixed triangulation of particles, connected by harmonic springs. The action in Eq. (2.8) is the continuum limit of the energy associated with these springs. The particle positions can be stretched to approximate any particular simply-connected surface with free boundaries $\vec{r}(\zeta^1,\zeta^2)$; there is, however, a significant energetic cost associated with large deviations from the surfaces preferred by the underlying connectivity or "background metric". To carry out the functional integral (2.8) on a computer, one would first integrate over all particle positions for a fixed triangulation, and then sum over different triangulations.

Tethered surfaces, discussed more completely in Refs. 23 and 24, are an example of the string partition function (2.8), specialized to a single "flat" triangulation, where every particle is connected to exactly six nearest neighbors. Although the

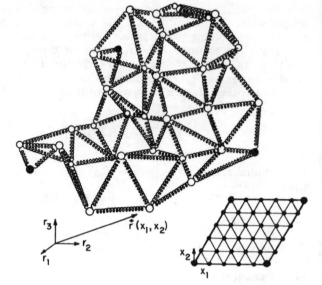

Fig. 6: Lattice of Gaussian springs with a fixed connectivity.

springs are usually replaced by a square well tethering potential, the particles behave as if they were connected by springs at long wavelengths for entropic reasons.[23,24] The "background metric" is

$$g_{0,ab} = \delta_{ab} \tag{2.10}$$

and the partition function is

$$\mathcal{Z}_0 = \int \mathcal{D}\vec{r}(x_1, x_2)e^{-F_0} \tag{2.11}$$

where

$$F_0 = \frac{1}{2} K \int d^2x \left(\left[\frac{\partial \vec{r}}{\partial x_1} \right]^2 + \left(\frac{\partial \vec{r}}{\partial x_2} \right)^2 \right) . \tag{2.12}$$

As it stands, we now have a model for "phantom" polymerized membranes, without self-avoiding interactions between distant particles. To obtain a model for a real self-avoiding membrane, we replace the free energy F_0 by

$$F = \frac{1}{2} K \int d^2x \left(\frac{\partial \vec{r}}{\partial \mathbf{x}} \right)^2 + \frac{1}{2} v \int d^2y \int d^2y' \, \delta^d[\vec{r}(\mathbf{y}) - \vec{r}(\mathbf{y}')] . \tag{2.13}$$

The second term assigns a positive energetic penalty v whenever two elements of the surface occupy the same position in the three-dimensional embedding space.

To make analytic progress with the statistical mechanics associated with (2.13), it is useful to generalize (2.13), and consider *manifolds* $\vec{r}(\mathbf{x})$ with a D-dimensional flat internal space embedded in a d-dimensional external space.[24-27] The associated free energy is

$$F = \frac{1}{2} K \int d^Dx \left(\frac{\partial \vec{r}}{\partial \mathbf{x}} \right)^2 + \frac{1}{2} v \int d^Dy \int d^Dy' \, \delta^d[\vec{r}(\mathbf{y}) - \vec{r}(\mathbf{y}')] , \tag{2.14}$$

or

$$F = \frac{1}{2} \int d^D x \left(\frac{\partial \vec{R}}{\partial x} \right)^2 + \frac{1}{2} v K^{d/2} \int d^D y \int d^D y' \, \delta^d [\vec{R}(\mathbf{y}) - \vec{R}(\mathbf{y}')] \ , \qquad (2.15)$$

where we have introduced the d-dimensional rescaled variable,

$$\vec{R}(x^1, x^2) = \sqrt{K} \, \vec{r}(x^1, x^2) \ . \qquad (2.16)$$

When $v = 0$, we have a free field theory, and it is easy to show that the mean squared distance between points with internal coordinates $\mathbf{x}_A$ and $\mathbf{x}_B$ is

$$\langle |\vec{r}(\mathbf{x}_A) - \vec{r}(\mathbf{x}_B)|^2 \rangle \underset{\mathbf{x}_{AB} \to \infty}{\simeq} \frac{2 d S_D}{(2-D)K} \left[|\mathbf{x}_{AB}|^{2-D} - a^{2-D} \right] \qquad (2.17)$$

where $S_D = 2\pi^{D/2}/\Gamma(D/2)$ is the surface area of a D-dimensional sphere, $\mathbf{x}_{AB} = \mathbf{x}_A - \mathbf{x}_B$ and a is a microscopic cutoff. If we take $\mathbf{x}_A$ and $\mathbf{x}_B$ to be close to opposite sides of the manifold (in the internal space), Eq. (2.17) becomes a measure of the squared radius of gyration. When $D = 1$, we are dealing with a linear polymer chain and we see that the size R_G increases as the square root of the linear dimension $L \sim |\mathbf{x}_{AB}|$, i.e., $R_G \sim L^{1/2}$. The same argument, however, shows that the characteristic membrane size R_G increases only as the square root of the logarithm of the linear dimension L for $D = 2$,

$$R_G \sim \frac{1}{K} \, \ell n^{1/2}(L/a) \ . \qquad (2.18)$$

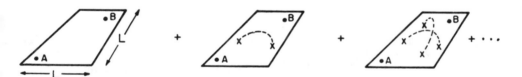

Fig. 7: Graphical representation of the perturbative calculation of the mean square distance in the embedding space between the points $\vec{r}(\mathbf{x}_A)$ and $\vec{r}(\mathbf{x}_B)$.

To see how self-avoiding corrections affect Eq. (2.17), we can carry out perturbation theory in the excluded volume parameter. Each term can be represented as in Fig. 7, where the dotted lines represent self-avoiding interactions between different pieces of the manifold. Dimensional analysis using the rescaled free energy Eq. (2.15) shows that this perturbation theory becomes singular in the limit of large internal linear dimension L: The correction to Eq. (2.17) must take the form

$$\langle |\vec{r}(\mathbf{x}_A) - \vec{r}(\mathbf{x}_B)|^2 \rangle \simeq \frac{2 d S_D}{(2-D)K} \, |\mathbf{x}_{AB}|^2 \left[1 + \text{const.} \times v K^{d/2} L^{2D-(2-D)\frac{4}{2}} + \cdots \right] \ .$$

$$(2.19)$$

Whenever

$$2D > (2 - D)\,\frac{d}{2} \tag{2.20}$$

the corrections to the free field result (1.31) diverge as $L \to \infty$, signalling a breakdown of perturbation theory. If this inequality is reversed, however, we expect self-avoidance to be asymptotically irrelevant in large systems. This is the case for polymers $(D = 1)$ when $d > 4$. Note, however, that the perturbative correction in (2.19) is always large for membranes, i.e., for $D = 2$.[28]

Fig. 8: Different regimes in the (d, D)-plane for self-avoiding tethered surfaces.

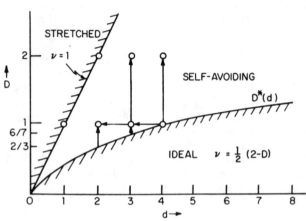

Figure 8 shows the critical curve

$$D^*(d) = \frac{2d}{4 + d} \tag{2.21}$$

which separates ideal from self-avoiding behavior in the (d, D)-plane. Also shown is the line $D = d$, along which the manifold becomes fully stretched due to self-avoidance. The critical line $D^*(d)$, of course, passes through the point $(d^* = 4, D^* = 1)$, which is the basis for epsilon expansions of polymers,[8] but in fact *any* *point* on this line is an equally good expansion candidate. We could, for example, stay in three dimensions $(d = 3)$, and change the manifold dimensionality D. Self-avoidance dominates for solid elastic cubes $(D = 3)$, but is less important for elastic surfaces $(D = 2)$. It produces relatively small corrections to the Gaussian result for linear manifolds $(D = 1)$, and becomes formally negligible when $D < D^* = 6/7$! This idea forms the basis for a $6/7 + \epsilon$ expansion for tethered surfaces,[24-27] as will be discussed in more detail in the lecture of Duplantier. The result is that the radius of gyration scales with the linear dimension according to

$$R_G \sim L^\nu \tag{2.22}$$

where

$$\nu = \frac{2 - D}{2} + 0.469\left(D - \frac{6}{7}\right) + \mathcal{O}\left(D - \frac{6}{7}\right)^2 . \tag{2.23}$$

This novel epsilon expansion gives excellent results for linear polymers in three dimensions ($\epsilon = 1/7$, $\nu = 0.567$), but is not very accurate for polymerized membranes ($\epsilon = 8/7$, $\nu = 0.536$).

We know from computer simulations[24] that $\nu \approx 0.8 \pm 0.03$ in self-avoiding membranes. A good approximation for ν in D-dimensional manifold embedded in d-dimensions follows from applying the Flory approximation to Eq. (2.14). Proceeding as in the case of linear polymers we find

$$F \approx \frac{1}{2} \, K R_G^2 L^{D-2} + \frac{1}{2} \, v \left(\frac{L^D}{R_G^d} \right)^2 R_G^d \quad . \tag{2.24}$$

Minimization with respect to R_G leads to $R_G \sim L^\nu$, with[23,24]

$$\nu = \frac{2+D}{2+d} \tag{2.25}$$

i.e., $\nu = 4/5$ for membranes ($D = 2$, $d = 3$).

3. Normal-Normal Correlations in Liquid Membranes

The analysis in the previous subsection is restricted to floppy membranes, without an appreciable bending rigidity. The conformations of untethered liquid membranes, however, are dominated by such bending energies. Before discussing how bending energies affect tethered membranes, we first review results for liquid membranes.

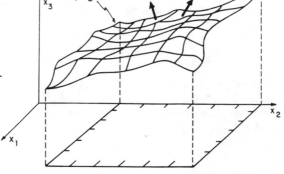

Fig. 9: Liquid membrane broken up into plaquettes each characterized by a unit normal.

Figure 9 shows a fragment of a liquid membrane which we assume is approximately parallel to the (x_1, x_2)-plane so that we can use a Monge representation for its position,

$$\vec{r}(x_1, x_2) = (x_1, x_2, f(x_1, x_2)) \quad . \tag{3.1}$$

The unit normal at any point is easily shown to be

$$\vec{n}(x_1, x_2) = (-\partial_1 f, -\partial_2 f, 1)/\sqrt{1 + [\vec{\nabla} f]^2} \quad . \tag{3.2}$$

If we think of the membrane as composed of rod-shaped amphiphillic molecules, it is natural to associate these normals with the local rod axis. Upon partitioning the membrane into segments as shown in Fig. 1, we can write down a lattice model of the bending energy,

$$F_b = -\tilde{\kappa} \sum_{\langle i,j \rangle} \vec{n}_i \cdot \vec{n}_j \ , \tag{3.3}$$

in analogy with a one-Frank-constant elastic energy for nematic liquid crystals.[29] Here the sum is over nearest-neighbor segments, $\vec{n}_i$ is the normal associated with the i-th segment, and $\tilde{\kappa}$ is a microscopic bending rigidity.

Equation (1.3) resembles the energy of a classical Heisenberg ferromagnet on a two-dimensional lattice.[30] The normals are like spin vectors, and the rigidity $\tilde{\kappa}$ is like a Heisenberg exchange constant. Rotational symmetry is broken at $T = 0$ by a "ferromagnetic" flat surface, with a uniformly aligned normal field. It is well-known, however, that long-range order is destroyed at any *finite* temperature in the two-dimensional Heisenberg model by spin wave fluctuations.[31] In surfaces, we might also expect long range-order to be destroyed, in this case by surface undulations. The analogy with spin waves is not perfect, however, because the normals are constrained to be part of a surface, which forces them to be expressible as in Eq. (3.2). For small undulations, this restriction means that "spin waves" in the normals must be purely longitudinal; transverse "spin waves" would tear the surface.

To determine how undulations affect correlations in the normals it is useful to take the continuum limit of Eq. (3.3). To leading order in a expansion in gradients of $f(x_1, x_2)$, we can neglect the factor $\sqrt{g} = \sqrt{1 + |\vec{\nabla} f|^2}$ in the measure as well as the $|\vec{\nabla} f|^2$ term in the denominator of (3.2) and find

$$F_b \approx \frac{1}{2} \kappa \int d^2 x |\nabla \vec{n}|^2 \approx \frac{1}{2} \kappa \int d^2 x \left[(\partial_1^2 f)^2 + 2(\partial_1 \partial_2 f)^2 + (\partial_2 f)^2 \right]$$

$$\approx \frac{1}{2} \kappa \int d^2 x \left[(\nabla^2 f)^2 - 2 \det(\partial_i \partial_j f) \right] \tag{3.4}$$

where κ is proportional to $\tilde{\kappa}$. The last two terms of (3.4) are just the mean curvature and Gaussian curvature pieces of the Helfrich[32] bending energy of a liquid membrane. The Gaussian curvature is a perfect derivative, which we can see by writing the second term as

$$2 \det(\partial_i \partial_j f) = -\epsilon_{im} \epsilon_{jn} \partial_m \partial_n \left[(\partial_i f)(\partial_j f) \right] \ . \tag{3.5}$$

Upon neglecting the contribution from this surface term, we can write

$$F_b \approx \frac{1}{2} \kappa \int d^2 x (\nabla^2 f)^2 \ . \tag{3.6}$$

Following DeGennes and Taupin[33], we can now estimate fluctuations in the normals. The angle $\theta(x_1, x_2)$ which the normal $\vec{n}(x_1, x_2)$ makes with respect to the $\hat{x}_3$ axis is given by

$$\vec{n} \cdot \hat{x}_3 = \cos \theta = 1/\sqrt{1 + |\vec{\nabla} f|^2} \ . \tag{3.7}$$

If there is a broken symmetry such that the normals point on average along the $\hat{x}_3$ axis, fluctuations in $\theta^2 = |\vec{\nabla} f|^2$ should be small at low temperatures. Because Eq. (3.6) is a quadratic form, we can calculate $\langle \theta^2 \rangle$ by passing to Fourier space and using the equipartition theorem,

$$\langle \theta^2(x_1, x_2) \rangle \approx k_B T \int \frac{d^2 q}{(2\pi)^2} \frac{1}{\kappa q^2} \approx \frac{k_B T}{\kappa} \ln(L/a) \ . \tag{3.8}$$

Just as in many other systems with continuous symmetries in two dimensions,[34] there is a logarithmic divergence with system size L, signalling the breakdown of long-range order in the normals. More sophisticated calculations by Peliti and Leibler[35] show that the renormalized wave-vector-dependent rigidity $\kappa_R(q)$ is softened by these fluctuations,

$$\kappa_R(q) = \kappa - \frac{3k_B T}{4\pi} \ln(1/qa) \ . \tag{3.9}$$

Note that if we replace κ by $\kappa_R(q)$ in Equation (3.8), this only makes the divergence worse. The renormalization group calculations of Ref. 35 are consistent with exponential decay of the normal-normal correlation function.

$$\langle \hat{n}(\mathbf{x}) \cdot \hat{n}(0) \rangle \propto e^{-x/\xi} \tag{3.10}$$

with a correlation length which diverges at low temperatures

$$\xi \approx a e^{4\pi\kappa/3k_B T} \ . \tag{3.11}$$

The low temperature behavior of the two-dimensional Heisenberg model[31] is very similar.

4. Tethered Surfaces with Bending Energy[36,37]

Figure 10 shows a surface in which bending energy and tethering are present simultaneously. If $\vec{r}_i$ denotes the position of the i-th vertex, and $\vec{n}_\alpha$ is the normal to the α-th triangular plaquette, a microscopic model Hamiltonian would be

$$H = -\tilde{\kappa} \sum_{\langle \alpha, \beta \rangle} \vec{n}_\alpha \cdot \vec{n}_\beta + \sum_{\langle i,j \rangle} V(|\vec{r}_i - \vec{r}_j|) \ , \tag{4.1}$$

where $V(r)$ is a tethering potential between nearest neighbor vertices. We have just seen that liquid membranes, subject only to bending energy, crumple at finite temperatures, in the sense that long-range order in the normals is destroyed. As we saw in Sec. II, tethered surfaces, subject only to the constraint of fixed bonding connectivity, also crumple, like polymers in a good solvent.[23,24] We shall now argue that

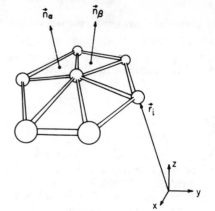

Fig. 10: Unit normals defined on the triangular plaquettes of a tethered surface.

there is a fundamental incompatibility when these two energies are simultaneously
present which stabilizes a flat phase at sufficiently low temperatures.

As in our discussion of liquid membranes, we start with a locally flat surface
and ask if a state with long-range order in the normals is stable to thermal fluctuations. We assume that the tethering potential induces nonzero elastic constants in
approximately planar membranes, so that there is elastic stretching energy, as well
as the bending energy characteristic of liquid membranes.

As shown in Figure 11, a three-component displacement field is necessary to
describe the deformation of an initially flat membrane. If $\vec{r}_0(x_1, x_2) = (x_1, x_2, 0)$
describes the undistorted membrane at $T = 0$, an arbitrary membrane configuration
for $T > 0$ is given by

$$\vec{r}(x_1, x_2) = \vec{r}_0 + \begin{pmatrix} u_1(x_1, x_2) \\ u_2(x_1, x_2) \\ f(x_1, x_3) \end{pmatrix} \quad . \tag{4.2}$$

A small line element $d\vec{r}_0 = (dx_1, dx_2, 0)$ in the undistorted membrane is mapped
by this transformation into a line element,

$$d\vec{r} = \begin{pmatrix} (1 + \partial_1 u_1)dx_1 + (\partial_2 u_1)dx_2 \\ (\partial_1 u_2)dx_1 + (1 + \partial_2 u_2)dx_2 \\ (\partial_1 f)dx_1 + (\partial_2 f)dx_2 \end{pmatrix} \quad . \tag{4.3}$$

As usual in discussions of continuum elastic theory,[38] we describe the stretching of
this line element by a strain matrix $u_{ij}(x_1, x_2)$,

$$d^2 r = d^2 r_0 + 2u_{ij}dx_i dx_j \quad , \tag{4.4}$$

where using Eq. (4.3) we find

$$u_{ij} = \frac{1}{2}[\partial_i u_j + \partial_i u_i] + \frac{1}{2}(\partial_i f)(\partial_j f) + \frac{1}{2}(\partial_i u_k)(\partial_j u_k) \quad . \tag{4.5}$$

To lowest order in gradients of u and f we can neglect the term $\frac{1}{2}(\partial_i u_k)(\partial_j u_k)$ and simply write

$$u_{ij} \approx \frac{1}{2}\left[\partial_i u_j + \partial_j u_i + (\partial_i f)(\partial_j f)\right] \quad . \tag{4.6}$$

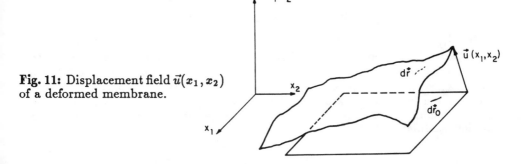

Fig. 11: Displacement field $\vec{u}(x_1, x_2)$ of a deformed membrane.

The free energy of a nearly flat tethered membrane is a sum of bending the stretching energies,

$$F[f, u] = \frac{1}{2}\,\kappa \int d^2x (\nabla^2 f)^2 + \frac{1}{2}\int d^2x [2\mu u_{ij}^2 + \lambda u_{kk}^2] \tag{4.7}$$

where the elastic stretching energy has been expanded in powers of the strain matrix, and μ and λ are elastic constants.[38] The vertical membrane displacement f in Eq. (4.6) introduces an important element of "frustration" into (4.7). To see this, imagine that we are given a vertical displacement field $f(x_1, x_2)$ which has a particularly low bending energy, and hence would lead to a low overall energy if the membrane were a liquid. If this is to be a low energy configuration for tethered membranes as well, it must be possible to choose phonon displacements such that u_{ij} vanishes for the given f. Note that the term $(\partial_i f)\partial_j f$ acts like a matrix vector potential in Eq. (4.6). It will in general be impossible to choose the two independent phonon displacement fields $u_1(x_1, x_2)$ and $u_2(x_1, x_2)$ to cancel all *three* distinct components of this symmetric matrix. We conclude that there must be many low energy configurations of a liquid membrane which will be energetically unfavorable when we introduce stretching energy.

To treat stretching energy more quantitatively, it is useful to eliminate the quadratic phonon field field in (4.7) and define

$$\tilde{F}(f) = -k_B T \ln\left\{ \int \mathcal{D}\vec{u}(x_1, x_2) e^{-F[f, u]/k_B T} \right\} \quad . \tag{4.8}$$

To carry out the functional integral in (4.8), it is essential to separate u_{ij} into its $q = 0$ and $q \neq 0$ Fourier components

$$u_{ij}(\mathbf{x}) = u_{ij}^0 + A_{ij}^0 + \sum_{q \neq 0} \left\{ \frac{1}{2} i[q_i u_j(\vec{q}) + q_i u_i(\vec{q})] + A_{ij}(\vec{q}) \right\} e^{iq \cdot x} \ . \tag{4.9}$$

Here, $A_{ij}(q)$ is the qth Fourier component of the "vector potential" $A_{ij}(x) = (\partial_i f)(\partial_j f)$,

$$A_{ij}(\mathbf{q}) = \int d^2 x e^{-i\mathbf{q}\cdot\mathbf{x}} (\partial_i f)(\partial_j f) \ , \tag{4.10}$$

while A_{ij}^0 is the corresponding component for $\vec{q} = 0$. Although there are only two independent phonon degrees of freedom $\vec{u}_i(\mathbf{q})$ in the in-plane strain matrix at nonzero wavevectors, the uniform part of the in-plane strain matrix u_{ij}^0 has in fact *three* independent components, reflecting the three independent ways of macroscopically distorting a flat two-dimensional crystal.[39] The $q \neq 0$ part of A_{ij} can be decomposed, as can any two-dimensional symmetric matrix, into transverse and longitudinal parts,[40]

$$A_{ij}(\mathbf{x}) = \frac{1}{2} [\partial_i \phi_j(\mathbf{x}) + \partial_j \phi_i(\mathbf{x})] + P_{ij}^T \Phi(\mathbf{x}) \ , \tag{4.11}$$

where P_{ij}^T is the transverse projection operator, $P_{ij}^T = \delta_{ij} - \partial_i \partial_j / \nabla^2$. By applying the transverse projector to both sides of (4.11), we find that

$$\Phi(\mathbf{x}) = \frac{1}{2} P_{ij}^T (\partial_i f)(\partial_i f) \ . \tag{4.12}$$

The functional integral can now be efficiently performed by integrating over the shifted variables

$$\tilde{u}_{ij}^0 = u_{ij}^0 + A_{ij}^0 \tag{4.13a}$$

$$\tilde{u}_i = u_i + \phi_i \ , \tag{4.13b}$$

which leads to an effective free energy

$$F_{\text{eff}} = \frac{1}{2} \kappa \int d^2 x (\nabla^2 f)^2 + \frac{1}{2} K_0 \int' d^2 x \left[\frac{1}{2} P_{ij}^T (\partial_i f)(\partial_i f) \right]^2 \tag{4.14}$$

where the prime on the integral means that the $q = 0$ part of P_{ij} has been integrated out, and

$$K_0 = \frac{4\mu(\mu + \lambda)}{2\mu + \lambda} \ . \tag{4.15}$$

To obtain a physical interpretation of the peculiar form assumed by the stretching energy in (4.14), we note first that the Laplacian of the square root of the integrand is just the Gaussian curvature,

$$-\nabla^2 \left[\frac{1}{2} P_{ij}(\partial_i f)(\partial_i f) \right] = \det \left(\frac{\partial^2 f}{\partial x_i \partial x_j} \right) = S(\mathbf{x}) \ . \tag{4.16}$$

Thus, the parts of a membrane with a nonzero Gaussian curvature act as source terms in the Laplace equation (4.16), leading inevitably to a large positive contribution to the stretching energy. The elastic coupling K_0 penalizes all membrane distortions which are not "isometric," $i.e.$, those which a nonzero Gaussian curvature. There are still many low energy configurations available to the membrane, however, as one can verify by crumpling a piece of paper, which has essentially infinite in-plane elastic constants.

The effect of the nonlinear stretching energy on the renormalized wave-vector-dependent rigidity

$$\kappa_R^{-1}(\mathbf{q}) \equiv q^4 \langle |f(\mathbf{q})|^2 \rangle \tag{4.17}$$

can be calculated perturbatively in K_0.[36] Figure 12 summarizes the relevant Feynman graphs. The slashes on the interaction vertex denote derivatives of $f(x_1, x_2)$. Note that the "tadpole" graphs vanish identically because we have integrated out the $\mathbf{q} = 0$ part of the interaction. The first two terms in the perturbation series for κ_R are

$$\kappa_R(\mathbf{q}) = \kappa + k_B T K_0 \int \frac{d^2 k}{(2\pi)^2} \frac{[\hat{q}_i P_{ij}^T(\mathbf{k}) \hat{q}_j]^2}{\kappa |\mathbf{q} + \mathbf{k}|^4} . \tag{4.18}$$

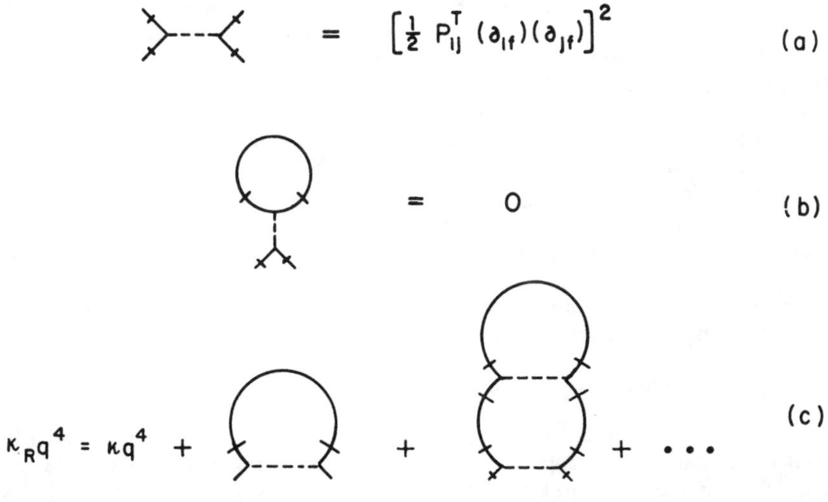

$$\left[\tfrac{1}{2} P_{ij}^T (\partial_{if})(\partial_{if}) \right]^2 \tag{a}$$

$$= \quad 0 \tag{b}$$

$$\kappa_R q^4 = \kappa q^4 + \quad + \quad + \cdots \tag{c}$$

Fig. 12: Graphical rules for calculating the renormalized rigidity: (a) interaction vertex, (b) a graph which vanishes because the $\vec{q} = 0$ part of the interaction has eliminated by integrating out the in-plane strain field, and (c) most divergent terms in the perturbation series for κ_R.

In contrast to the weak logarithmic singularity for liquid membranes displayed in Eq. (3.9), the integral in Eq. (4.18) exhibits a strong $1/q^2$ divergence for small q. The correction to the bare rigidity is positive, showing that the stretching energy $stiffens$ the resistance of the membrane to undulations. Summing up the series of badly diverging diagrams displayed in Fig. 12 leads to a self-consistent equation for $\kappa_R(q)$,

$$\kappa_R(\mathbf{q}) = \kappa + k_B T K_0 \int \frac{d^2 k}{(2\pi)^2} \frac{\left[\hat{q}_i P_{ij}^T(\mathbf{k})\hat{q}_j\right]^2}{\kappa_R(\mathbf{q}+\mathbf{k})|\mathbf{q}+\mathbf{k}|^4} \qquad (4.19)$$

which has the solution, valid for small q,

$$\kappa_R(\mathbf{q}) \sim \sqrt{k_B T K_0} \ q^{-1} \ . \qquad (4.20)$$

We can now repeat the analysis of fluctuations of the surface normals carried out for liquid membranes in Sec. II. Upon inserting the renormalized rigidity (4.20) into Eq. (3.8), we find that the fluctuations in θ^2 are now finite,

$$\langle \theta^2(\mathbf{x}) \rangle = k_B T \int \frac{d^2 q}{(2\pi)^2} \frac{1}{\kappa_R(q)q^2} \simeq \sqrt{\frac{k_B T}{K_0}} \int \frac{d^2 q}{(2\pi)^2} \frac{1}{q} < \infty \ . \qquad (4.21)$$

The mode-coupling analysis which led to this important result *assumed* no significant renormalization of the elastic coupling K_0.[36] A recent renormalization group calculation of Aronovitz and Lubensky[41] suggests that the elastic constants will in fact exhibit weak singularities as $q \to 0$, but that κ_R still diverges strongly enough to make (3.8) finite. Equation (4.21) shows that tethering stabilizes long-range order in the normals at low temperatures, and suggests the existence of the finite temperature crumpling transition. There is now strong evidence from computer simulations[37] for this crumpling transition, at least for tethered surfaces without self-avoidance. These arguments strongly suggest a crumpling transition in real self-avoiding membranes as well. The possibility of a crumpling transition due to *long-range forces* was suggested in the paper on liquid membranes by Peliti and Leibler.[12] Although it is possible to rewrite the stretching energy in Eq. (2.14) terms of a long-range interaction between *Gaussian* curvatures,[1] this interaction is rather different from the long-range forces considered in Ref. 35. There are, moreover, no true microscopic long-range forces in a tethered surface: they arise here only to compensate for integrating out the underlying physical phonon field.

5. Defects and Hexatic Order in Membranes

In our discussion of tethered surfaces, it has been convenient to consider polymerized membranes, tied together with permanent covalent bonds. It is interesting to consider instead nominally crystalline membranes bound together by weaker, van der Waals forces. This situation arises, for example, in unpolymerized lipid bilayers at sufficiently low temperatures. If these materials were constrained to be flat, their low temperature crystalline phase would eventually become unstable upon heating to a proliferation of unbound dislocations. As first elucidated in a famous argument by Kosterlitz and Thouless,[42] a dislocation with Burger's vector $\vec{b}$ (see Fig. 13) in a flat two dimensional crystal of radius R has an energy of order $K_0 b^2 \ln(R/a)$, where K_0 is given by Eq. (4.15) and a is the lattice spacing. This energy cost suppresses the formation of dislocations at low temperatures. However, there is also an entropy of roughly $2k_B \ln(R/a)$ associated with a dislocation, since it can be located at $(R/a)^2$ possible positions. Above the critical melting temperature $k_B T_M \sim K_0 b^2$, the entropy term dominates, dislocations proliferate, and the crystal melts into a hexatic phase.[21] Figure 13 also shows another type of defect, the disclination, which has an energy of order $K_0 R^2 s^2$. Here s is the disclination charge, defined as the angle in radians of the wedge which must be removed or added to a perfect crystal to make the defect. Disclination energies diverge so rapidly with system size that these defects are extremely unlikely in equilibrated 2d crystals.

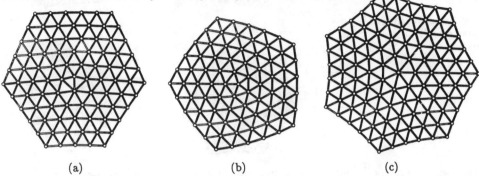

(a)   (b)   (c)

Fig. 13: Defects of interest in polymerized crystalline membranes: (a) dislocation with Burger's vector b, (b) $+2\pi/6$ disclination and (c) $-2\pi/6$ disclination.

In this section we sketch what happens when crystals containing defects are allowed to buckle into the third dimension.[36,43] As shown in Figure 14a, an initially flat disclination in a triangular lattice (obtained by removing a 60° wedge of material from a perfect crystal) will prefer to buckle into an approximately conical shape in a large enough crystal. A similar, hyperbolic buckling (Fig. 14b) occurs in a sufficiently large crystal containing a negative disclination, obtained by adding a 60° wedge of material. This buckling occurs because the system finds it energetically preferable to trade in-plane elastic energy for bending energy. Suppose for simplicity that the in-plane elastic constants are very large, as in an ordinary piece of writing paper. It is easy to check by inserting disclinations into pieces of paper that there are then essentially no elastic distortions in the buckled state. The only remaining energy is now the bending energy, i.e., the first term of Eq. (4.7). It is not hard to show that, for small s, the vertical displacement in polar coordinates (r, θ) is,[36]

$$f(r, \theta) \approx \sqrt{\frac{s}{\pi}} \, r \quad , \tag{5.1}$$

and

$$f(r, \theta) \approx \sqrt{\frac{2|s|}{3\pi}} \, r \cos 2\theta \quad , \tag{5.2}$$

for positive and negative disclinations respectively. In both cases $\nabla^2 f \propto 1/r$, and the bending energy in Eq. (4.7) diverges logarithmically, $F \sim \kappa \ln(R/a)$. The disclination energy still diverges with system size, but it has been reduced considerably from the quadratic divergence characteristic of flat membranes. The logarithmic

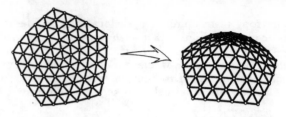

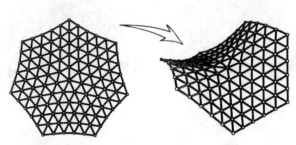

Fig. 14: Buckling of (a) positive and (b) negative disclinations.

dependence on system size is quite general, and not restricted to the large in-plane elastic constant, small disclination charge approximation considered here.[43]

The dislocation in Fig. 13 can be regarded as a tightly bound pair of oppositely charged disclinations, in the sense that its core consists of two displaced points of local five- and seven-fold symmetry. When the dislocation is allowed to buckle, as in Figure 15, we might expect the two (logarithmically divergent) strain fields to cancel at large distances leading to a *finite* dislocation energy.[36] The underlying elasticity equations for buckled membranes are nonlinear,[38] however, so we cannot really apply the superposition principle in this way. Numerical calculations have recently been carried out[43] using the model Hamiltonian (4.1), with

$$V(\vec{r}) = \frac{1}{2} \left[|\vec{r}|^2 - a^2 \right] \quad , \tag{5.3}$$

where a is the lattice constant. The energy of an isolated dislocation at the origin in a system of size of R is shown in Fig. 16 for a variety of ratios of the elastic parameter K_0 to the bending rigidity $\tilde{\kappa}$. The energy initially increases logarithmically with distance, but eventually breaks away and increases more slowly after a critical buckling radius $R_c(K_0/\tilde{\kappa})$. Arguments given in Ref. 43 suggest that the energy $E(R)$ approaches a constant in large R,

$$E(R) \underset{R \to \infty}{\approx} E_D \left[1 - \frac{cR_c}{R} \right] \tag{5.4}$$

where c is a constant of order unity.

The finiteness of the dislocation energy has important consequences for the statistical mechanics of these membranes.[36] The entropy term in the Kosterlitz-Thouless argument now dominates at all nonzero temperatures. Instead of logarithmically bound dislocation pairs, one now has a finite density of unbound dislocations, with density

$$n_D \approx a^{-2} e^{-E_D/k_B T} \quad . \tag{5.5}$$

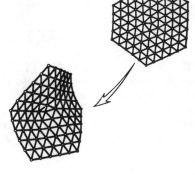

Fig. 15: Buckling of a dislocation. Note the $\pm 2\pi/6$ disclination pair in the dislocation core.

Translational order will be broken up at length scales greater than the translational correlation length

$$\xi_T \approx n_D^{-1/2} \approx a e^{E_D/2k_B T} \ . \tag{5.6}$$

The resulting phase will be a hexatic, with extended bond orientational order.[21] Because disclinations retain a logarithmically divergent energy even after buckling, hexatic membranes should be stable against disclination unbinding into an isotropic liquid over a range of temperatures. Note that hexatics replace crystals as the inevitable low temperature phase in equilibrated membranes. Similar conclusions would apply even if the energy increased indefinitely for $R > R_c$ in Fig. 16, provided only that the rate of increase is less than logarithmic.

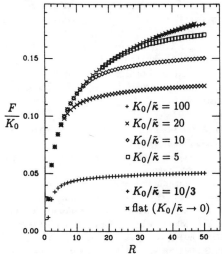

Fig. 16: Dislocation energy as a function of membrane radius R. The energy ceases to increase logarithmically beyond a critical buckling radius.

The properties of undulating hexatic membranes are intermediate between the liquid and tethered membranes discussed so far in this Chapter. The free energy Eq. (2.7) must now be replaced by[36]

$$F_H = \frac{1}{2}\,\kappa \int d^2x (\nabla^2 f)^2 + \frac{1}{2}\,K_A \int d^2x \left[\partial_i\theta + \frac{1}{2}\,\epsilon_{jk}\partial_k\big((\partial_i f)(\partial_j f)\big)\right]^2 \tag{5.7}$$

where $\theta\ (x_1, x_2)$ is the local bond angle field defined within the membrane, and K_A is the hexatic stiffness constant.[21] Gradients of θ are accompanied by a "vector potential"

$$A_i = \frac{-1}{2}\ \epsilon_{jk}\partial_k\left((\partial_i f)(\partial_j f)\right) \tag{5.8}$$

in Eq. (5.7). Just as in Eq. (4.6), this vector potential reflects frustration: the bond angle field cannot return to itself when parallel transported around a region of nonzero Gaussian curvature (The Gaussian curvature is given by the curl of (5.8)). The statistical mechanics associated with Eq. (5.8) was worked out by David *et al.*[44] Remarkably, there is a low temperature "crinkled" phase with a radius of gyration R_G controlled by a *continuously variable* Flory exponent

$$R_G \sim L^{\nu(K_A)} \tag{4.9}$$

where L is a characteristic membrane linear dimension and

$$\nu(K_A) = 1 - \frac{k_B T}{2\pi K_A} + \mathcal{O}(k_B T)^2\ . \tag{4.10}$$

Both disclination unbinding transitions and finite crumpling transitions are possible in hexatic membranes.

These notes are adapted from lectures first given at the Fifth Jerusalen Winter, January, 1988. For closely related papers, and more information about the statistical mechanics of membranes and interfaces, see *Statistical Mechanical of Membranes and Surfaces*, edited by D.R. Nelson, T. Piran, and S. Weinberg (World Scientific, to be published). The work described here was carried out in collaboration with L. Peliti, Y. Kantor, M. Kardar, S. Seung, and M. Paczuski. It is a pleasure to acknowledge these fruitful collaborations, as well as stimulating interactions with F. David and S. Leibler. This work was supported by the National Science Foundation, through grant DMR85-14638 and through the Harvard Materials Research Laboratory.

1. M. B. Green, J. H. Schwarz, and E. Witten, *Superstring Theory 1 and 2* (Cambridge University Press, Cambridge, 1987).

2. J. Weeks, in *Ordering in Strongly Fluctuating Condensed Matter Systems*, ed. T. Riste (Plenum, New York, 1980).

3. For an introduction useful to physicists, see B. A. Dubrovin, A. T. Fomenko, and S. P. Novikov, *Modern Geometry — Methods and Applications* (Springer, New York, 1984).

4. J. V. Jose, L. P. Kadanoff, S. Kirkpatrick, and D. R. Nelson, *Phys. Rev.* B16, 1217 (1977).

5. J. Fröhlich, in *Applications of Field Theory to Statistical Mechanics*, ed. L. Garido, Lecture Notes in Physics, Vol. 216 (Springer-Verlag, Berlin, 1985).

6. P. G. deGennes, *Rev. Mod. Phys.* 57, 827 (1985).

7. E. Brézin, B. I. Halperin, and S. Leibler, *Phys. Rev. Lett.* 50, 1387 (1983); R. Lipowsky and D. M. Kroll and R. K. P. Zia, *Phys. Rev.* B27, 4499 (1983).

8. Y. Oono, *Adv. Chem. Phys.* 61, 301 (1985).

9. P. G. deGennes, *Scaling Concepts in Polymer Physics* (Cornell U. Press, Ithaca,1979).

10. E. Evans and R. Skalak, *Mechanics and Thermodynamics of Biomembranes* (CRC Press, Boca Raton, 1980).

11. P. G. deGennes and C. Taupin, *J. Phys. Chem.* 86, 2294 (1982).
12. *Physics of Complex and Supermolecular Fluids*, eds. S. A. Safran and N. A. Clark (John Wiley and Sons, New York, 1987).
13. *Physics of Amphiphillic Layers*, eds. J. Meunier, D. Langevin, and N. Boccara (Springer-Verlag, Berlin, 1987).
14. H. Fendler and P. Tundo, *Acc. Chem. Res.* 17, 3 (1984).
15. Blumstein, R. Blumstein, and T. H. Vanderspurt, *J. Colloid Interface Sci.* 31, 236 (1969).
16. M. J. Azis, E. Nygren, J. F. Hays, and D. Turnbull, *J. Appl. Phys.* 57, 2233 (1985).
17. Zallen, *The Physics of Amorphous Solids* (Wiley, New York, 1983).
18. J. Larche, J. Appell, G. Porte, P. Bassereau, and J. Marignan, *Phys. Rev. Lett.* 56, 1700 (1986).
19. C. R. Safinya, D. Roux, G. S. Smith, S. K. Sinha, P. Dimon, and N. A. Clark, *Phys. Rev. Lett.* 57, 2718 (1986).
20. D. R. Nelson and L. Peliti, *J. Physique* 48, 1085 (1987); S. Seung and D. R. Nelson, *Phys. Rev. A* (in press).
21. D. R. Nelson and B. I. Halperin, *Phys. Rev.* B19, 2457 (1979); D. R. Nelson, *Phys. Rev.* B27, 2902 (1983).
22. A. M. Polyakov, *Phys. Lett.* 103B, 207 (1981).
23. Y. Kantor, M. Kardar, and D. R. Nelson, *Phys. Rev. Lett.* 57, 791 (1986).
24. Y. Kantor, M. Kardar, and D. R. Nelson, *Phys. Rev. A35*, 3056 (1987).
25. Kardar and D. R. Nelson, *Phys. Rev. Lett.* 58, 1289 (1987); and *Phys. Rev. A* (in press).
26. J. A. Aronovitz and T. C. Lubensky, *Europhys. Lett.* 4, 395 (1987).
27. D. Duplantier, *Phys. Rev. Lett.* 58, 2733 (1987).
28. There are logarithmic corrections to the result of naive dimensional analysis in this case. See the Appendix of Ref. 24.
29. P. G. de Gennes, *The Physics of Liquid Crystals* (Clarendon, Oxford, 1974).
30. A. M. Polyakov, *Nucl. Phys.* B268, 406 (1986).
31. A. M. Polyakov, *Phys. Rev. Lett. 59B*, 79 (1975).
32. W. Helfrich, *Z. Naturforsch.* 28C, 693 (1973).
33. P. G. de Gennes and C. Taupin, *J. Phys. Chem.* 86, 2294 (1982).
34. D. R. Nelson, in *Phase Transitions and Critical Phenomena*, Vol. 7, edited by C. Domb and J. Lebowitz (Academic, New York, 1983).
35. L. Peliti and S. Leibler, *Phys. Rev. Lett.* 54, 690 (1985); see also W. Helfrich, *J. Physique* 46, 1263 (1985).
36. D. R. Nelson and L. Peliti, *J. Physique* 48, 1085 (1987).
37. Y. Kantor and D. R. Nelson, *Phys. Rev. Lett.* 58, 2774 (1987); *Phys. Rev. A36*, 4020 (1987).
38. L. D. Landau and E. M. Lifshits, *Theory of Elasticity* (Pergammon, New York, 1970).
39. For an analogous treatment of compressible spin models in d-dimensions, see J. Sak, *Phys. Rev.* B10, 3957 (1974).
40. See, e.g., S. Sachdev and D. R. Nelson, *J. Phys.* C17, 5473 (1984).
41. J. A. Aronovitz and T. C. Lubensky, *Phys. Rev. Lett.* (in press).
42. J. M. Kosterlitz and D. J. Thouless, *J. Phys.* C5, 124 (1972); *J. Phys.* C6, 1181 (1973).
43. S. Seung and D. R. Nelson, *Phys. Rev. A* (in press).
44. F. David, E. Guitter, and L. Peliti, *J. Phys. (Paris)* 48, 2059 (1987).

FLUCTUATIONS IN FLUID AND HEXATIC MEMBRANES

LUCA PELITI

Dipartimento di Scienze Fisiche and Unitá GNSM-CISM
Universitá di Napoli, Mostra d'Oltremare, Pad. 19
I-80125, NAPOLI (Italy)

Amphiphilic membranes are characterized by a very small value of their surface tension. As a consequence, as it was pointed out long ago by Helfrich[1], the free energy associated with their configuration is essentially determined by curvature. This also implies that shape fluctuations (undulations) are much more violent in these membranes than they are in usual interfaces. I shall discuss in this lecture the effects of such fluctuations in fluid membranes, whose only relevant degrees of freedom are the geometrical ones; I shall then briefly dwell on the effects of orientational order in hexatic[2,3] and smectic[4] membranes and I shall close by reporting the results of a recent work by David[5] on the effect of the incompressibility constraint.

The geometry[6] of a surface embedded in d-dimensional Euclidean space is defined by the mapping $\vec{\sigma} \to \mathbf{X}(\vec{\sigma})$, where $\vec{\sigma} = (\sigma^1, \sigma^2)$ is the inner two-dimensional coordinate, and $\mathbf{X} = (X^1, \ldots, X^d)$ is the corresponding location in bulk space. This geometry is locally described by two tensors: the metric tensor g_{ij} defines the square distance in bulk space between two points whose coordinates differ by ds:

$$ds^2 = g_{ij} d\sigma^i d\sigma^j, \qquad i, j = 1, 2. \tag{1}$$

The curvature tensor $\mathbf{K}_{ij}$ defines how the surface locally deviates from its tangent plane. It is most conveniently defined in terms of the covariant derivatives D_i:

$$\mathbf{K}_{ij} = D_i D_j \mathbf{X} = \partial_i \partial_j \mathbf{X} - \Gamma_{ij}^k \partial_k \mathbf{X}, \tag{2}$$

where Γ_{ij}^k is a Christoffel symbol. Each component of this tensor is a vector in d-dimensional space, locally normal to the tangent plane. Therefore, for $d = 3$ (and for an orientable surface), it may be represented just by a scalar. In the general case, one can build two Euclidean scalars out of it. The first is given by

$$H^2 = \mathbf{K}_i^i \cdot \mathbf{K}_j^j. \tag{3}$$

In $d = 3$ it is just the square of the sum of the inverse curvature radii. The second is given by

$$2K = H^2 - \mathbf{K}_i^j \cdot \mathbf{K}_j^i. \tag{4}$$

This quantity is just the scalar curvature and may be expressed purely in terms of the metric tensor g_{ij}.

The point to be kept in mind for fluid membranes is that all their physical properties should depend just on their geometry, and not on the particular mapping $\vec{\sigma} \to \mathbf{X}(\vec{\sigma})$ we chose to represent it. By exploiting this argument, Helfrich[1] showed that the free energy of an undulating membrane (neglecting membrane-membrane interactions) should be of the form

$$\mathcal{H} = \int d\vec{\sigma}\sqrt{g}\left[r_0 + \left(\frac{1}{2}\right)\kappa_0 H^2 + \overline{\kappa}_0 K + \text{higher order terms}\right]. \qquad (5)$$

We have defined $g = Det(g_{ij})$. The factor $d\vec{\sigma}\sqrt{g}$ is just the area element on the membrane. In $d = 3$ there may appear, for membranes whose two sides are different, a term linear in H. The "higher order terms" involve higher derivatives of X. The coefficients r_0, κ_0, and $\overline{\kappa}_0$ are respectively called the (bare) *surface tension, rigidity, and Gaussian rigidity*. The third term does not play a role for small fluctuations of isolated membranes, since it may be shown that the integral $\int d\vec{\sigma}\sqrt{g} \cdot K$ is a topological invariant. In usual situations, the first term is dominant. But it may be argued that the coefficient r_0 is rather small for fluctuating amphiphilic membranes.

For an *almost planar* membrane we may choose the coordinates $\mathbf{X}$ and $\vec{\sigma}$ such that $\mathbf{X} = (\sigma^1, \sigma^2, \mathbf{h})$, where $\mathbf{h}$ is a small, $(d-2)$-dimensional vector. We can then expand Eq. (5) in powers of $\mathbf{h}$, to obtain:

$$\mathcal{H} \simeq \int d\vec{\sigma}\left(\frac{1}{2}\right)\left[r_0(\partial_i\mathbf{h})^2 + \kappa_0(\partial_i\partial_i\mathbf{h})^2\right]. \qquad (6)$$

This tells us that the fluctuations of $\partial_i h$ are Gaussian, and are correlated up to distances of the order of $(\kappa_0/r_0)^{1/2}$. This distance diverges as r_0 goes to zero, what makes us expect that this limit should lead us to some sort of critical theory. Helfrich[7] has drawn attention to this fact, what has encouraged a few investigators[8-10] to look into the properties of such a model. Probably the clearest picture is obtained from the $d \to \infty$ calculation of Ref. 10, which I shall summarize here. We consider the membrane spanning a square frame of side L, and being immersed in a solution, thus fixing the chemical potential of the amphiphilic molecules, and therefore the coefficient r_0. Since the membrane area per molecule is more or less constant at equilibrium, the *true area* S of the membrane is conveniently measured just by the number of molecules forming it. In general, we have $S \propto L^2$, with a proportionality constant A which depends on r_0, κ_0, and the temperature T. We adopt units in which $kT = 1$. Then there appears to be a critical line $r_0 = r_c(\kappa_0)$ in the (r_0, κ_0) plane at which the constant A diverges. On this line S increases faster than L^2, and one is tempted to associate with it a fractal dimension d_f larger than 2. On the other hand, if one defines an effective rigidity by computing the energy needed to bend the frame, one finds that it decreases with increasing L—and actually (in the $d \to \infty$ limit) vanishes as L diverges. Both these phenomena were anticipated by Helfrich[7] and were confirmed by renormalized perturbation theory.[8-9] One of the main consequences of the reduction of the effective rigidity is that the fluctuations in the tangents to the membrane are correlated—also along the critical line—only up to a finite *persistence length* ξ. In practice the rigidity does not have any effects on fluctuations of wavelength larger than ξ. As a consequence, the membrane is completely crumpled, and in the absence of self-avoidance, its fractal dimension equals infinity. This effect may also be seen as a restoration of Euclidean symmetry, in accordance with the Mermin-Wagner theorem: this symmetry is broken by the existence of the frame, but is restored very far away from it, near the center of the membrane, where no orientation of the tangent plane is preferred.

These considerations lead to two very natural questions: (i) can we envisage a situation in which Euclidean symmetry is spontaneously broken, i.e., where the membrane remains flat in the thermodynamical limit? (ii) what is the correct description of the membrane at distances larger than ξ? Förster and Polyakov[9] have answered the second question by showing that the effective theory for these membranes (if we do not into account self-avoidance effects) is the celebrated Liouville field theory which was first introduced to describe the bosonic string.[11] This also means that all the known pathologies of the Liouville field theory also plague the theory of fluctuating membranes. In practice this should not worry us too much, since self-avoidance effects and topological fluctuations are surely more important for real world membranes than the Liouville field instabilities. It is true that they are, if possible, even less understood.

Stanislas Leibler and myself attempted to answer the first question by pointing out[8] that two appeared as the lower critical dimension for Euclidean symmetry breaking, and that hypothetical membranes with more than two inner dimensions should exhibit a finite-temperature *crumpling transition*, such that at low temperatures the tangents should be on the average parallel to the frame. We also expressed the conjecture that long-range interactions along the membrane should make such a transition possible also for two-dimensional systems.

A way by which long-range forces can originate is by the coupling with internal degrees of freedom.[2] They can be related to additional in-plane order. Positional order leads to the consideration of crystalline membranes (actually of polymerized, or tethered ones, since crystalline order cannot be sustained in a fluctuating membranes, if the core energy of a dislocation is finite). These systems are discussed in the lectures by D. R. Nelson.[12] A weaker form of order is the orientational one presented by hexatic or smectic-C membranes. In plane hexatics, the bonds connecting each molecule to its six nearest neighbors have a preferred orientation. If we introduce the angle θ between this bond and, say, the x-axis, the correlation $\langle \exp[6i\theta(0)] \exp[6i\theta(\vec{r})] \rangle$ decays like $|r|^{-\eta}$, where η is a temperature-dependent exponent. This is described by a free energy $\mathcal{H}_h$ associated with orientational rigidity:

$$\mathcal{H}_h = \int d\mathbf{r} \left(\frac{1}{2}\right) K_A (\nabla\theta)^2. \tag{7}$$

If the membrane is not flat, it is natural to associate with the local bond directions a unit tangent vector field n_i, and to assume that the orientational energy is minimum when this field is mapped onto itself by parallel transport. This leads us to postulate the following form of $\mathcal{H}_h$ for a corrugated membrane:

$$\mathcal{H}_h = \int d\vec{\sigma}\sqrt{g} \left(\frac{1}{2}\right) K_A D_i n^j D^i n_j. \tag{8}$$

Now, whenever the scalar curvature K does not vanish, it is not possible to find a tangent field ni which is mapped onto itself by parallel transport. In fact, if a tangent vector is carried along a closed path, encircling an area $\sum$ where the scalar curvature is K, it comes back rotated by an angle equal to $\sum K$. This frustration effect may be taken into account by introducing a "vector potential" $\vec{A}$, whose rotor is proportional to the scalar curvature K. For almost planar surfaces we are thus led to

$$\mathcal{H}_h = \int d\vec{r} \left(\frac{1}{2}\right) K_A (\nabla\theta + \vec{A})^2. \tag{9}$$

When one integrates over the Gaussian field θ, this produces a long-range effective interaction between the scalar curvatures K at different points, which contributes to a stiffening of the membrane. It is possible[3] to treat by renormalized perturbation theory this effective Hamiltonian in the limit of large stiffness constant K_A. One obtains as a result a corrugated version of the low temperature phase of the Kosterlitz-Thouless transition: in this phase the membrane is self-similarly crumpled, with characteristic exponents which depend on temperature via K_A. At some critical temperature, presumably, defects which disrupt the orientational order nucleate, and the membrane goes back to the fluid state.

In smectic-C membranes the molecules are tilted, and there is also a kind of orientational order along the membrane. The situation is different from the case of hexatics, since there turning the local bond direction by $\pi/3$ left the physical situation invariant. This prevented any low order coupling between the orientational field ni and the geometry. In smectic-C, on the other hand, nothing prevents terms directly coupling the field n_i and the curvature tensor K_{ij} from arising, like for instance

$$K_{ik}K_j^k n^i n^j. \tag{10}$$

These terms represent a stronger coupling between geometrical and orientational degrees of freedom than those present in hexatic membranes. Nevertheless they do not induce long-range interactions. For instance, for membranes in $d = 3$, they introduce a tendency for the field n_i to be aligned along the direction of the principal curvature axes. If the membrane is now subject to a small, localized deformation, this will not modify these directions far away, hence also the corresponding modification of the field ni will be localized. We conclude that smectic-C membranes will behave like ordinary ones.

François David[5] has recently shown that a similar effect also prevents the incompressibility constraint from modifying the behavior of fluid membranes. For almost flat membranes this constraint does indeed introduce long-range interactions between the extrinsic curvatures at different points. But when the effect of geometry fluctuations is taken into account, one finds that these interactions are screened over a length which is much smaller than the persistence length ξ that we have mentioned before. Therefore the doubts that have sometimes been expressed on the validity of the field theory for incompressible membranes do not seem to be justified.

1. W. Helfrich, Z. Naturforsch. 28c 693 (1973).
2. D. R. Nelson and L. Peliti, J. Phys. France 48 1085 (1987).
3. F. David, E. Guitter, and L. Peliti, J. Phys. France 48 2059 (1987).
4. J. Prost and L. Peliti, to be published.
5. F. David, Europhys. Lett. (in press).
6. a useful reference is: B. A. Dubrovin, A. T. Fomenko, and S. P. Novikov, *Modern Geometry*, Vol. 1 (New York, Springer, 1984).
7. W. Helfrich, J. Phys. France 46, 1263 (1985).
8. L. Peliti and S. Leibler, Phys. Rev. Lett. 54 1690 (1985).
9. D. Förster, Phys. Lett. A114, 115 (1986); A. M. Polyakov, Nucl. Phys. B268 406 (1986).
10. F. David, Europhys. Lett. 2 577 (1986).
11. A. M. Polyakov, Phys. Lett. B103 207 (1981).
12. D. R. Nelson, this School.

UNBINDING TRANSITION AND MUTUAL ADHESION IN GENERAL OF DGDG MEMBRANES

W. HELFRICH and M. MUTZ

Fachbereich Physik
Freie Universität Berlin
Arnimallee 14
D-1000 Berlin 33, F.R.G.

ABSTRACT. The glycolipid DGDG (digalctosyl diglyceride) undergoes an unbinding transition driven by temperature. Observing single and bundled membranes under the phase contrast microscope, we have discovered it in highly swollen samples containing NaCl (30-100 mM). Adhesion induced by lateral tension is also seen. We infer from its large contact angles that the membranes are much rougher than predicted on the basis of their thermal undulations.

Digalactosyl diglyceride (DGDG) is common in plant membranes and obtained by extraction. The molecule has a polar head consisting of two ring sugars in a row, otherwise it is like lecithin. We purchased DGDG from Sigma (Munich) and used it without further purification.

DGDG swelled in pure water, forming within a day vesicles and other large membrane structures. The membranes are fluid and stable at all practical temperatures. Salt concentrations sometimes as low 10 mM prevent swelling at room temperature. However, swelling sets in if the temperature is sufficiently raised.

We swelled DGDG at elevated temperatures in 30-100 mM NaCl solutions. Subsequent cooling caused separate membranes to coalesce into bundles. Vesicles, which probably were multilamellar, developed contact areas. All this happened at the same temperature ($\pm 1/2$°C) and was reversible. We think it is evidence for an unbinding transition of stacked DGDG membranes.

Most of the observations were made with a sample containing 100 mM NaCl. The transition temperature was 24°C at the beginning, but dropped to 22°C within two days. It was crossed many times to trigger mutual adhesion or separation of the membranes.

Binding and unbinding proceeded very slowly but without apparent hysteresis. The progress of binding in an array of eight parallel membranes is shown in Figs. 1 to 3. The water is seen to be driven into pockets as adhesion spreads with a sharp front. The reverse process (not shown) is much less dramatic, the water returning in a diffuse way. The evolution of adhesion between multilamellar vesicles is also visible in Figs. 1-3.

The transition temperature, apart from its decrease with time, varied strongly among the samples. It was detected between 20 and 36°C in five samples, in others it was not found at all down to 6°C.

We believe that we did not see the binding transition of two membranes, in accordance with the expectation that its temperature should be lower for the pair than for the stack. Some pictures suggest (see Figs. 1-3) that four membranes adhere as readily as eight. This may be an artefact since adhesion produces lateral tensions as soon as it is stopped by an obstruction, e.g., a tight water pocket.

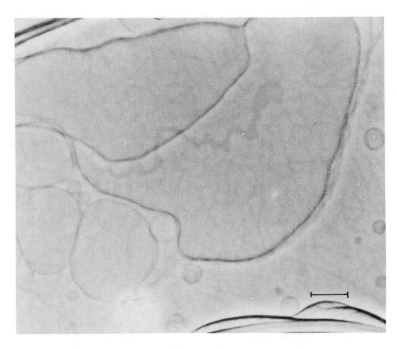

Figure 1

Very low lateral tensions are known to induce adhesion of lecithin membranes that separate otherwise.[1] It would be attractive to interpret the unbinding transition in terms of a competition of a attractive van der Waals and repulsive undulation forces. A renormalization group treatment of this type and based on a realistic ansatz for the direct forces has recently been presented by Lipowsky and Leibler.[2] However, we doubt that our transition can be satisfactorily explained by undulation theory.

Let us briefly (and superficially) check if the membranes may be expected to unbind. The most relevant material parameter are the Hamaker constant $A_h = (3$ or $7) \cdot 10^{-14}$ erg, measured by Marra[3] for 0.2 M salt or pure water, respectively and the bending rigidity $k_c \approx 1 \cdot 10^{-12}$ erg, being measured by us. We also write down the half-space approximation for the van der Waals interaction energy of two membranes per unit area

$$g_{vdW} = -\frac{A_h}{12\pi z^2}$$

and a formula for the enery of pure steric interaction due to thermal undulations, in a stack and per unit area,[4]

$$g_{st} = \frac{3\pi^2 (k_B T)^2}{128 k_c \bar{z}^2}$$

Here, z is a uniform and $\bar{z}$ a mean spacing. Inserting $k_B T = 4 \cdot 10^{-14}$ (room temperature) and the lower value of A_h, which favors separation, and equating

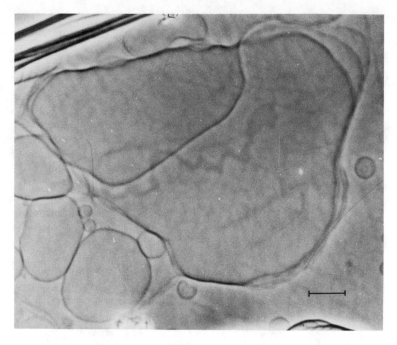

Figure 2

$z=\bar{z}$, we find steric repulsion to be half as strong as van der Waals attraction. We feel the comparison is inconclusive (and remains so if the direct forces are treated more realistically).

Adhesion induced by ultralow lateral tensions has become a rather common phenomenon in studies of swollen phospholipids.[1,5] An example obtained with DGDG in pure water (no binding transition) is shown in Fig. 4. The contact angle of symmetric adhesion (two single membranes with equal angles) were found to be 45 - 60°C, those of a single membrane adhering to a stack ca. 85°C. At some places a rounding of the membrane next to the contact area reveals a lateral tension below 10^{-3} dyn/cm.[1,6] Note that E. Evans[7] measured the mutual adhesion energy of highly stressed DGDG membranes to be 0.22 erg/cm^2.

Considering adhesion induced by ultralow tensions, let us now show the insuffiency of undulation theory by using the following three formulas: First, Young's equation

$$g_a = (1 - cos\Psi)\sigma_f \tag{1}$$

relating the energy of adhesion per unit area, g_a, to the lateral tension of the free membrane and to the contact angle Ψ of a single membrane adhering to a stack. Second, the energy of stretching, g_f, per unit area of the free membrane[6]

$$g_f = \frac{k_B T}{8\pi k_c}\sigma_f \tag{2}$$

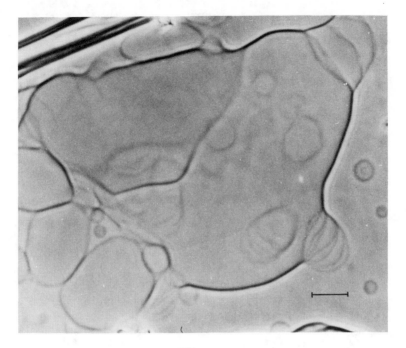

Figure 3

which is the energy needed to flatten undulations. Third, the self-evident relationship

$$g_a = g_f - g_s \qquad (3)$$

expressing the adhesion energy by the stretching energies of the membrane where
it is free, g_f, and where it adheres to the stack, g_s. Above the unbinding transition
we have to expect $g_s > 0$ so that

$$g_a < g_f$$

and, because of (2),

$$g_a < \frac{k_B T}{8 \pi k_c} \sigma_f$$

Inserting the above numbers we arrive at $g_a < 1.6 \cdot 10^{-3} \sigma_f$. On the other hand, the
contact angle of single membranes adhering to a stack of membranes beeing almost
90°C, we read from Young's equation (1) $g_a \approx \sigma_f$. We think that the enormous
discrepancy can only be removed by postulating an optically unresolved additional
roughness of the membranes. It would have to absorb much more real membrane
area than the $< 5\%$ absorbed by undulations.[6] Eqs. (1-3) will then have to be
modified.[1]

We did a preliminary experiment to check the hypothesis of a large reservoir
of hidden area in unstressed DGDG membranes. Two unilamellar spherical vesicles
with regular fluctuations were observed when the sample was cooled from 70 to
10°C and heated up to 70°C again. Despite an area expansivity of 0.02 per °C,[7]

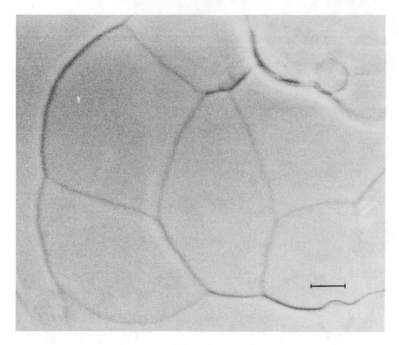

Figure 4

as measured on stretched membranes, i.e. an expected total change of real area by ca. 12%, the vesicle neither became elliptical when heated nor ruptured when cooled, although the volume appeared to be constant. The fluctuation amplitudes of the higher-order spherical harmonic were larger at the higher temperature but this change seemed again too small to absorb an excess area of 12%.

What could be the structure of the postulated membrane roughness? The answer might be a tendency of the bilayer to form local "saddles." In fact, it can be shown[8] in a molecular model that the energy per unit area of Gaussian curvature K, if expanded in K, should have a negative K^2 term, which is a precondition for such saddles. (Note that $K = c_1 c_2$ where c_1 and c_2 are the principle curvatures) Unlike local "hats" the saddles may have little effect on the bending rigidity. However, they could give rive to a considerable roughness if they cooperate.

1. R. M. Servuss and W. Helfrich, to be published.
2. R. Lipowsky and S. Leibler, Phys. Rev. Lett. 56, 2541 (1986).
3. J. Marra, J. Colloid Interface Sci. 109, 11 (1986).
4. W. Helfrich, Z. Naturforsch. 33a, 305 (1078).
5. M. Muts, unpublished.
6. W. Helfrich and R.M. Servuss, Nuovo Cimento D 3, 137 (1984).
7. E. Evans and D. Needham, J. Phys. Chem. 91, 4219 (1987).
8. W. Helfrich, Liquid Crystals (in press).

SCALING PROPERTIES OF INTERFACES AND MEMBRANES

REINHARD LIPOWSKY

Institut für Festkörperforschung der KFA Jülich
Postfach 1913, D-5170 Jülich, FRG

1. Outline and Summary

This review deals with three topics: (i) *Fluctuations* of interfaces and membranes (Sec. 2); (ii) *Interactions* of interfaces and membranes (Sec. 3); and (iii) *Dynamics* of interfaces and growth of wetting layers (Sec. 4). All three topics are intimately related: fluctuations renormalize the interactions, and the interactions act as a driving force for the dynamics.

I will focus on interfaces and membranes which are *rough* but not (yet) crumbled. The scaling properties of such surfaces are governed by the roughness exponent, ζ (Sec. 2.2). For *interfaces*, this exponent depends on the nature of the two phases separated by the interface which can be fluids, periodic crystals, quasicrystals, or random systems (Sec. 2.3). For *membranes*, ζ depends on the internal membrane structure which may be fluid, crystalline, or hexatic (Sec. 2.4). In all cases, the fluctuations give a singular contribution to the surface free energy, see Sec. 2.5.

The effective *interaction* of surfaces represents a *unifying concept* by which one can understand many different phenomena, see Sec. 3.1 and 3.2. This point of view is not entirely new: there exists a huge literature on interactions between *planar* surfaces (Sec. 3.3). However, it has been realized only recently that these interactions are often renormalized by *fluctuations*. Several scaling regimes for the renormalized interaction must be distinguished, see Sec. 3.4. These different regimes can be studied in a systematic way, starting from an effective Hamiltonian for the interacting surfaces (Sec. 3.5). So far, the most useful method has been a *functional renormalization group* (RG), see Sec. 3.6. As a result, one finds RG fixed points for the critical behavior associated with the unbinding of surfaces (Sec. 3.7 and 3.8).

Unbinding occurs for repulsive interactions which drive the surfaces apart. The dynamics of this process is discussed in Sec. 4. Several growth modes for interfaces are considered: adhesive growth (Sec. 4.1 and 4.2), diffusion- limited growth (Sec. 4.3), and activated growth resulting from quenched impurities (Sec. 4.4). These growth modes are relevant for the thickening of wetting layers or thin films.

2. Fluctuations of Interfaces and Membranes

2.1. INTRODUCTION

An interface or domain wall represents the contact region between two bulk phases of matter. This contact region has an intrinsic width which is usually microscopic and set by the bulk correlation lengths of the two phases. At finite temperatures, $T > 0$, the interface undergoes thermally–excited fluctuations which lead to a certain interfacial roughness. The interface may also be roughened by quenched impurities which provide an effective random potential. These fluctuations change the area of the interface and, thus, are controlled by *tension*.

Membranes are sheets of amphiphilic molecules which form spontaneously when these molecules are dissolved in water or in mixtures of water and oil. Such membranes play an essential role in biophysics since they provide the spatial organization of cells and organelles. For $T > 0$, membranes are also deformed by thermal fluctuations but typical undulations are *bending* modes controlled by *curvature*.

In this section, I will discuss the properties of a *single* interface or membrane. First, the roughness of surface fluctuations is defined in a precise way and the basic *roughness exponent*, ζ, is introduced (Sec. 2.2). The possible values of ζ are then summarized in Sec. 2.3 and Sec. 2.4 for interfaces and membranes, respectively. Finally, the free energy arising from the surface fluctuations is discussed in Sec. 2.5. This free energy has a scaling form which is characterized by the *decay exponent*, τ.

2.2. SCALE INVARIANCE OF SURFACES AND THE ROUGHNESS EXPONENT ζ

Consider a 2–dim surface segment with linear size $\sim L$ which is embedded in 3–dim space. At $T > 0$, the surface segment will make bumps in order to increase its configurational entropy. The bumpy surface can be characterized by two length scales, $L_\parallel$ and $L_\perp$. The scale $L_\parallel$ gives the area $\sim L_\parallel^2$ of a typical surface configuration when *projected* onto the planar reference state; the scale $L_\perp$ measures the average distance of the surface from this reference state. In general, I will consider (d-1)–dim surfaces with projected area $\sim L_\parallel^{d-1}$ which fluctuate in d–dim space. Two examples in $d = 1 + 1$ are: (a) domain boundaries within an adsorbed monolayer, and (b) steps or ledges on a crystal surface.

The two length scales $L_\parallel$ and $L_\perp$ are not independent but satisfy the scaling relation $L_\perp \sim L_\parallel^\zeta$ where $\zeta \geq 0$ is the *roughness exponent*. [1,2] Several cases must be distinguished: (i) A *smooth* surface with $\zeta = 0$; in this case, the size of $L_\perp$ is set by a microscopic length scale (or cutoff), a; (ii) A *marginally rough* surface with $\zeta = 0(\sqrt{\log})$ which corresponds to

$$L_\perp \sim [\ln(L_\parallel/a)]^{1/2} \qquad \text{for} \qquad L_\parallel \gg a . \tag{2.1}$$

(iii) A *rough* surface with $0 < \zeta < 1$ and

$$L_\perp \sim L_\parallel^\zeta \qquad \text{for} \qquad L_\parallel \gg a . \tag{2.2}$$

(iv) A rough surface with $\zeta \geq 1$; in this case, the scaling relation between $L_\perp$ and $L_\parallel$ holds only on *intermediate* length scales:

$$L_\perp \sim L_\parallel^\zeta \qquad \text{for} \qquad \xi_p \gg L_\parallel \gg a \tag{2.3}$$

where ξ_p is the *persistence length* of the surface. This length which was introduced in the context of membranes[3] gives the correlation length of the surface normals. For $L_\parallel \gg \xi_p$, the normals decorrelate and the surface is *crumpled*.[3] For $\zeta < 1$, one always has $\xi_p = \infty$.

Thermal fluctuations provide an *entropic* mechanism for the roughening of surfaces. Alternatively, the surface can be roughened by an external potential. In this case, it tries to minimize its *energy* by adapting its shape to the minima of the external potential. The latter mechanism underlies, e.g., the roughening by quenched random impurities.

So far, a finite surface segment has been considered. Now, take the thermodynamic limit, $L_\parallel \to \infty$. Then, the roughness exponent ζ can be extracted from the difference correlation function, $\Delta C(x) \equiv < [\ \ell(\vec{x}) - \ell(\vec{0})\ ]^2 >$, for the variable $\ell(\vec{x})$ which measures the distance of the fluctuating surface from its planar reference state with coordinate $\vec{x} = (x_1, \ldots, x_{d-1})$. Indeed, the scaling relation (2.2) implies[1]

$$\Delta C(x) \sim x^{2\zeta} \quad \text{or} \quad \Delta C(x) \approx b^{2\zeta}\Delta C(x/b) \quad \text{for} \quad x \gg a \ . \tag{2.4}$$

Thus, the bumpy surface is invariant under the rescaling transformation $x \to x/b$ and $\ell \to \ell/b^\zeta$.

2.3. ROUGHNESS OF INTERFACES

1. Fluid phases and periodic crystals — An interface separating two fluid phases is characterized by the roughness exponent $\zeta = \zeta_o$ with

$$\begin{aligned}
\zeta_o &= 0(\sqrt{\log}) & d &= 3 \\
&= (3-d)/2 & 1 &< d < 3 \ .
\end{aligned} \tag{2.5}$$

The same value applies to an interface which separates a periodic crystal from another phase *provided* T exceeds the roughening temperature, T_R.

2. Ideal quasicrystals — In this case, the interface feels a quasiperiodic potential which stiffens the interface and *decreases* the value of ζ. A Fibonacci potential in $d = 2$, which approximates the ideal Penrose tiling, leads to the *non-universal* value [4,5]

$$\zeta < \zeta_o = 1/2 \quad \text{and} \quad \zeta \approx 2 \ln[(1 + \sqrt{5})/2]\, T/J \quad \text{for small T} \ . \tag{2.6}$$

For Harper's potential, the interface undergoes a roughening transition in $d = 2$.[4]

In general, the large scale properties of a rough interface with roughness exponent $0 < \zeta \le \zeta_o = (3-d)/2$ can be described by the *scale-dependent stiffness*

$$\tilde{\Sigma}_{\text{eff}}(L_\parallel) \sim L_\parallel^{\eta_\sigma} \quad \text{with} \quad \eta_\sigma = 2(\zeta_o - \zeta) \ge 0 \ . \tag{2.7}$$

In fluids and in periodic crystals with $T > T_R$, the exponent $\eta_\sigma = 0$ and $\tilde{\Sigma}_{\text{eff}}(L_\parallel = \infty) \equiv \tilde{\Sigma}$ is finite. A finite $\tilde{\Sigma}$ also applies to random systems with $\zeta < 1$.

3. Random quasicrystals — Random tilings can be obtained from the ideal tiling by a random rearrangement of the tiles thereby abandonning the matching rules.[6,7] Then, the interfacial roughness is *enhanced* compared to a periodic system, and[8]

$$\zeta = 2/3 \quad \text{in} \quad d = 1+1 \quad \text{for} \quad T > 0 \ . \tag{2.8}$$

For $T = 0$, the interface is marginally rough and smooth in $d = 1+1$ and $d = 2+1$, respectively.[8]

4. Quenched random impurities — Two types of impurities must be distinguished; (i) Random fields which couple directly to the order parameter density, and (ii) Random bonds which couple to the energy density. For *random fields*, a scaling argument[9] leads to [10-12]

$$\zeta = 0 \qquad \text{for} \quad d > 5$$
$$= (5 - d)/3 \qquad \text{for} \quad 2 < d < 5 \ . \tag{2.9}$$

For *random bonds*, the corresponding scaling argument fails but

$$\zeta = 2/3 \quad \text{in} \quad d = 1 + 1 \quad . \tag{2.10}$$

is known exactly both for $T = 0$ and for $T > 0$.[13-16]

2.4. ROUGHNESS OF MEMBRANES

1. Fluid membranes — For fluid membranes such as lipid bilayers in the L_α phase, [17] the elastic bending energy is governed by the mean curvature.[18] This leads to $\zeta = \zeta_o$ with[19]

$$\zeta_o = 0(\sqrt{\log}) \qquad \text{for} \quad d = 5$$
$$= (5 - d)/2 \qquad \text{for} \quad 1 < d < 5 \ . \tag{2.11}$$

Thus, in $d = 3$, $\zeta = 1$,[20] and the persistence length, ξ_p, is finite[3].
2. Crystalline and polymerized membranes — Lipid bilayers usually exhibit L_β phases at low T in which the lipid molecules form a lattice or network.[17] A similar situation occurs for membranes composed of polymerizable lipids which provide models for biological systems.[21] Then, the bending modes are coupled to the internal degrees of freedom.[22] At low T, these membranes exhibit a *scale–dependent bending rigidity*:[22,23]

$$\kappa_{\text{eff}}(L_\parallel) \sim L_\parallel^{\eta_\kappa} \qquad \text{with} \quad 2/3 \leq \eta_\kappa \leq 1 \quad \text{in} \quad d = 2 + 1 \ . \tag{2.12}$$

In general, a scale–dependent rigidity implies

$$\zeta = \zeta_o - \eta_\kappa/2 \tag{2.13}$$

with ζ_o as given by (2.11).
3. Hexatic membranes — The crystalline membrane could melt into a hexatic one if the free energy of a single dislocation is finite as a result of buckling.[22] At low T, hexatic membranes acquire a scale–independent rigidity[24] which implies

$$\eta_\kappa = 0 \quad \text{and} \quad \zeta = 1 \quad \text{in} \quad d = 2 + 1 \ . \tag{2.14}$$

2.5. SCALING OF SURFACE FREE ENERGY

The surface fluctuations give a singular contribution, Δf, to the surface free energy. Consider again a surface segment with projected area $\sim L_\parallel^{d-1}$, and assume that the surface is 'clamped' to a reference plane along one edge. The opposite edge is free and will make transverse excursions $\sim L_\perp$ from the reference plane. These largest humps have wavenumber $\sim 1/L_\parallel$ and should contain a thermal free energy $\sim T$ as suggested by the equipartition theorem. This gives an entropic contribution, Δs, to the free energy per unit (projected) area which scales as

$$-T\Delta s \sim T/L_\parallel^{d-1} \ . \tag{2.15}$$

On the other hand, the elastic free energy per unit area, Δe, behaves as

$$\Delta e \sim \tilde{\Sigma}_{\text{eff}} \ (L_\perp/L_\parallel)^2 \sim L_\parallel^{\eta_\sigma}(L_\perp/L_\parallel)^2 \tag{2.16}$$

for interfaces, and as

$$\Delta e \sim \kappa_{\text{eff}} \ (L_\perp/L_\parallel^2)^2 \sim L_\parallel^{\eta_\kappa}(L_\perp/L_\parallel^2)^2 \tag{2.17}$$

for membranes. Therefore, Δe has the general scaling form

$$\Delta e \sim (L_\perp/L_\parallel^n)^2 \tag{2.18}$$

with

$$\begin{aligned} n &= 1 - \eta_\sigma/2 && \text{for interfaces} \\ &= 2 - \eta_\kappa/2 && \text{for membranes} \ . \end{aligned} \tag{2.19}$$

Now, the free energy per unit area, Δf, is taken to be $\Delta f = \Delta e - T\Delta s$.[25,8] For a marginally rough surface with $\zeta = 0(\sqrt{\log})$, the elastic free energy dominates and

$$\Delta f \sim \ln(L_\parallel/a)/L_\parallel^{2n} \sim L_\perp^2 \exp\{-2nL_\perp^2/a_\perp^2\} \tag{2.20}$$

where $a_\perp$ is a microscopic length scale which must be distinguished from a[26]. For a rough surface with $\zeta > 0$ (and $a \ll L_\parallel \ll \xi_p$), Δf exhibits the power law behavior

$$\Delta f \sim 1/L_\parallel^{\zeta\tau} \sim 1/L_\perp^\tau \tag{2.21}$$

The *decay exponent*, τ, depends on ζ, d, and n. Two cases must be distinguished:

(i) For thermally–excited surface fluctuations, one usually has $\zeta = (2n+1-d)/2$. Then, the entropic contribution and the elastic free energy have the same order of magnitude, and

$$\tau = 2(n/\zeta - 1) = (d-1)/\zeta \ . \tag{2.22}$$

(ii) For *interface* fluctuations induced by a random potential, the roughness exponent satisfies $\zeta > (2n+1-d)/2$ with $n = 1$, and the elastic free energy scales as the energy gain resulting from the potential. In this case, the elastic free energy dominates and[1]

$$\tau = 2(1/\zeta - 1) < (d-1)/\zeta \ . \tag{2.23}$$

3. Interactions of Interfaces and Membranes

3.1. INTRODUCTION

In the previous section, a single interface or membrane has been considered. In real systems, one usually has two or several surfaces rather than a single one. In some cases, the surfaces are *oriented* in such a way that they are, on average, *parallel*. Condensed matter physics provides several examples for such a behavior where the surfaces are 1– or 2–dimensional:

(a) Wetting, surface melting, and surface–induced disorder[27] — In this case, the contact region of two bulk phases, α and γ, contains a thin film or layer of a third phase, β. This film is bounded by two interfaces which have the same average orientation. Surface melting[28–31], edge melting[32], and surface–induced disorder[33–35] have recently been studied by a variety of experimental methods.

(b) Equilibrium shape of crystals — This shape is determined, to a large extent, by the behavior of the steps or ledges which separate flat terraces of the crystal surface.[36] The average orientation of these steps is determined by the index of the crystal surface.

(c) Commensurate–Incommensurate transitions — Near such a transition, the system consists of commensurate domains which are separated by domain walls. In many cases, these walls (or lines) are parallel and form a striped phase, see Ref. 2 and references therein.

(d) Adhesion of vesicles or biological cells — A vesicle composed of an amphiphilic membrane can adhere to another surface (which may be a second membrane). Within the region of contact, the membrane and the surface are roughly parallel.[17]

(e) Lamellar phases of membranes — Such phases occur (i) in binary systems of water and lipids, and (ii) in oil–water mixtures containing amphiphilic molecules. They are composed of a stack of membranes which have the same average orientation. For recent experimental studies, see Refs. 37 and 38.

In all of these examples, the surfaces experience a mutual interaction. In this section, I discuss the nature and the form of this interaction. First, the *total* interaction, $V_{TI}(\ell)$, of two surfaces with separation ℓ is defined in Sec. 3.2. In the *absence* of surface fluctuations, this interaction reflects the intermolecular forces and is called the *direct* interaction, $V_{DI}(\ell)$, see Sec. 3.3.

Surface fluctuations lead to a nontrivial *renormalization* of $V_{DI}(\ell)$. Several scaling regimes (or universality classes) must be distinguished (Sec. 3.4). In general, one must include fluctuations on *all* length scales. This can be done starting from a systematic theory, see Sec. 3.5. So far, the most powerful method has been a nonperturbative functional renormalization group (RG). The results of this RG will be described in Sec. 3.6. Finally, the critical behavior associated with the unbinding of surfaces will be discussed in Sec. 3.7 and 3.8.

3.2. EXTERNAL PRESSURE AND TOTAL INTERACTION OF SURFACES

Consider two roughly parallel surfaces (or surface segments) and assume that one can change the separation, ℓ, of these surfaces by an external pressure, H. I will use the sign convention that *increasing* pressure leads to *decreasing* surface separation.

In the context of wetting, a thin film of a metastable phase, β, intrudes between two coexisting bulk phases, α and γ. Then, the free energy per unit volume, $\hat{f}_\beta$, of the β phase exceeds the free energy per unit volume, $f_\alpha = f_\gamma = f_{\alpha\gamma}$, of the coexisting phases α and γ, and the external pressure is given by $H = \hat{f}_\beta - f_{\alpha\gamma}$.[27] Thus, for wetting, H can be experimentally controlled via the temperature, T, and via the chemical potentials of the molecular species. For membranes, H can be controlled by an osmotic or hydrostatic pressure, or by changing a relative humidity.[39]

The external pressure H is balanced by a disjoining pressure, $-\partial V_{TI}/\partial \ell$, arising from the total interaction, $V_{TI}(\ell)$, of the surfaces.[40] Thus, the mean separation, $\bar{\ell}$, is determined by

$$H = -\partial V_{TI}/\partial \ell \qquad \text{for} \qquad \ell = \bar{\ell}. \tag{3.1}$$

In an experiment (or in a computer simulation), one can measure the mean surface separation, $\bar{\ell}$, as a function of H. Then, the total interaction, $V_{TI}(\ell)$, can be obtained via a Legendre transformation from H to ℓ.[27] First, one inverts the relation $\ell = \bar{\ell}(H)$ in order to obtain $H = \bar{H}(\ell)$. When this is inserted into (3.1), one has $-\partial V_{TI}/\partial \ell = \bar{H}(\ell)$ which determines $V_{TI}(\ell)$ up to a constant. This shows that the total interaction, $V_{TI}(\ell)$, is a well-defined quantity which contains the same information as $\ell = \bar{\ell}(H)$. Now, from a *theoretical* point of view, one would like to *predict* the behavior of $\ell = \bar{\ell}(H)$ and, thus, of $V_{TI}(\ell)$. This can be done in two steps: (i) First, one ignores surface fluctuations and studies the *direct* interaction, $V_{DI}(\ell)$, between planar surfaces [41]; (ii) Secondly, one takes surface fluctuations into account which renormalize $V_{DI}(\ell)$. *This renormalized interaction represents the theoretical prediction for $V_{TI}(\ell)$.*

3.3. DIRECT INTERACTIONS OF PLANAR SURFACES

The direct interactions, $V_{DI}(\ell)$, depend on the microscopic forces acting between the molecules. Two contributions have been known for a long time:[42] (i) *van der Waals interactions* — In $d = 3$, this interaction decays as $\sim 1/\ell^2$ and $1/\ell^3$ for nonretarded and retarded forces, respectively. (ii) *Electrostatic interactions* — Interfaces or membranes often contain electric charges. If the surface separation ℓ is large compared to the Debye screening length, λ_E, the associated direct interaction decays as $\sim \exp(-\ell/\lambda_E)$. For $a \ll \ell \ll \lambda_E$, V_{DI} exhibits a power law behavior.

More recently, the direct interaction between surfaces has been studied within Landau or van der Waals theories. As a result, one finds several cases depending on the nature of the phase, β, between the surfaces: (iii) *Exponential interactions* — If the intervening β phase is composed of small molecules and has a microscopic correlation length, ξ_β, $V_{DI} \sim \exp(-\ell/\xi_\beta)$.[43-45] (iv) *Interactions induced by critical fluctuations* — If the intervening β phase is critical or near–critical, scaling theory implies $V_{DI} \sim 1/\ell^{d-1}$.[46] (v) *Polymer–induced interactions* — When the surfaces are covered by adsorbed polymers, one has additional direct interactions which decay like a power of ℓ.[47]

3.4. DIFFERENT SCALING REGIMES FOR THE UNBINDING OF SURFACES

Now, consider a *fluctuating* surface which is bound, via attractive direct interactions, to another surface. In such a situation, the roughness of the fluctuating surface is restricted by the presence of the second surface. The size, $\xi_\perp$, of this roughness is set by the transverse extension of the largest bumps. For smaller bumps with roughness $L_\perp \ll \xi_\perp$, the fluctuations are essentially unrestricted, and their lateral extension, $L_\parallel$, scales as in (2.1)–(2.3). Therefore, the largest bumps can be characterized by a longitudinal correlation length, $\xi_\parallel$, which satisfies $\xi_\perp \sim [\ln(\xi_\parallel/a)]^{1/2}$ for $\zeta = 0(\sqrt{\log})$ and $\xi_\perp \sim \xi_\parallel^\zeta$ for $\zeta > 0$.

The two length scales, $\xi_\parallel$ and $\xi_\perp$, play essentially the same role for the bound surfaces as the scales $L_\parallel$ and $L_\perp$ for a finite surface segment, compare Sec. 2.2. Therefore, the excess free energy, $V_{FL}(\xi_\perp)$, of these largest bumps can be estimated as in Sec. 2.5:[25]

$$V_{FL}(\xi_\perp) \sim 1/\xi_\perp^\tau \quad \text{with} \quad \tau = 2(n/\zeta - 1) \tag{3.2}$$

for a rough surface with $\zeta > 0$, and

$$V_{FL}(\xi_\perp) \sim \xi_\perp^2 \ \exp[-2n\xi_\perp^2/a_\perp^2] \tag{3.3}$$

for a marginally rough surface, compare (2.21) and (2.20).

This excess free energy can be interpreted as a *fluctuation–induced repulsion* acting between the surfaces. It can be used in order to identify different scaling regimes for the *unbinding* of surfaces. In the process of unbinding, the mean separation, $\bar{\ell}$, becomes large compared to microscopic length scales, and the surfaces probe the tail of their direct interaction, V_{DI}. Therefore, each scaling regime contains a certain *class* of direct interactions.

1. *Rough surfaces with $\zeta > 0$* — In this case, one must distinguish four scaling regimes: a *mean–field* (MF) regime and three different *fluctuation* (FL) regimes. The two length scales, $\bar{\ell}$ and $\xi_\perp$, behave as[48]

$$\bar{\ell} \gg \xi_\perp \quad \text{in MF regime} \quad \text{and} \quad \bar{\ell} \sim \xi_\perp \quad \text{in FL regimes} . \tag{3.4}$$

(i) The MF regime is characterized by[1,25]

$$V_{DR}(\ell) \gg V_{FL}(\ell) \sim 1/\ell^\tau \quad \text{for large } \ell \tag{3.5}$$

where V_{DR} represents the *repulsive* part of V_{DI}. (ii) The *weak–fluctuation* (WFL) *regime* is defined by[1,25]

$$V_{DR}(\ell) \ll V_{FL}(\ell) \sim 1/\ell^\tau \ll V_{DA}(\ell) \quad \text{for large } \ell \tag{3.6}$$

where V_{DA} represents the *attractive* part of V_{DI}. In addition, one has two nontrivial fluctuation regimes: (iii) the *intermediate fluctuation* (IFL) regime with[49]

$$|V_{DI}(\ell)| \sim V_{FL}(\ell) \sim 1/\ell^\tau \quad \text{for large } \ell \quad , \tag{3.7}$$

which contains, in fact, three different subregimes; and (iv) the *strong–fluctuation* (SFL) regime characterized by[1,25]

$$|V_{DI}(\ell)| \ll V_{FL}(\ell) \sim 1/\ell^\tau \quad \text{for large } \ell \quad . \tag{3.8}$$

2. Marginally rough surfaces with $\zeta = 0(\sqrt{\log})$ — In this case, the two length scales, $\bar{\ell}$ and $\xi_\perp$, satisfy

$$\bar{\ell} \gg \xi_\perp^2 \quad \text{in MF regime} \quad \text{but} \quad \bar{\ell} \sim \xi_\perp^2 \quad \text{in FL regimes} . \qquad (3.9)$$

The MF regime is now characterized by $V_{DR}(\ell) \gg V_{FL}(\sqrt{\ell}) \sim \exp(-c\ell/a)$ for large ℓ. Thus, one enters the FL regimes for a short–ranged repulsion $V_{DR} \sim \exp(-c\ell/a)$.

For $V_{DI}(\ell) \approx -W/\ell^r + U\exp(-\ell/a_1)$, the linear renormalization group (RG) introduced in Ref.[50] leads to two subregimes depending on the size of the microscopic length scale, a_1.[51] For $V_{DI}(\ell) \approx -W\exp(-\ell/a_1) + U\exp(-\ell/a_2)$, this linear RG yields three subregimes[50,52] as has been confirmed, to a certain extent, by MC simulations of interface models[53]. Exponential interactions should apply to 3–dim lattice models with short–ranged interactions. However, MC simulations of such models have not produced, so far, any evidence for the FL regimes.[54,55]

3.5. EFFECTIVE HAMILTONIAN FOR INTERACTING SURFACES

Consider two surface segments which are, on average, parallel and are described by two coordinates, $\ell_1(\vec{x})$ and $\ell_2(\vec{x})$. For thermally–excited fluctuations their elastic free energy per unit area is given by $\frac{1}{2}K_1(\nabla^n\ell_1)^2$ and $\frac{1}{2}K_2(\nabla^n\ell_2)^2$, respectively, with n as in (2.19) and $\zeta = (2n+1-d)/2$. Furthermore, the two surfaces experience a direct interaction, $V_{DI}(\ell_1 - \ell_2)$. Then, their separation $\ell \equiv \ell_1 - \ell_2 \geq 0$ is governed by the effective Hamiltonian

$$\mathcal{H}\{\ell\} = \int d^{d-1}x \ \{\frac{1}{2}K(\nabla^n\ell)^2 + V(\ell)\} \qquad (3.10)$$

with $K = K_1 K_2/(K_1 + K_2)$ and $V(\ell) \equiv H\ell + V_{DI}(\ell)$ for $\ell > 0$.

The field ℓ is restricted to positive values since the surfaces considered here cannot intersect. Therefore, the model (3.10) should be supplemented with the *hard wall* condition

$$V(\ell) = \infty \quad \text{for} \quad \ell < 0 . \qquad (3.11)$$

Then, all configurations with negative ℓ–values have a vanishing Boltzmann weight, $\exp\{-\mathcal{H}/T\}$. For $\ell > 0$, the interaction $V(\ell)$ has the generic form $V(\ell) = H\ell + V_{DI}(\ell)$ with $V_{DI}(\ell) \approx 0$ for large ℓ. For effective pressure $H > 0$, the term $H\ell$ provides an exponential cutoff in the weight for large ℓ. For $H = 0$, configurations with arbitarily large ℓ have a finite weight. Therefore, the surfaces are always *unbound* at $H = 0$ as long as the number of surface modes is finite ($L_\parallel < \infty$). In order to get a bound state of the surfaces, one must *first* perform the thermodynamic limit ($L_\parallel \to \infty$) for $H > 0$, and subsequently let H approach zero.[27]

The model (3.10) can be studied with a variety of theoretical methods. [27] So far, the only approach which has been useful for general d and n is a functional renormalization group (RG)[56,19,25] which represents an extension of Wilson's approximate recursion relations[57].

3.6. FUNCTIONAL RENORMALIZATION

1. Strong-fluctuation (SFL) regime — The functional RG acts as a *nonlinear* map on the direct interaction, $V(\ell) = V_{DI}(\ell)$, of interfaces or membranes. When applied to the model given by (3.10), this RG leads to the recursion relation[56,19,25]

$$V^{(N+1)}(\ell) = R[V^{(N)}(\ell)] \tag{3.12}$$

with

$$R[V(\ell)] = -\tilde{v}b^{d-1}\ln\left\{\int_{-\infty}^{\infty}\frac{d\ell'}{\sqrt{2\pi}\tilde{a}_\perp}\exp\left[-\frac{1}{2}(\ell'/\tilde{a}_\perp)^2 - G(\ell,\ell')\right]\right\} \tag{3.13}$$

and

$$G(\ell,\ell') = [V(b^\zeta\ell - \ell') + V(b^\zeta\ell + \ell')]/2\tilde{v} . \tag{3.14}$$

The parameters $\tilde{a}_\perp$ and $\tilde{v}$ which depend on the rescaling factor $b > 1$ represent a length and a free energy scale. The length scale $\tilde{a}_\perp$ is determined by the requirement that the RG transformation is exact to first order in V. This leads to

$$\tilde{a}_\perp^2 = a_\perp^2(b^{2\zeta} - 1)/2\zeta \qquad \text{with} \qquad a_\perp^2 = c_d(T/K)a^{2\zeta} , \tag{3.15}$$

$c_d = 2/(4\pi)^{(d-1)/2}\Gamma(\frac{d-1}{2})$, and $\zeta = (2n + 1 - d)/2$ as before. The scale $\tilde{a}_\perp$ has a simple interpretation: it is the roughness of the small–scale fluctuations with wavelengths $a \leq L_\parallel \leq ba$. The free energy scale, $\tilde{v}$, is not determined by the linearized RG. Wilson's original decomposition of phase space leads to the choice

$$\tilde{v} = v(1 - b^{1-d})/(d - 1) \qquad \text{with} \qquad v = c_d T/a^{d-1} . \tag{3.16}$$

For rough surfaces with $\zeta = (2n + 1 - d)/2 > 0$, the above recursion relation leads to two fixed points, $V_o^*(\ell)$ and $V_c^*(\ell)$, [56,19,25] which have a Gaussian tail $\sim \exp(-\ell^2)$. These two fixed points govern the behavior within the SFL regime as defined by (3.8). The fixed point $V_o^*(\ell)$ is purely repulsive and describes the unbound state of the surfaces. All direct interactions, $V(\ell) = V_{DI}(\ell)$, which lead to completely separated surfaces are mapped onto this fixed point. On the other hand, the fixed point $V_c^*(\ell)$ has an attractive well and describes the continuous unbinding transition within the SFL regime. All direct interactions which correspond to such a transition point are mapped onto V_c^*. Thus, within the SFL regime, the fixed points V_o^* and V_c^* have a domain of attraction with codimension zero and one, respectively.

2. Intermediate (IFL) regime — Now, consider interactions $V(\ell) = V_{DI}(\ell)$ which belong to the IFL regime and decay as $\sim 1/\ell^\tau$ for large ℓ. This regime is governed by a *whole line* of nontrivial RG fixed points as found from the functional RG in the infinitesimal rescaling limit $b \to 1 + \triangle t$. [58] In this limit, the RG transformation (3.12) becomes[25]

$$\partial V/\partial t = (d - 1)V + \zeta\ell\partial V/\partial\ell + \frac{1}{2}v \ \ln[1 + (a_\perp^2/v)\partial^2 V/\partial\ell^2] \tag{3.17}$$

with $a_\perp$ and v as in (3.15) and (3.16).

It is convenient to use the dimensionless variables $z \equiv \sqrt{2\zeta}\ell/a_\perp$ and $U(z) \equiv 2\zeta\ V(a_\perp z/\sqrt{2\zeta})/v$. Then, the flow equation becomes[58]

$$\partial U/\partial t = \zeta[\tau U + zU' + \ln(1 + U'')] . \tag{3.18}$$

The fixed points, $U^*(z)$, of this RG transformation satisfy

$$\tau U^* + z\partial U^*/\partial z + \ln[1 + \partial^2 U^*/\partial z^2] = 0 . \tag{3.19}$$

Therefore, the rescaled fixed point equation depends only on *one* parameter, namely τ.

This fixed point equation has solutions which are singular at $z = 0$ and behave as[58]

$$U^*(z) \approx \sigma/z^\tau + \frac{\tau + 2}{\tau}\ln(z) \quad \text{with} \quad \sigma > 0 \tag{3.20}$$

for *small* z. For *large* z, all solutions to (3.19) decay, and U^* is then governed by the linear equation, $\tau U^* + z\partial U^*/\partial z + \partial^2 U^*/\partial z^2 = 0$. This implies[58]

$$U^*(z) \approx \rho(\sigma)/z^\tau + \bar{\rho}(\sigma)z^{\tau-1}\exp(-z^2/2) \tag{3.21}$$

for large z where the amplitudes ρ and $\bar{\rho}$ are uniquely determined by σ. This line of RG fixed points, $U^*(z|\sigma)$, leads to non-universal critical behavior and essential singularities in agreement with exact calculations[49] for $(d,n) = (2,1)$. The corresponding RG flow is unusual and has a parabolic character.

3. Fluctuation regimes for marginally rough surfaces — For the marginal case $\zeta = (2n + 1 - d)/2 = 0$, the flow equation (3.17) reduces to

$$\partial V/\partial t = 2nV + \frac{1}{2}v\ \ln[1 + (a_\perp^2/v)\partial^2 V/\partial \ell^2] \tag{3.22}$$

Then, the equation for the fixed points, $V^*(\ell)$, can be written as $\partial^2 V^*/\partial \ell^2 = -\partial\Phi(V^*)/\partial V^*$ where $\Phi(V^*)$ has a unique minimum at $V^* = 0$.[25] This implies that one has *no* (stationary) unbinding fixed points for $\zeta = 0$. Instead, a line of *drifting* fixed points, $V(\ell,t) = V^\dagger(\ell - gt)$, is found which move with constant velocity, g, under the RG.[56,25] They arise because the redundant perturbation, $\partial V^*/\partial \ell$, associated with a shift of the ℓ-coordinate, becomes marginal for $\zeta = 0$.

In the absence of a nontrivial fixed point, the critical behavior must be determined by a matching procedure: one applies the RG up to a matching point at which the renormalized interaction, $V^{(N)}(\ell)$ or $V(\ell,t)$, can be analyzed by mean-field theory. Such a matching procedure can be done analytically if one *linearizes* the RG transformation. If the recursion relation (3.13) is linearized as it stands, one obtains a linear RG for $V(\ell)$[52,25] which is completely equivalent to a normal ordering of $V(\ell)$[59,48,27]. On the other hand, one may first incorporate the hard wall condition (3.11) into (3.13), and *subsequently* linearize. The latter procedure leads to the modified linear RG introduced in Ref.[50].

3.7. CRITICAL BEHAVIOR AT UNBINDING TRANSITIONS

The critical behavior associated with the unbinding of surfaces can be characterized by critical exponents. First, consider the case of *complete* unbinding which occurs for a repulsive total interaction, $V_{TI} \geq 0$, as the effective pressure, H, goes to zero.[60,1,61] Then,

$$\bar{\ell} \sim H^{-\psi^c} \quad , \quad \xi_\perp \sim H^{-\nu_\perp^c} \quad , \quad \text{and} \quad \xi_\parallel \sim H^{-\nu_\parallel^c} . \tag{3.23}$$

with critical exponents $\psi^c, \nu_\perp^c$, and $\nu_\parallel^c$. (The superscript c stands for 'complete'). Likewise, the surface free energy, $f_s(H) \equiv H\bar{\ell} + V_{TI}(\bar{\ell})$ with $\bar{\ell} = \bar{\ell}(H)$ contains the singular part, $f_s \sim H^{2-\alpha^c}$, and

$$\partial^2 f_s / \partial H^2 = \partial\bar{\ell}/\partial H \sim H^{-\alpha^c} . \tag{3.24}$$

The four critical exponents $\psi^c, \nu_\perp^c, \nu_\parallel^c$, and α^c are not independent but satisfy the scaling relations

$$\alpha^c = \psi^c + 1 \quad , \quad \nu_\perp^c = \zeta\nu_\parallel^c \quad , \quad \text{and} \quad 2n\nu_\parallel^c = 2 - \alpha^c + 2\psi^c \tag{3.25}$$

with n as in (2.19). Therefore, there is *only one independent* critical exponent, say ψ^c.[60] For a direct interaction, $V_{DI}(\ell) \sim 1/\ell^p$, one has

$$\begin{aligned} \psi^c &= 1/(1+p) & \text{for} & \quad p < \tau \\ &= 1/(1+\tau) & \text{for} & \quad p \geq \tau. \end{aligned} \tag{3.26}$$

This holds both for thermally–excited[60,61] and for impurity–induced[1,2,62] fluctuations with τ as given by (2.22) and (2.23), respectively.

For $H = 0$, the surfaces may be bound (in the thermodynamic limit, $L_\parallel = \infty$), i.e., the total interaction, V_{TI}, may have an attractive part. The strength of this attraction depends on various parameters, and can vanish as these parameters are changed. Then, the surfaces undergo a continuous or discontinuous unbinding transition.

At a *discontinuous* transition, the mean separation, $\bar{\ell}$, jumps from a finite value to infinity.[63] Such a transition can occur for direct interactions which satisfy $V_{DI}(\ell) \gg V_{FL}(\ell) \sim 1/\ell^\tau$ for large ℓ, and which have an attractive part at smaller values of ℓ. In contrast, a discontinuous transition is impossible in the SFL regime with $|V_{DI}(\ell)| \ll V_{FL}(\ell) \sim 1/\ell^\tau$.[48,58] In the IFL regime with $|V_{DI}(\ell)| \sim V_{FL}(\ell)$, discontinuous transitions are still possible but acquire very unusual scaling properties.[48,49,64]

A *continuous* transition is *critical* if the singular behavior depends on *two* scaling fields.[65] This applies to *all* transitions in the SFL regime.[58]. At a critical unbinding transition, one has $\bar{\ell} = y^{-\psi}\Omega_\ell(H/y^\Delta)$ where y is an appropriate scaling field. Similar scaling forms with corresponding exponents $\nu_\perp, \nu_\parallel$, and $2 - \alpha$ hold for $\xi_\perp, \xi_\parallel$, and f_s. For $y = 0$, one has $\bar{\ell} \sim H^{-\psi/\Delta}$ and, thus, $\psi^c = \psi/\Delta$ etc. It then follows from (3.25) that $\nu_\perp = \zeta\nu_\parallel$ and $2n\nu_\parallel = 2 - \alpha + 2\psi$.

3.8. HOW FAR IS IT TO ASYMPTOTIA?

In the previous subsections, the *asymptotic* behavior associated with the unbinding of surfaces has been discussed. If one wants to study these critical effects in experiments or numerical simulations, one must worry about: (i) *Crossover behavior*; (ii) *Finite size effects*; and (iii) *Time scales* for equilibration. The first two topics are discussed in this subsection while equilibration will be considered in Sec. 4.

1. *Crossover behavior* — As an example, consider complete wetting (or edge melting) in $d = 2$ for a direct interaction, $V_{DI}(\ell) \approx C_p/\ell^p$ with $p > \tau = 2$. Then, the unbinding of the 1–dim interfaces is driven by thermally–excited fluctuations for $T > 0$. Indeed, their mean separation behaves as[60]

$$\bar{\ell} \approx cs^{1/3}/h^{1/3} \quad \text{for small} \quad h \equiv H/C_p \quad \text{with} \quad s \equiv T^2/\bar{\Sigma}C_p$$

and interfacial stiffness $\bar{\Sigma} = \bar{\Sigma}(T)$. If the 1–dim interface feels a periodic potential, one expects[66] $\bar{\Sigma} \sim \exp(J/T)$ and, thus, $s \sim \exp(-J/T)$ for small T where J is the step energy. At $T = 0$, on the other hand, the unbinding is entirely controlled by V_{DI}, and $\bar{\ell} \sim 1/h^{1/(p+1)}$ for small h as follows from minimization of $V(\ell) \approx H\ell + C_p/\ell^p$.

Thus, one has a characteristic *crossover* at low T which can be described by the scaling form

$$\bar{\ell} = h^{-1/(1+p)} \,\Omega(s/h^\phi) \quad \text{with} \quad \phi = (p-2)/(p+1) \tag{3.27}$$

where the shape function $\Omega(x) \approx cx^{1/3}$ for large x. It seems that such a crossover has been observed, for $p = 3$, in recent experiments[31] on edge melting.

2. *Finite size effects* — Now, assume that the surfaces are embedded in a finite system which has a linear extension, $N_\parallel$ and $N_\perp$, in the direction parallel and perpendicular to the surfaces. The anisotropy of the surface fluctuations, characterized by $\xi_\perp \sim \xi_\parallel^\zeta$ leads to different finite size effects in these two directions.

In the parallel direction, finite size effects set in once $\xi_\parallel \simeq N_\parallel$. This is very important for the interpretation of computer simulations. It leads, in fact, to quasi-critical behavior as $H \to 0$ for finite $N_\parallel$.[67] In the perpendicular direction, finite size effects are always present for $\xi_\perp \simeq N_\perp$ since the boundaries then act as a confining potential on the surfaces. However, such effects may set in even for $\xi_\perp \ll N_\perp$. This happens in the context of wetting. In this case, a finite value of $N_\perp$ leads to a shift of the coexistence curve and the process of wetting is truncated at $H = H_* > 0$. Quite generally, one has $H_* \sim 1/N_\perp$.[68] Since the limit $H = 0$ is no longer accessible, the mean separation $\bar{\ell}$ of the surfaces can no longer diverge. For a non-conserved order parameter, the maximal value of the equilibrium separation is given by[68]

$$\max(\bar{\ell}) \sim 1/H_*^{\psi^c} \sim N_\perp^{\psi^c} \quad . \tag{3.28}$$

with ψ^c as in (3.26). This behavior applies, e.g., to wetting by one–component fluids (where it is called capillary condensation) and to surface melting. For a conserved order parameter, the truncation of $\bar{\ell}$ sets in somewhat earlier, and $\max(\bar{\ell}) \sim N_\perp^{\psi^c/(1+\psi^c)}$.

4. Interface Dynamics and Growth of Wetting Layers

4.1. INTRODUCTION

In *equilibrium*, interacting surfaces feel two forces: an external pressure, H, which is balanced, for $H \geq 0$, by the disjoining pressure, $-\partial V_{TI}/\partial \ell$, arising from the interactions.[40] Then, the total force,

$$\hat{H}(\ell) \equiv H + \partial V_{TI}/\partial \ell \tag{4.1}$$

acting on the surfaces, vanishes as in (3.1). Now, consider an *unbalanced* situation *away* from equilibrium with $\hat{H}(\ell) \neq 0$. Then, the surfaces will move apart for $\hat{H}(\ell) < 0$ or will come closer together for $\hat{H}(\ell) > 0$.

In the following, I will focus on the *thickening of wetting layers* or on the *growth of thin films*. In this case, the total force, $\hat{H} < 0$, corresponds to an effective *undersaturation*. This undersaturation *scales* as

$$\hat{H} \sim \partial V_{TI}/\partial \ell \sim -1/\ell^{1/\psi^c} \qquad \text{for} \quad H = 0 \tag{4.2}$$

with $\psi^c = 1/(1 + p)$ or $\psi^c = 1/(1 + \tau)$ as in (3.26). Thus, for $H = 0$, the driving force for the dynamics is either determined by the direct repulsion $V_{DR} \sim 1/\ell^p$ or by the fluctuation–induced repulsion $V_{FL} \sim 1/\xi_\parallel^\tau$.

Several growth modes for the thickening of wetting layers will be discussed: (i) Adhesive growth for rough interfaces;[69] (ii) Adhesive growth for smooth interfaces; (iii) Diffusion–limited growth;[70] and (iv) Activated growth in the presence of quenched impurities. In all cases, the wetting layers are taken to be close to thermal and *chemical* equilibrium. This has to be distinguished from the dynamics of dry spreading[71] where the total volume of the wetting film is fixed.

4.2. ADHESIVE GROWTH FOR ROUGH INTERFACES

Consider a liquid phase, β, which is adsorbed from a vapor phase, α, onto a solid substrate, γ. Then, a wetting layer builds up and the mean separation, $\bar{\ell}$, of the $(\alpha\beta)$ interface from the solid wall steadily increases with time, t. The deposition rate from the vapor phase is proportional to the vapor pressure while the evaporation rate depends on the binding energies within the condensed phase. The growth rate is then proportional to the undersaturation: $\partial \ell/\partial t \sim -\hat{H}(\ell)$, which implies[69]

$$\bar{\ell}(t) \sim t^\theta \quad \text{with} \quad \theta = \psi^c/(1 + \psi^c) \qquad \text{for} \quad H = 0 \tag{4.3}$$

where ψ^c is given by (3.26) and $\tau = (d - 1)/\zeta$. On the other hand, for small $H > 0$, the equilibrium thickness $\bar{\ell}(\infty) \sim 1/H^{\psi^c}$. It then follows that the equilibration time, t_{eq}, scales as $t_{eq} \sim 1/H^{1 + \psi^c}$ for small $H > 0$. For a marginally rough interface in $d = 3$ and $V_{DI}(\ell) \sim \exp(-\ell/a_1)$, the mean separation $\bar{\ell}(t) \sim \ln(t)$, and the equilibration time $t_{eq} \sim 1/H$.[72]

This growth mode also applies to wetting in lattice models with nonconserved dynamics provided $T > T_R^{\alpha\beta}$ where $T_R^{\alpha\beta}$ is the roughening temperature of the $(\alpha\beta)$ interface. Indeed, the growth law (4.3) has been confirmed by MC simulations of SOS–models in $d = 2$ and $d = 3$[73], of a 3–state chiral Potts model in $d = 2$[74], and of an Ising model in $d = 3$[75]. For a 2–dim model on a periodic lattice (with short–ranged interactions), the growth law (4.3) becomes $\bar{\ell}(t) \approx A_\ell t^{1/4}$ with amplitude $A_\ell \sim (T^2/\tilde{\Sigma})^{1/4} \sim \exp(-J/T)$ since the stiffness $\tilde{\Sigma} \sim \exp(J/T)$ for low T. Therefore, $\bar{\ell}(t)$ exhibits strong crossover behavior at low T which has indeed been observed in one of the MC studies.

What about the length scales $\xi_\parallel$ and $\xi_\perp$? Scaling implies that $\xi_\parallel(t) \approx b \xi_\parallel(b^{-z}t)$ and, thus,[69]

$$\xi_\parallel(t) \sim t^{\theta_\parallel} \quad \text{with} \quad \theta_\parallel = 1/z \qquad (4.4)$$

which will be taken as a *definition* of the dynamic critical exponent, z. On the other hand, one may again define a roughness exponent, ζ, via $\xi_\perp(t) \sim \xi_\parallel(t)^\zeta$ as before. Then,

$$\xi_\perp(t) \sim t^{\theta_\perp} \quad \text{with} \quad \theta_\perp = \zeta/z . \qquad (4.5)$$

The growth of $\bar{\ell}$ as given by (4.3) is rather slow. Then, one may assume *local equilibrium* such that the length scales $\bar{\ell}, \xi_\parallel$, and $\xi_\perp$ are related via $\xi_\parallel \sim \bar{\ell}^{\nu_\parallel/\psi}$ and $\xi_\perp \sim \xi_\parallel^\zeta$ where the exponents $\nu_\parallel/\psi = \nu_\parallel^c/\psi^c$ and ζ have the same values as in equilibrium. This implies that $\theta_\parallel = \theta \nu_\parallel^c/\psi^c = 1/(2 - \eta_\sigma), \theta_\perp = \zeta/(2 - \eta_\sigma)$ and $z = 2 - \eta_\sigma$ with $\eta_\sigma \geq 0$. The local equilibrium assumption is indeed confirmed by a systematic study of the Langevin equation $\partial \ell/\partial t = -C \delta \mathcal{H}/\delta \ell + f$ with effective Hamiltonian $\mathcal{H}\{\ell\}$ as in (3.10) and Gaussian white noise, f.[69]

It is possible, however, that the value of ζ is determined by the *dynamics* rather than by the statics. Such a behavior has been found for a *single* interface which grows by *ballistic deposition* with a local growth rule.[76,77] For some growth rules, the interfacial coordinate, $\ell(x,t) = gt + h(x,t)$, evolves according to[76] $\partial h/\partial t = g(\nabla h)^2 + CK\nabla^2 \ell + f$. This leads to $z = z_1$ and $\zeta = \zeta_1$ with $z_1 + \zeta_1 = 2$.[78] In $d = 1 + 1$, one has $z_1 = 3/2$ and $\zeta_1 = 1/2$.[76] For $d > 2$, the exponents are expected to satisfy the bounds $1/d \leq \zeta_1 \leq 1/2$ and $3/2 \leq z_1 \leq 2 - 1/d$.[76,79]

In the present context, the velocity, g, is not constant but depends on time: $g = \partial \ell/\partial t \sim t^{\theta-1}$. Therefore, the fluctuations, $h(\vec{x},t) = \ell(\vec{x},t) - \bar{\ell}(t)$, should evolve according to

$$\partial h/\partial t = B t^{\theta-1}(\nabla h)^2 + CK\nabla^{2n}\ell + f . \qquad (4.6)$$

where f is again a Gaussian white noise. A scaling analysis of this equation shows that the nonlinear term $\sim (\nabla h)^2$ is *irrelevant* for $\theta < \theta_* \equiv z_1/2n$. The adhesive growth law (4.3) implies $\theta < 1/2$ for $\psi^c < 1$, while $\theta_* \geq 3/4$ follows from the bound $z_1 \geq 3/2$. Thus, local equilibrium should be generally valid for (4.3).

On the other hand, if $\theta > \theta_*$ is enforced by $H < 0$ or by an external potential, one may enter the ballistic deposition regime which is then characterized by $\zeta = \zeta_1$ and $z = z_1/\theta$. Even in this regime, $\xi_\perp$ is always small compared to $\bar{\ell}$ since $\theta_\perp = \zeta/z = \theta\zeta_1/(2 - \zeta_1) \leq \theta/3$.

4.3. ADHESIVE GROWTH FOR SMOOTH INTERFACES

Now, consider a wetting layer in $d = 3$ consisting of a periodic crystal or an ideal quasicrystal. At low T, the interfaces bounding this layer are *smooth* (with $\zeta = 0$), and the approach to complete wetting proceeds via multilayering. Each new layer starts from 2–dim nucleation clusters with critical radius $\simeq \Sigma_s/a(-\hat{H})$ and free energy $\triangle F \simeq \Sigma_s^2/a(-\hat{H})$ where Σ_s is the step free energy per unit length. Therefore, the growth is activated, and $\partial \ell/\partial t \sim \exp(-\triangle F/T) \sim \exp[C/\hat{H}(\ell)]$. For $H = 0$ and $V_{DI} \sim 1/\ell^p$, this leads to the logarithmic growth law

$$\bar{\ell}(t) \sim [\ln(t)]^{\psi^c} \quad \text{with} \quad \psi^c = 1/(1+p) \tag{4.7}$$

in $d = 3$. For small $H > 0$, the equilibrium separation $\bar{\ell}(\infty) \sim 1/H^{\psi^c}$ which implies the equilibration time $t_{eq} \sim \exp[C/H]$ for small $H > 0$.

The above analysis applies to multilayering in the 3–dim Ising model as has been studied in a MC simulation.[54] In this case, $V_{TI}(\ell) \sim \exp(-c\ell/a)$ which implies $\bar{\ell}(t) \sim \ln[\ln(t)]$ for $H = 0$, and $t_{eq} \sim \exp[C/H]$ as before. Thus, observation of more than the first few layers requires an exponentially large time.

In real solid films, *defects* have a dramatic effect on the growth rate. In epitaxial growth, *mismatch dislocations* lead to an effective interaction, $V_{DI}(\ell) \sim -1/\ell$, which is *attractive* and, thus, prevents complete wetting at $H = 0$.[80] On the other hand, for non–epitaxial growth, the growth rate can be greatly enhanced by the presence of *screw dislocations* which act as a source for steps. Classical theories for spiral growth predict that the rate is $\sim$ (supersaturation)^2.[81] In the present context, this leads to $\partial \ell/\partial t \sim [\hat{H}(\ell)]^2$. For $H = 0$ and $V_{DI}(\ell) \sim 1/\ell^p$, this implies the power law growth

$$\bar{\ell}(t) \sim t^\theta \quad \text{with} \quad \theta = \psi^c/(\psi^c + 2) \quad \text{and} \quad \psi^c = 1/(1+p) \tag{4.8}$$

in $d = 3$.

4.4. DIFFUSION–LIMITED GROWTH

Next, consider a binary mixture or alloy of two molecular species, A and B, which separates into two phases, α and β, below its consolute point. Assume that the B–rich phase β forms a wetting layer which intrudes beween the α phase and a solid wall, γ. At complete wetting, the β layer will grow into the α phase, and the region of α phase adjacent to the $(\alpha\beta)$ interface becomes depleted of B molecules. If there is no hydrodynamic flow, further growth can only occur by diffusion through the depleted region.

The thickness of the depleted region is set by the diffusion length, $\delta(t) \sim t^{1/2}$. Mass conservation implies that $\ell(t) \sim [X_\alpha - X(\ell)]\delta(t)$ where X_α is the concentration of B molecules deep in the α phase while $X(\ell)$ is the concentration in front of the $(\alpha\beta)$ interface. In local equilibrium, one has $X_\alpha - X(\ell) \sim$ undersaturation of the relative chemical potential. Then, the thickness, ℓ of the wetting layer evolves according to $\ell(t) \sim -\hat{H}(\ell)/\delta(t)$, which leads to[70]

$$\bar{\ell}(t) \sim t^\theta \quad \text{with} \quad \theta = \psi^c/2(1 + \psi^c) \quad . \tag{4.9}$$

where ψ^c is given by (3.26) with $\tau = (d-1)/\zeta$. This is confirmed by a systematic study of the interfacial motion using the Green's function formalism.[70]

An interface which moves as a result of bulk diffusion can be unstable with respect to the Mullins–Sekerka instability.[82] It turns out, however, that the interfacial motion is *stable* as long as the growth exponent $\theta < 1/2$ which applies to the growth given by (4.9).[70] On the other hand, for effective pressure $H < 0$, the above scaling analysis leads to $\bar{\ell}(t) \sim \delta(t) \sim t^{1/2}$, and the interface develops 'fingers' as a result of the Mullins–Sekerka instability. This fingering could accelerate the interfacial motion and θ could become larger than $1/2$.

4.5. ACTIVATED GROWTH IN THE PRESENCE OF QUENCHED IMPURITIES

Finally, consider the same wetting geometry as in the previous subsections but assume that the $(\alpha\beta)$ interface feels a random potential arising from the presence of quenched impurities. In such a situation, the interface gets caught in *metastable* states and its dynamics is then controlled by the size of activation barriers. [83,13]

In order to make an interfacial fluctuation of longitudinal and transverse extension, $\xi_\parallel$ and $\xi_\perp \sim \xi_\parallel^\zeta$, one must overcome free energy barriers, $\triangle F \sim \xi_\parallel^{d-1}(\xi_\perp/\xi_\parallel)^2$. On the other hand, if the interface moves out by such a hump, it will typically gain a free energy $\sim (-\hat{H})\xi_\perp\xi_\parallel^{d-1}$. Therefore, the interface must nucleate humps which are characterized by a critical size $\xi_\parallel = \xi_{\parallel c} \sim (-\hat{H})^{-1/(2-\zeta)}$ corresponding to an activation free energy $\triangle F \sim (-\hat{H})^{-(d-3+2\zeta)/(2-\zeta)}$. This leads to $\partial\ell/\partial t \sim \exp[-\triangle F/T] \sim \exp[-C/(-\hat{H}(\ell))^x]$, and

$$\bar{\ell}(t) \sim [\ln(t)]^{\psi^c/x} \quad \text{with} \quad \chi = (d - 3 + 2\zeta)/(2 - \zeta) \tag{4.10}$$

where ψ^c is given by (3.26) with $\tau = 2(1/\zeta - 1)$.

Acknowledgements. I thank Michael E. Fisher, Wolfgang Helfrich, Christopher Henley, Daniel Kroll, Stanislas Leibler, Heiner Müller-Krumbhaar, Thomas Nattermann, Theo Nieuwenhuizen, Erich Sackmann, Jacques Villain, and Barbara Zielinska for stimulating interactions, and Gene Stanley and Nicole Ostrowsky for the invitation to this Advanced Study Institute.

1. R. Lipowsky and M. E. Fisher, Phys. Rev. Lett. **56**, 472 (1986).

2. M. E. Fisher, J. Chem. Soc., Faraday Trans. 2 **82**, 1569 (1986).

3. P. G. De Gennes and C. Taupin, J. Phys. Chem. **86**, 2294 (1982).

4. C. L. Henley and R. Lipowsky, Phys. Rev. Lett. **59**, 1679 (1987).

5. A. Garg and D. Levine, Phys. Rev. Lett. **59**, 1683 (1987); A. Garg, Phys. Rev. B (in press).

6. V. Elser, Phys. Rev. Lett. **54**, 1730 (1985); C. L. Henley, J. Phys. A **21**, 1649 (1988).

7. Interfaces in a 'quasiglass' have been studied by T. L. Ho, J. A. Jasscsak, Y. H. Li, and W. F. Saam, Phys. Rev. Lett. **59**, 1116 (1987).

8. R. Lipowsky and C. L. Henley, Phys. Rev. Lett. **60**, 2394 (1988).

9. Y. Imry and S.-K. Ma, Phys. Rev. Lett. **35**, 1399 (1975).

10. G. Grinstein and S.-K. Ma, Phys. Rev.B**28**, 2588 (1983).

11. J. Villain, J. Physique Lett. **43**, L551 (1982).

12. T. Nattermann, J. Phys. C**16** 4113 (1983).

13. D. A. Huse and C. L. Henley, Phys. Rev. Lett. **54**, 2708 (1985).

14. M. Kardar and D. R. Nelson, Phys. Rev. Lett. **55**, 1157 (1985).

15. D. A. Huse, C. L. Henley, and D. S. Fisher, Phys. Rev. Lett. **55**, 2924 (1985).

16. T. Nattermann and W. Rens, Phys. Rev. B (in press).

17. See, e.g., *Physics of Amphiphilic Layers*, eds. J. Meunier, D. Langevin, and N. Boccara [Springer Proc. in Physics, Vol. 21] (Springer, 1987).

18. W. Helfrich, Z. Naturforschung 28c, 693 (1973).

19. R. Lipowsky and S. Leibler, Phys. Rev. Lett. 56, 2541 (1986), and p. 98 in Ref. 17.

20. W. Helfrich, Z. Naturforschung 33a, 305 (1978).

21. E. Sackmann, P. Eggl, C. Fahn, H. Bader, H. Ringsdorf, and M. Schollmeier, Ber. Bunsenges. Phys. Chem. 89, 1198 (1985).

22. D. R. Nelson and L. Peliti, J. Physique 48, 1085 (1987).

23. F. David and E. Guitter, Europhys. Lett. 5, 709, (1988); J. A. Aronovits and T. C. Lubensky, Phys. Rev. Lett. 60, 2634 (1988).

24. F. David, E. Guitter, and L. Peliti, J. Physique 48, 2059 (1987).

25. R. Lipowsky and M. E. Fisher, Phys. Rev. B 36, 2126 (1987).

26. For thermally-excited fluctuations, $a_\perp$ is given by (3.15) with $\zeta = 0$ and $d = 2n + 1$.

27. A review is in R. Lipowsky, Habilitations–Thesis, University of Munich, 1987 (Juel–Spes–438, KFA Juelich, W–Germany); see also R. Lipowsky, Ferroelectrics 73, 69 (1987).

28. J. Krim, J. P. Coulomb, and J. Bousidi, Phys. Rev. Lett. 58, 583 (1987).

29. B. Pluis, A. W. Denier van der Gon, J. W. M. Frenken, and J. F. van der Veen, Phys. Rev. Lett. 59, 2678 (1987).

30. D.-M. Zhu and J. G. Dash, Phys. Rev. Lett. 60, 432 (1988).

31. K. C. Prince, U. Breuer, and H. P. Bonzel, Phys. Rev. Lett. 60, 1146 (1988).

32. See, e.g., D.-M. Zhu, D. Pengra, and J. G. Dash, Phys. Rev. B 37, 5586 (1988).

33. E. G. McRae and R. A. Malic, Surf. Sci. 148, 551 (1984).

34. S. F. Alvarado, M. Campagna, A. Fattah, W. Uelhoff, Z. Phys. B 66, 103 (1987).

35. H. Dosch, L. Mailaender, A. Lied, J. Peisl, F. Grey, R. L. Johnson, and S. Krummacher, Phys. Rev. Lett. 60, 2382 (1988).

36. See, e.g., C. Rottmann, M. Wortis, J. C. Heyraud, and J. J. Metois, Phys. Rev. Lett. 52, 1009 (1984).

37. P. Bassereau, J. Marignan, and G. Porte, J. Physique 48, 673 (1987).

38. D. Roux and C. R. Safinya, J. Physique 49, 307 (1988).

39. V. A. Parsegian, N. Fuller, and R. P. Rand, Proc. Natl. Acad. Sci. USA 76, 2750 (1979).

40. In the classical literature, the term 'disjoining pressure' usually stands for $-\partial V_{DI}/\partial \ell$ arising from the direct interactions in the *absence* of surface fluctuations.

41. More precisely, the direct interactions act between surfaces with constant separation. For direct interactions of curved surfaces, see M. P. Gelfand and R. Lipowsky, Phys. Rev. B 36, 8725 (1987).

42. See, e.g., J.N. Israelachvili, *Intermolecular and Surface Forces* (Academic, 1985).

43. B. Widom, J. Chem. Phys. 68, 3878 (1978).

44. R. Lipowsky, D. M. Kroll, and R. K. P. Zia, Phys. Rev. B 27, 4499 (1983); and R. Lipowsky, J. Appl. Phys. 55, 2485 (1984).

45. E. Bresin, B. I. Halperin, and S. Leibler, J. Physique 44, 775 (1983).

46. M. E. Fisher and P. G. De Gennes, C. R. Acad. Sci. 287, 207 (1978).

47. See, e.g., P. G. De Gennes, Macromolecules 15, 492 (1982).

48. D. M. Kroll, R. Lipowsky, and R. K. P. Zia, Phys. Rev. 32, 1862 (1985); and R. K. P. Zia, R. Lipowsky, and D. M. Kroll, Am. J. Phys. 56, 160 (1988).

49. R. Lipowsky and T. M. Nieuwenhuizen, J. Phys. A 21, L89 (1988).

50. E. Bresin, B. I. Halperin, and S. Leibler, Phys. Rev. Lett. 50, 1387 (1983).

51. One finds $\bar{\ell}/\ell_o \approx (\xi_\perp/a_\perp)^2 \approx \ln(\xi_\parallel/a)$ with $\ell_o = (2n + a_\perp^2/2a_1^2)a_1$ and $\ell_o = 2\sqrt{n}a_\perp$ for $a_\perp < 2\sqrt{n}a_1$ and $a_\perp > 2\sqrt{n}a_1$, respectively.

52. D. S. Fisher and D. A. Huse, Phys. Rev. B 32, 247 (1985).

53. G. Gompper and D. M. Kroll, Phys. Rev. B 37, 3821 (1988), and Europhys. Lett. 5, 49 (1988).

54. K. Binder and D. P. Landau, Phys. Rev. B 37, 1745 (1988).

55. G. Gompper and D. M. Kroll, Phys. Rev. B 38 (in press).

56. R. Lipowsky and M. E. Fisher, Phys. Rev. Lett. 57, 2411 (1986).

57. K. G. Wilson, Phys. Rev. B 4, 3184 (1971).

58. R. Lipowsky, KFA – Juelich preprint.

59. D. M. Kroll and R. Lipowsky, Phys. Rev. B 26, 5283 (1982).

60. R. Lipowsky, Phys. Rev. Lett. 52, 1429 (1984), and Phys. Rev. B 32, 1731 (1985).

61. S. Leibler and R. Lipowsky, Phys. Rev. Lett. 58, 1796 (1987), and Phys. Rev. B 35, 7004 (1987).

62. M. Huang, M. E. Fisher, and R. Lipowsky, University of Maryland preprint.

63. In the context of wetting, discontinuous transitions were first found by J. Cahn, J. Chem. Phys. 66, 3667 (1977); and by C. Ebner and W. F. Saam, Phys. Rev. Lett. 38, 1486 (1977).

64. Similar transitions occur in the necklace model for three or more random walkers, see M. E. Fisher and M. P. Gelfand, University of Maryland preprint.

65. In the context of wetting, continuous transitions were first found by D. B. Abraham, Phys. Rev. Lett. 44, 1165 (1980); and D. Sullivan, J. Chem. Phys. 74, 2604 (1981). For membranes, continuous transitions were first found by R. Lipowsky and S. Leibler, Ref. 19.

66. M. E. Fisher and D. S. Fisher, Phys. Rev. B 25, 3192 (1982).

67. D. M. Kroll and G. Gompper, KFA–Juelich preprint.

68. R. Lipowsky and G. Gompper, Phys. Rev. B 29, 5213 (1984).

69. R. Lipowsky, J. Phys. A 18, L585 (1985).

70. R. Lipowsky and D. A. Huse, Phys. Rev. Lett. 57, 353 (1986).

71. P. G. De Gennes, Rev. Mod. Phys. 57, 827 (1985).

72. The linear RG gives $\bar{\ell}(t) \approx (\ell_o/2)\ln(t)$ with ℓ_o as in Ref. /51/ for $n = 1$.

73. Z. Jiang and C. Ebner, Phys. Rev. B 36, 6976 (1987).

74. M. Grant, K. Kaski, and K. Kankaala, J. Phys. A 20, L571 (1987).

75. K. K. Mon, K. Binder, and D. P. Landau, Phys. Rev. B 35, 3683 (1987).

76. M. Kardar, G. Parisi, and Y.-C. Zhang, Phys. Rev. Lett. 56, 889 (1986).

77. J. Krug and H. Spohn, University of Munich preprint.

78. J. Krug, Phys. Rev. A 36, 5465 (1987).

79. D. E. Wolf and J. Kertess, Europhys. Lett. 4, 651 (1987).

80. D. A. Huse, Phys. Rev.B 29, 6985 (1984).

81. See, e.g., H. Müller-Krumbhaar, in *Current topics in materials science*, Vol. 1, ed. E. Kaldis (North-Holland, 1978).

82. W. W. Mullins and R. F. Sekerka, J. Appl. Phys. 35, 444 (1963).

83. J. Villain, Phys. Rev. Lett. 52, 1543 (1984).

MIKE SHLESINGER AND JOSSI KLAFTER

SURFACTANTS IN SOLUTION: AN EXPERIMENTAL TOOL TO STUDY FLUCTUATING SURFACES

DIDIER ROUX

Centre de Recherche Paul Pascal
Domaine Universitaire, 33405 Talence Cedex, France

ABSTRACT. Surfactants in solution lead to a variety of structures which can be considered as phases of surfaces. One can distinguish between "rigid" surfaces where the bending elastic constant of the surface is much larger than $k_B T$, in this case the structures of the phases result mainly from a balance between elastic energy of the film and interactions. In other interesting cases, the elasticity moduli are of order $k_B T$ and the structure and stability of the phases result from a competition between thermal fluctuations and elastic energy. These systems can be considered as examples of fluctuating surfaces.

1. Introduction and Generalities

1.1. SURFACES OF SURFACTANTS

A surfactant, or amphiphilic molecule is made of two parts which have opposing natures: one is water soluble (hydrophilic) and the other is oil soluble (hydrophobic). The hydrophilic and hydrophobic parts of the molecule are linked together by a chemical bond, which gives to the surfactant molecules their peculiarities.[1] Indeed, such molecules prefer in general to stay at the interface between hydrophilic-hydrophobic regions or if it is not possible they aggregate in order to build up such interfaces. The consequence of this property is easily illustrated by adding increasing amount of a typical surfactant (such as SDS, Sodium Dodecyl Sulfate) to water. For small amount of surfactant, the molecules lie practically all at the interface between the water and its vapor. As a consequence the liquid vapor surface tension decreases until it levels off to remain nearly constant when the concentration of bulk surfactant goes beyond the CMC (Critical Micellar Concentration). Above the CMC, the molecules of surfactant aggregate as small objects (micelles) which are usually globular in shape. These micelles stay in solution in the bulk of the liquid and more surfactant molecules added will form new micelles rather than staying at the vapor-liquid interface. Upon the addition of increasing amount of surfactant, phase transitions to more organized structures occur.

The complete evolution of the phase behavior of a binary system of surfactant plus water is described by a phase diagram, a typical example is shown on Fig. 1. High concentration of surfactant leads usually to liquid crystalline phases where a long range order appears in one or more direction of the space. The most common phases are respectively the lamellar phase (smectic A, La phase) where the surfactant molecules build planes which are equally spaced and separated with water, and the hexagonal phase where they aggregate in infinite cylinders arranged on a two dimensional hexagonal lattice (see Fig. 1).

1.2. FILM ELASTICITY AND TENSION

There are two levels of description of an aggregate of surfactants in solution: the first one corresponds to a microscopic description of the surfactant aggregate, the second level is more phenomenological and consists of considering essentially the interface between the hydrophobic and hydrophilic regions to be a thin elastic surface.

There are in the literature several attempts to identify the important parameters at the microscopic level that are responsible for the structure of the phase observed for a given surfactant. A remarkably simple description based on geometrical arguments has been proposed by Israelachvili, Mitchell and Ninham.[2]

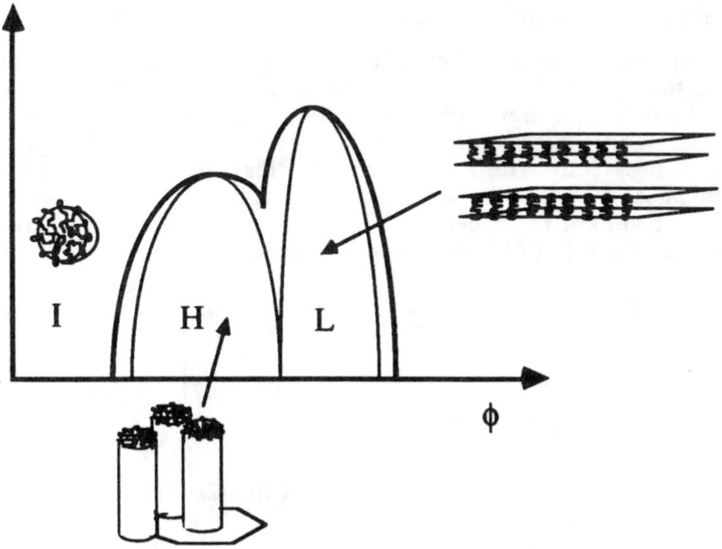

Fig. 1: Typical phase diagram of a binary system (surfactant + water). Three phases are shown: the micellar isotropic liquid phase (I), the hexagonal phase (H) made of infinite cylinders on a two dimensional hexagonal lattice, and the lamellar phase (L) consisting of a stack of layers.

The description in terms of elasticity of the interface is due to Helfrich[3] and has been intensively used in the last years to understand the properties of microemulsions and membranes. Consider a surface of total area S made with a large number of surfactants molecules of area per polar head a. The membrane is considered in a liquid state, which means that each surfactant molecule is free to move in the two dimensional space defined by the surface (but obviously cannot leave this surface). Assuming that the surfactant molecule has an equilibrium area per polar head a_0, the elastic energy per unit area E_{def} is

$$E_{\text{def}} = \frac{1}{2}\Gamma_0 \left[\frac{(a - a_0)^2}{a_0^2} \right] + \frac{1}{2}k_c(H - H_0)^2 + kK. \tag{1}$$

The first term is related to the deviation of the surface from its equilibrium position, Γ_0 is a coefficient and has the units of an energy per unit area.[4] The two last terms correspond to the elastic response to a curvature of the surface: H is the mean curvature of the surface at a given point $H = 1/R_1 + 1/R_2$ where R_1 and R_2 are the principle radii of curvature of the surface, $K \equiv (1/R_1) \cdot (1/R_2)$ is the Gaussian curvature. H_0 is the spontaneous curvature and reflects the dissymmetry of the film, k_c and $\bar{k}$ are the elastic constants of mean and Gaussian curvatures. This energy corresponds to a quadratic expansion for small deviations (stretching and curvature) around the flat surface with its equilibrium area.

In the limit of large Γ_0 and with no constraint on the limits of the surface (the deformations are considered at constant total area S), the system equilibrates at $a = a_o$ and the elastic energy of the film is mainly determined by the bending part. We can formally distinguish between two extreme cases: either the bending constants (k_c and $\bar{k}$) are large compared to $k_B T$ (thermal fluctuations are not important) or they are of the order of $k_B T$ and thermal fluctuations will play an important role.

1.3. LARGE BENDING CONSTANT (LOW TEMPERATURE LIMIT)

As an example of the effect of the bending energy on the structure of the aggregates lets take a ternary system: oil, water and surfactant. Following previous authors,[5,6] and forgetting about entropy and interactions, we can compare the stability of three types of aggregates: spheres, infinite cylinders and infinite planes as a function of the concentrations for a given system. This is probably the simplest way to consider the phase transition between the micellar, lamellar and hexagonal structures. If R_s is the radius of the sphere and R_c the radius of the cylinder, the elastic energy per unit area are respectively for each phases:

$$\begin{cases} \text{spheres}: & \Delta f_s = 2k_c \left[\frac{1}{R_s^2} - \frac{H_o}{R_s} \right] \\[2mm] \text{cylinders}: & \Delta f_c = k_c \left[\frac{1}{2 \cdot R_c^2} - \frac{H_o}{R_c} \right] \quad \text{with} \quad \Delta f = \frac{F}{S} - \frac{k_c H_0^2}{2} \\[2mm] \text{planes}: & \Delta f_1 = 0. \end{cases}$$

In order to express these energies as a function of the concentrations we use the two conserved quantities:

— volume fraction of object $\Phi = \Phi_o + \Phi_s$, $\Phi = n \cdot v/V$
— total area: $n \cdot S = n_s \cdot a_0$.

Here n is the number of objects, v the volume of objects, V the total volume of the phase and n_s the number of surfactant. Φ_o is the oil volume fraction. Introducing v_s the volume of a surfactant molecule ($\Phi_s = n_s \cdot v_s/V$) we get: $R_s = 3 \cdot \Phi/\Phi_s \cdot (v_s/a_o)$ and $R_c = 2 \cdot \Phi/\Phi_s \cdot (v_s/a_o)$. The free energies are then functions of the surface to volume reduced ratio

$$x = a_o/v_s/H_o \cdot \left(\frac{\Phi_s}{\Phi} \right)$$

$$\begin{cases} \text{spheres}: & \Delta f_s = k_c H_o^2 \left[2x^2/9 - 2x/3 \right] \\[1mm] \text{cylinders}: & \Delta f_c = k_c H_o^2 \left[x^2/8 - x/2 \right] \\[1mm] \text{planes}: & \Delta f_1 = 0. \end{cases}$$

We can then define the range of stability of each phases and get the following phase diagram (Fig. 2). Obviously this is far from being a rigorous calculations, in order to have a more accurate determination of the stabilities of each phases, the entropy of mixing should be added to the energy and the interactions between

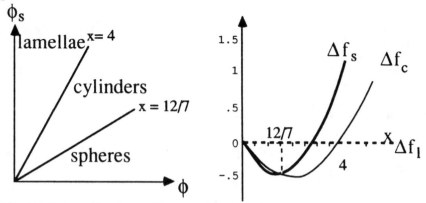

Fig. 2: Schematic phase diagram showing the zones of stability for the spheres, the cylinders and the lamellae in the approximation of large bending constant and neglecting entropy and interactions.

objects too. Then a correct thermodynamic analysis should be done including a tangent construction.[6] However this very simple approach illustrates the effect of surfactant concentration which may lead to changes of structures as it increases (spheres → cylinders → planes, see Fig. 2).

The very simple example given above is related to a more general problems, namely to find out the shape of surfaces respecting geometrical constraints such as having a fixed curvature. The experimental evidence for cubic phases of connected surfaces[7] corresponding to some kind of Schwarz's surfaces (mean curvature equals to zero but finite Gaussian curvature), is an example of the complexity of the problem which is actually the subject of a lot of interest.[8,9]

1.4. WEAK BENDING CONSTANTS (HIGH TEMPERATURE LIMIT)

Let us now consider the case where k_c is of order $k_B T$. Then thermal fluctuations can no longer be neglected and they have very important consequences. The notion of persistence length, first introduced by de Gennes and Taupin[10] is very useful to understand in a simple way the effect of thermal fluctuations. Let consider a free membrane, at finite temperature this membrane is crumpled[11,12] and it is possible to define a typical length below which the membrane is nearly flat and above which the membrane makes a two-dimensional random walk. This length (ξ_k) is similar to the persistence length of a linear polymer[13] but instead of being proportional to the bending constant is an exponential function of it,

$$\xi_k \propto \exp[\alpha(k_c/k_B T)]. \tag{2}$$

Here α is a numerical coefficient (2π in the de Gennes-Taupin calculation). We can now understand better why thermal fluctuations are only relevant when $k_c \approx k_B T$. Indeed, a numerical evaluation of formula (2) leads to $\xi_k \approx 1000$Å for $k_c = k_B T$ and $\xi_k \approx 50\mu$m for $k_c = 2k_B T$.

In the following we will examine several experimental consequences of the effect of thermal fluctuations on surfactant films. Two main examples will be detailed, the consequence of fluctuations on the interactions between membranes in lyotropic lamellar phases and on the stability and structure of microemulsions. Before going in more details let us address the question of what are the microscopic parameters which govern the bending energy. This question is of very important practical interest since, as we will see, the value of the bending constant controls the stability and the structure of many interesting phases.

1.5. FACTORS INFLUENCING THE BENDING ENERGY

Besides the stretching coefficient, three parameters are involved in Eq. (1), the bending constants k_c and $\bar{k}$ and the spontaneous curvature H_0. Very little is known experimentally about and the parameters controlling its value; more experimental and theoretical work is needed in order to identify its role. Much more has been done on k_c and H_0.

It has been recognized early that H_0 is partially controlled by the shape of the surfactant molecule.[2] As shown in Fig. 3, the molecular shape depends mainly on the polar head area to chain area ratio. Very naturally, Israelachvili et al[2] have argued that cylindrical like molecules would prefer a lamellar environment ($H_0 = 0$) when conical like molecules would rather prefer a curved (micellar or cylindrical) environment ($H_0 > 0$ or $H_0 < 0$). Besides the "steric" shape, interactions between molecules may also modify their effective shape. For example, electrostatic interactions between polar heads will tend to favor inverted shapes, screening these interactions with salt may change continuously H_0. This effect is of practical importance for microemulsion systems.

a   b   c

Fig. 3: Illustration of the molecular structure of surfactant on the elastic properties of the film. (a) The shape of the molecule influences mainly the spontaneous radius of curvature H_0. (b) A mixture of surfactant (open circle head) and cosurfactant (filled circle head) makes a flexible film compare to a pure surfactant one, this is due to a thinning of the film adding small tail amphiphiles. (c) a nonionic surfactant makes also a flexible film due to an increase of the area per polar head.

Typical surfactant films such as lipids make quite rigid films ($k_c \approx 20 k_B T$).[14] In certain favorable cases, flexible films can be obtained with very special surfactant molecules, such as nonionic surfactant ($C_i E_j$, see Fig. 3c) consisting of very short copolymers; one chain is hydrophilic and the other hydrophobic. More usually, flexible films can be obtained in adding to a regular surfactant a cosurfactant. This cosurfactant is often a short length alcohol molecule such as pentanol or hexanol (see Fig. 3b). In order to better understand the factors influencing the bending constant, three theoretical approaches have been followed. The surfactant film can be modeled as an elastic sheet[15] or each surfactant molecule is assumed to be related to its neighbors by springs in order to model surfactant-surfactant interactions.[16] Another approach uses polymer scaling laws to calculate the scaling behavior of the elastic constant.[17] Recently a direct calculation of the effect of mixing surfactant and cosurfactant molecules has been published.[18] All these different theories agree on the result that the elastic constant scales with the film thickness L and with the area per polar head a of the surfactant,

$$k_c \propto \frac{L^n}{a^p}, \tag{3}$$

where n and p are numbers ($n \approx$ 2-3 and $p \approx$ 5). This is consistent with the experimental observations that alcohol molecules are needed to reduce k_c (in decreasing L) or that nonionic surfactant which have a rather large a ($a \sim 45\text{Å}^2$ compared to $a \sim$25-30Å^2 for regular surfactant) leads to flexible films.

2. Interactions Between Membranes: Undulation Forces

2.1. GENERALITIES

2.1.1. _Classical Interactions_. The problem consists of determining the interactions between two membranes of thickness L separated by a distance r. In the absence of layers undulations (we will come back to this point in the next paragraph), the free energy f per unit area is given by the sum of three terms,

$$f = \frac{F}{S} = f_{\text{vdW}} + f_{\text{elec}} + f_{\text{hyd}}. \tag{4}$$

Neglecting the retardation effects, the van der Waals interaction is simply given by a double integration of the $1/r^6$ potential which leads to the following formula[19],

$$f_{\text{vdW}} = -\frac{A}{12\pi} \left[\frac{1}{r^2} + \frac{1}{(r+2L)^2} - \frac{2}{(r+L)^2} \right], \tag{5}$$

this interaction is attractive and varies as $1/r^2$ for small distances and as $1/r^4$ for larger r ($r > 200\text{Å}$). A is the Hamaker constant and is typically of order $k_B T$.

The next term in Eq. (4) is due to electrostatic interactions and is relevant mainly for charged surfactant swelled with water. This interaction can be calculated from the one-dimension Poisson-Boltzmann equation.[20] Two limiting cases have to be considered, on one hand the water in between the membranes contains only the counter-ions of the charged surfactant, on the other hand salt has been added such that the Debye length is smaller than the distance r between membranes. In the first case an exact solution has been calculated by Parsegian[20] and the expansion for large r reads[21]

$$F_{\text{elec}} = \frac{\pi^2 k_B T}{4 L_e r} \left[1 - \frac{a}{L_e r} + \left[\frac{a}{L_e r} \right]^2 + \cdots \right]. \tag{6}$$

L_e is a length ($L_e = \pi e^2/\epsilon k_B T \approx 22$Åat room temperature and for the water dielectric constant $\epsilon = 80$. One should notice that the dominant term in this Eq. (6) varies as $1/r$ (very long range) and is not dependent on the area a per charged polar head (this corresponds to the so-called ionic condensation[22] The limit of large addition of ions and counterions (salt) is simply given by an exponential decay of characteristic length the Debye screening length λ_D

$$f_{elec} = E_0 \exp[-r/\lambda_D]. \tag{7}$$

More recently it has been shown that it exists a short range interaction coming from the organization of the water around the polar head.[23] This interaction may be represented empirically as $f_{hyd} = F_0 e^{-r/\lambda}$, where λ is a microscopic length (typically 2-3Å) and F_0 is a constant, this interaction is only important for length smaller than 10Å.

2.1.2. *Undulation forces.* Besides these classical interactions Helfrich[24] has proposed a novel interaction arising explicitly from thermal fluctuations. Let consider a "free" membrane at finite temperature, as we saw the membranes undulate under thermal fluctuations. Now let consider that this membrane approaches an impenetrable wall (or even another membrane) at a distance smaller than the persistence length ξ_k. The movements of the membrane is then restricted leading to an entropy loss which is responsible for an effective interaction. One may also say that there exists a pressure of the membrane against the wall due to the thermal fluctuations, this picture is fully equivalent to the perfect gas pressure[25] or to the entropic pressure of a constrained polymer.[13] This undulation interaction can be calculated by several means leading to the same powerlaw dependence of the free energy per unit area ($1/r^2$). A calculation using the Landau-de Gennes elastic energy for a smectic leads to:

$$F_{und} = \frac{3\pi^2}{128} \cdot \frac{(k_B T)^2}{k_c} \cdot \frac{1}{r^2}. \tag{8}$$

Notice that the amplitude of this interaction is inversely proportional to the bending constant k_c.

2.2. EXPERIMENTAL SYSTEM: THE LAMELLAR PHASE

Interactions between membranes can be studied using a very interesting material: the lamellar phase of surfactant. Indeed, as it has been described previously, this phase consists of stacks of membranes separated by solvent. The interlayer distance can be changed depending upon the amount of solvent added and on the stability of this phase toward other structures (micellar, hexagonal,...). The range of distances accessible experimentally varies a lot from one system to the other. In the absence of long range repulsive interactions (such as electrostatic or undulation) the dilution is limited to distances which remain small (10-20Å). When long range repulsive interactions dominate very high dilution are possible and distances as large as several hundreds or even thousands[26] of Angstroms can be obtained. These extremely diluted lamellar phases constitute a unique example of colloidal smectics and appeared to be extremely useful to study interactions on a very large range of distances.[27-30]

Several experimental technics have been used to directly measure the interactions between membranes. In a first paragraph, I will very briefly describe the principles of two methods which have been utilized in the last 20 years for measuring "classical" forces, then I will present more recent works where the interactions have been measured through the elastic properties of the smectic phase.

2.3. DIRECT EXPERIMENTAL MEASUREMENTS OF INTERACTIONS

2.3.1 *Osmotic pressure measurements*. In a series of very nice experiments Parsegian et al[31] have been able to measure interactions between membranes for charged and uncharged lipids. In these experiments, a lamellar phase of lipids bilayers is prepared with excess water. A polymer is added to the water, the polymer size is large enough to prevent the polymer to penetrate between the bilayer but it modifies the chemical potential of the water. The consequence of modifying the bulk water chemical potential (which is the same as the chemical potential of the water which is situated between the membranes) is to modify the equilibrium distance between layers in the lamellar phase. In other words, an external (osmotic) pressure is applied to the system. Knowing the water chemical potential μ_w and measuring the repeating distance d, curves $\mu_w(d)$ can be obtained and compared with theories. Measurements have shown that classical theories of van der Waals and electrostatic forces were quantitatively verified and that a novel short range interaction coming from the organization of the water around the polar head was needed to interpret the experimental data.

2.3.2. *Force measurement machine*. Another elegant technique has been extensively developed by Israelachvili and coworkers.[32] It consists of measuring directly the mechanical force between two plates (in fact half cylinders) on which has been previously deposed a bilayer of surfactant. This experimental "tour de force" has allowed to show that classical interactions can be effectively used to describe the behavior of the membrane-membrane forces. They were also able to directly measure the hydration force.[33]

2.4. MEASUREMENTS OF THE ELASTIC PROPERTIES OF SMECTICS

2.4.1. *Elastic constant of a smectic*. A very efficient method to measure interactions between membranes is to measure the elastic moduli of a lamellar phase. Indeed, a lamellar phase of surfactant is a smectic A liquid crystal and the elastic modulus of compression is in fact, directly related to the interactions between layers. The elastic free energy of a slightly deformed smectic reads[34]

$$F = \int_v \frac{1}{2} \left[B \left(\frac{\partial u}{\partial z} \right)^2 + K(\Delta_\perp u)^2 \right] dv. \tag{9}$$

Here B and K are respectively the compressional and the bending elastic constants, u is the layer displacement, taken along the axe z perpendicular to the layers, $\Delta_\perp$ is the Laplacian in the xy plane, v is the volume of the sample, B has the dimensions of an energy per unit length cube and K an energy per unit length. In a simple model where the smectic is described as a stack of interacting membranes of thickness L situated at a distance d_0 (the repeating distance d is given by $d = d_0 + L$), K is simply the curvature elastic constant of one membrane multiplied by the density of layers per unit length $(1/d)$,

$$K = k_c/d, \tag{10}$$

and B is the second derivative of the interaction potential per unit are a U

$$B = d\left(\frac{\partial^2 U}{\partial d^2}\right). \tag{11}$$

Assuming k_c to be constant as a function of d, and U a decreasing function of d ($U \sim d^{-n}$, we will see that for example $n = 1$ for electrostatic interaction and $n = 2$ for undulation), K and B appear to be both decreasing with d (K as $1/d$ and B as $1/d^{n+1}$). Considering the variation of two order of magnitude (at least) of d, K *may vary continuously of two orders of magnitude and B up to six orders of magnitude*. These large variations in the elastic constant of smectic have to be compared with the same kind of "renormalization" of the elastic constants by the basic size of the crystal in colloidal crystals such as those obtained with latex in water.[35] These systems may be considered as unique examples of colloidal smectic with the advantage compare to regular colloidal systems that it exists a continuous path from regular lyotropic smectic to dilute lamellar phases and consequently a continuous variation in the elastic constants (in a range of several orders of magnitude). It exists several methods to measure the elastic constant of a smectic. We will detail three methods which have been used recently in lyotropic systems to study the interactions between membranes. The first one is based on an accurate determination of the shape of the x-ray structure factor, which in the special case of smectics allow a determination of the elastic constant (based on the so-called Landau-Peierls effect). Another method is based on the consequence of applying a direct mechanical stress and we will present briefly some results obtained using dynamic light scattering.

   2.4.2. *Synchrotron X-ray studies.* X-ray measurements allow usually to access to structural properties of solids or liquid crystals. However in some interesting cases, such as smectic-A phase of liquid crystals or two-dimensional solids, thermodynamic properties can be obtained reliably from the analysis of the X-ray structure factor. Starting from the Landau-de Gennes elastic energy of smectics Eq. (9), it can be shown that thermal fluctuations diverge with the size of the sample. This divergence should destroy the long range ordering but due to the finite size of the sample and to the weakness of the effect (it diverges as the logarithm of the size) the system is marginally stable.This effect has an important consequence on the structure factor of a smectic. Indeed, the true Bragg peaks are replaced by power-law singularities[34] which signal the lack of true long range order. The asymptotic behavior of the structure factor has been calculated by Caillé[36] based on the elastic free energy of the smectic which has been previously defined in Eq. (9).

$$S(0, 0, q_\parallel) \propto \frac{1}{[q_\parallel - q_m]^{2-\eta_m}} \tag{12a}$$

$$S(q_\perp, 0, q_m) \propto \frac{1}{q_\perp^{4-2\eta_m}}, \tag{12b}$$

where q_m is the position of the m-th harmonic of the structure factor ($q_m = mq_0$, $m = 1, 2, ..$) and η_m is the power law exponent related to the elastic constants B and K

$$\eta_m = m^2 q_0^2 \frac{k_B T}{8\pi\sqrt{BK}}. \tag{13}$$

Consequently, the shape of the x-ray structure factor is directly related to the elastic constants of the smectic. More accurately, the structure factor is the Fourier transform of the correlation function[37]

$$G(\mathbf{R}) = G(\rho, z) \propto \left[\frac{1}{\rho}\right]^{2\eta} e^{\eta} \left[2\gamma + E_1\left(\frac{\rho^2}{4\lambda|z|}\right)\right]. \tag{14}$$

Here, γ is Euler's constant, $E_1(x)$ is the exponential integral function, $\mathbf{R}^2 = z^2 + \rho^2$ with $\rho^2 = x^2 + y^2$, η has been defined previously Eq. (13) with $m = 1$ and λ is a length related to the ratio between the elastic constants $\lambda = \sqrt{K/B}$. The structure factor is a function of the two parameters η and λ, which in principle allow the determination of the elastic constants K and B. Experimentally, the parameter η is more accurately measured than λ[28] and the experimental results will be discussed for η.

Assuming that K is directly related to k_c Eq. (10), η can easily be calculated for the different models of interactions between membranes. For example, assuming that the undulation forces dominate, one obtain for the undulation forces[28,29]

$$\eta_{und} = \frac{4}{3}\left[1 - \frac{L}{d}\right]^2. \tag{15}$$

In the case of electrostatic interactions (no salt added) Eq. (6) leads to

$$\eta_{elec} = \sqrt{\frac{k_B T L_e}{2k_c g d}}\left[1 - \frac{L}{d}\right]^{3/2}, \tag{16}$$

with $g = 1 - 3a_1/(d - L) + 6a_1^2/(d - L)^2$ and $a_1 = a/L_e$.

The evolution of η as a function of the dilution (d) has been measured for a series of systems.[28,29,30,38] The main conclusions of this work can be summarized as follows.

In the absence of long range electrostatic interactions (uncharged surfactant, water-salt dilution or oil-dilution) and for flexible membranes (systems where $k_c \approx k_B T$), the existence of long range entropically driven undulation forces has been demonstrated. Indeed, a universal behavior can be experimentally found for systems which are quite different. A summary of the results is shown on Fig. 4 where the variation of η as a function of the reduced distance $x = (1 - L/d)^2$ has been plotted. As it is clear from Fig. 4, the behavior of η is the same for a SDS-pentanol system diluted either with oil (o) or brine ($\triangle$), the same behavior has also been found for neutral membranes made of lipids (DMPC) and alcohol (pentanol) diluted with pure water.[30] When the elastic constant k_c is much larger than $k_B T$ on the contrary, "classical" interactions dominates and for an uncharged membrane made of pure lipids the repulsion between membrane is dominated by short range hydration force[31,38]. When long range electrostatic force exist, corresponding to a pure water dilution of a charged membrane, it has been demonstrated that electrostatic interactions dominate even for flexible membranes and the variation of η can be well described by Eq. (16).[29]

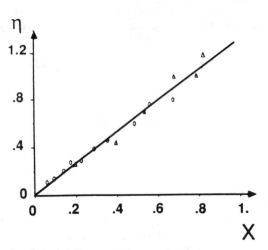

Fig. 4: Plot of the value of the exponent η versus the reduced distance squared $X = (1-L/d)^2$ for the data where the undulation forces are dominant. The circles corresponds to the oil dilution and the triangles to the water salt dilution. The straight line is the prediction of the Helfrich theory.

2.4.3. *Measurements of the elastic constants using mechanical constraint.* Recently an elegant method has been used to measure λ $(\lambda = \sqrt{K/B})$ as a function of the dilution for the same system that has been studied with high resolution X-ray (oil dilution, see previous paragraph). The method consists of constraining a lamellar phase to stay in a wedge consisting of a lens lying on a flat surface. The boundary constraints, namely homeotropic alignment (parallel to the glass) are not compatible with the constant layer spacing of the lamellar phase. To respect the limiting boundaries, edge dislocations of Burger vector B arise. It can be shown that the spatial arrangement of these dislocations results from a minimization of the elastic energy of the smectic and is a function of λ. By means of optical observation, the spatial location of the edge dislocations has been measured for a series of sample allowing measurements of λ as a function of the dilution. The results obtained are compatible with interactions between layers dominated by undulation forces.[39] Other mechanical technics have been used for measuring elastic properties of binary lyotropic systems,[40] but there is no systematic measurements as a function of the dilution .

2.4.4. *Dynamic Light scattering methods.* In principle light scattering on oriented samples of smectic A is one of the best method to determine elastic properties.[34] Order parameter fluctuations in a thermotropic smectic A are related to elastic constants of the liquid crystal, one expects six dynamical modes corresponding to propagative and diffusive movements.[41] For a binary lyotropic system one more mode is expected coming from the additional degree of freedom due to concentration fluctuations.[42] This mode is directly related to the compressional elastic constant B. Recent experiments[43] have demonstrated the feasibility of obtaining B using dynamic light scattering. These very promising technics has been used for measuring B in a series of systems for a large range of dilution.[43]

3. Structure and stability of microemulsions

3.1. STRUCTURE OF THE MICROEMULSION PHASE

A lamellar phase of membrane is a phase where the layers are stacked at a distance which even if it can be large remains smaller than the persistence length of the film. In certain cases, the lamellar phase may be unstable toward a phase in which the film is less constrained. Let start with an oil-water-surfactant ternary system with a water over oil ratio of one (same amount of water and oil). For large enough concentration of surfactant, this system may form a lamellar phase where the water and oil layers are separated with a surfactant film. With no surfactant the system will phase separate between a pure oil phase and a pure water phase. For intermediate surfactant concentration and with a suitable surfactant an isotropic, liquid and thermodynamically stable phase may form. This microemulsion phase has the structure of a sponge of oil where the water fills the pores; the film of surfactant making a skin between the water and oil domains. Figure 5 shows a schematic phase diagram of such a system. In the cases of practical interest, the microemulsion phase is destabilized at low surfactant concentration in a three phase equilibrium where it coexists with practically pure water and pure oil phases. On a theoretical point of view, this phase may be seen as a phase of fluctuating surface and the stability and structure of this phase understood using the statistical properties of a liquid surface.[12]

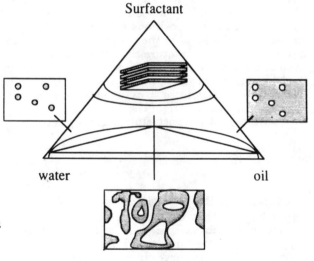

Fig. 5: Schematic phase diagram of a microemulsion system, at high concentration of surfactant, a liquid crystalline phase is stable (lamellar phase). At small concentration of surfactant a microemulsion isotropic liquid phase exists. The structure is micellar in the dilute regions (water or oil) and bicontinuous (sponge-like) when water and oil have similar concentrations.

3.2. THE DE GENNES-TAUPIN MODEL

Inspired by a previous model of Talmon and Prager,[44] de Gennes and Taupin[10] have proposed a simple microscopic model describing the microemulsion phase diagram . In this model and in a further development,[45] the space has been divided in a cubic lattice where basic cubes of size ξ have been randomly filled with water and oil. The surfactant film separates the microscopic water-oil interface (each time that a water domain neighbors an oil domain). Neglecting the surfactant concentration (i.e. the thickness of the film), the entropy of the film for size larger than ξ can be calculated as the entropy of mixing of the water and oil domains. If ϕ is the volume fraction of water and $1 - \phi$ the volume fraction of oil the energy density of mixing is simply

$$-\frac{TS}{V} = \frac{k_B T}{\xi^3}[\phi \ln \phi + (1 - \phi)\ln(1 - \phi)]. \tag{17}$$

The constraint of making a surfactant film staying at the water-oil interface lead to a fundamental relationship between the area to volume ratio (proportional to the surfactant concentration ϕ_s) and the water (or oil) volume fraction ϕ. Within the random mixing approximation one gets

$$z\phi(1 - \phi) = \phi_s \frac{1}{v_s} \xi, \tag{18}$$

where a is the area per polar head of the surfactant and v_s the surfactant molecular volume and z the number of nearest neighbor for the lattice (6 for the cubic lattice). Equation (18) is fundamental and one should point out that it comes only from random mixing approximation and geometrical constraints. Before going further in the description of the free energy of this model let us detail an important point. Considering that the stability of the microemulsion phase is driven by the physics of the amphiphilic film, we can understand that the film will be at its minimum of energy when the compositions will be such that $\xi = \xi_k$ and $a = a_0$. Indeed, when the lattice size is equal to the persistence length of the film it means that it does not cost any energy of curvature since the film is *spontaneously crumpled* at this length by thermal fluctuations. This remark is the key point of the de Gennes-Taupin model and is not obvious, we will come back in the next paragraph to this problem. The second constraint ($a = a_0$) is more obvious and results just from relaxing the stretching on the film Eq. (1).

With these two constraints, Eq. (18) becomes

$$\frac{\phi_s}{\phi(1 - \phi)} = \frac{6v_s}{a_0 \xi_k} = c^t. \tag{19}$$

The location of the points in the $\phi_s - \phi$ plane (phase diagram) is a parabola corresponding to what has been called the Shulman curve.[10,45] Outside this curve the system has to pay the price for not matching concentration and characteristic length. One should think that Eq. (1) describes the energetic elasticity of the film around the Shulman curve. Indeed, the stretching is well described within harmonic approximation but there is nothing favoring the persistence length in the bending part. The only way, at this stage, to incorporate the physics related to the persistence length is to fix the lattice size to be ξ_k and to keep only the stretching part of Eq. (1) in order to relax the constraint outside the Shulman curve. Within this approximation ($\xi = \xi_k$ and a varies), the free energy density reads

$$\frac{F}{V} = \frac{k_B T}{\xi_k^3}[\phi \ln \phi + (1 - \phi)\ln(1 - \phi)] + 3\frac{\Gamma_0}{\xi_k}\phi(1 - \phi)\left[\frac{a}{a_0} - 1\right]^2, \tag{20}$$

with a given by Eq. (18) replacing ξ by ξ_k.

For fixed ϕ_s and large Γ_0, this function has two minima corresponding to the Shulman curve ($a \approx a_0$) when ϕ_s is smaller than a critical value $\phi_s^c \approx 3v_s/(2\xi_k a_0)$.

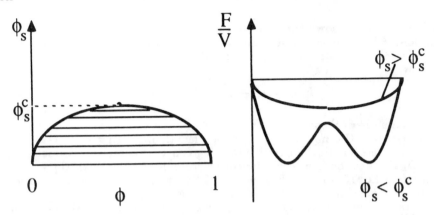

Fig. 6: Phase diagram obtained using the de Gennes-Taupin model. For concentration of surfactant smaller than a critical value $\phi_s^c \approx 3v_s/(2\xi_k a_0)$ there is a two phase region (where two microemulsion phases are in equilibrium. On the right a typical free energy density as a function of the volume fraction of water (ϕ) is shown below and above the critical point.

Consequently a phase separation occurs when $\phi_s < \phi_s^c$, a schematic phase diagram is shown on Fig. 6. The main result of the model is that the minimum amount of surfactant needed in order to get a stable microemulsion phase with the same amount of water and oil (ϕ_s^1) is inversely proportional to ξ_k (in this model $\phi_s^1 = \phi_s^c$). Since ξ_k is an exponential function of the bending elastic constant k_c, ϕ_s^1 is extremely sensitive to the film properties. Even if this model includes a large part of the physics, some important experimental behaviors are not well described, indeed this model does not lead to the three phase equilibrium (see Fig. 5).

3.3. IMPROVEMENTS OF THE BASIC MODEL

The reason why the three phase equilibrium between a microemulsion phase and phases of pure solvents (water and oil) were not found in the de Gennes model is directly related to the choice of fixing the lattice size at the persistence length ξ_k. More recently, B. Widom[46] realized that in order to correctly described the dilute phases of surfactant in water and oil, one need to let the lattice size to vary. Indeed, the typical aggregates size for these dilute phases is of the order of a molecular length rather than the film persistence length. In other words, the integrity of the film is destroyed for large dilution and a continuous behavior exists between a microemulsion phase (of large characteristic size) and a dilute phase of surfactant (of small characteristic size, micelles for example). Using a similar model, but relaxing the constraint on ξ and incorporating the bending curvature energy, Widom was able to find a three phase equilibrium.[46] In doing so, the model however loses the main result of the preceding version, namely the ξ_k dependence of the minimum amount of surfactant in the microemulsion phase (for half water and half oil). This can be understood since there is now nothing in this version of the model which favors the film bending to be of order its persistence length.

Another version of the model has been recently proposed.[47] This model keeps the cell size variation as proposed by Widom but incorporates in the bending energy part the physics of the film for length smaller than the persistence length. Indeed, it has recently been shown that the elastic constant k_c is renormalized by thermal fluctuations.[48,49,12] For a liquid membranes the entropy of undulations decreases the elastic energy needed to bend a film, the bending constant k_c is then scale dependent,

$$k_c(q) = k_c^0 + \alpha \frac{k_B T}{4\pi} \ln \left[\frac{q}{q_{max}} \right], \tag{21}$$

where α is a numerical constant.[48,49]

One recovers the notion of persistence length in setting $k_c(2\pi/\xi_k) = 0$ in Eq. (21). For this length the effective bending constant is zero (or a small number of order $k_B T$) since the film is spontaneously crumpled by the thermal fluctuations. Incorporating Eq. (21) in the free energy of the microemulsion phase and letting the lattice size varying, the system will now adjust the phase transition in order to favor the microemulsion phase for $\xi \approx \xi_k$. This model includes the Widom improvements while keeping the physics of the fluctuating film. In fact, as previously, the large wave length fluctuations are taken into account with the oil-water entropy of mixing (for $0 < q < 2\pi/\xi$) and the small wave length fluctuations are incorporated in the renormalization of the bending energy (for $2\pi/\xi < q < 2\pi/L$). In addition the relative stability of the phase toward the lamellar phases has been calculated and comparison with experimental results have been done.[50,51]

4. Conclusion

We have seen that phases of surfactant in solutions can be considered as phases of surfaces. When the bending elastic constant k_c is of order $k_B T$, thermal fluctuations may play an important in the structure and stability of these systems. Thermal fluctuations are responsible for undulations of the surface corresponding to a *two-dimensional Brownian motion* of the film. We have presented two extreme cases: on one hand in dilute lamellar phases the membrane is nearly flat and undulations correspond to small perturbations from the flat surface, on the other hand the film loses its long range orientation and a bicontinuous phase of fluctuating surfaces (microemulsion) is stabilized.

When membranes are stacked in a multilayer smectic phase, the undulations are responsible for a universal interaction arising from the restriction of the Brownian motion of each layer by the existence of the neighboring layers. This undulation interaction, first proposed by Helfrich[24] has been recently experimentally measured.[28-30,43] This system is an experimental example of *weakly fluctuating* surfaces.

The stability of microemulsion phases results from the gain of entropy of the *strongly fluctuating* film. From theoretical consideration it can be shown that the microemulsion exists when the respective concentration of solvents (water and oil) and surfactant leads to a characteristic length of local curvature close to the persistence length of the film. For larger concentration of surfactant the system rather than being too curved ($\xi \ll \xi_k$) prefers to form a lamellar phase, on the other hand, for too small concentration of surfactant the system prefers to phase separates into a microemulsion phase of characteristic length $\approx \xi_k$ and two dilute phases of surfactant in water and oil (stabilized by the entropy of mixing).

One may address the question of the stability of a very dilute lamellar phase. Indeed when a lamellar phase is swelled by a solvent (oil, for example) the distance between layers can become much larger than the thickness of the membrane. Once the repeating distance reaches a length of order the persistence length of the film, one might think that for the same reasons which lead to the stability of microemulsion, the lamellar phase should melt into a phase of *strongly fluctuating* surfaces. The structure of this phase should be spongy and bicontinuous as for a oil-water microemulsion with however the important difference that the same solvent (oil or water) should be on each side of the film consisting now of a bilayer of surfactant. Such a phase has been proposed recently[52] and recent experimental investigations seem to confirm its existence.[53]

The systems described in this course correspond all to fluid surfaces. This class of surfaces, as being developed in others lectures[12], is not the only one of practical interest. Polymeric, hexatic and crystalline phases have also to be studied. Much less experimental examples of these other types of surfaces exist, specially with weak enough bending constant in order to favor thermal fluctuations. New phase transitions such as the unbinding phase transition,[54] the crumpling phase transition[11,12] has been predicted and not actually observed. Future works should focus on identifying, preparing and studying new systems in order to contribute of a better understanding of the physics of fluctuating surfaces.

Acknowledgements: This presentation has benefited greatly from a series of works produced in collaboration with several scientists; I offer my special thanks to D. Andelman, A. M. Bellocq, A. Ben-Shaul, M. Cates, C. Coulon, W. Gelbart, F. Nallet, S. Safran and C. Safinya.

1. For general reviews on Surfactants in solution see (a) C. Tanford, in *The Hydrophobic Effect* (New York, 1973, 1980); *Physics of Amphiphiles, Micelles, Vesicles*; (b) *Microemulsions*, eds. M. Corti and V. Degiorgio (North-Holland, 1985).

2. J. N. Israelachvili, D. J. Mitchell and B. W. Ninham, J. Chem. Soc. Faraday Trans. 1, 72, 1525 (1976).

3. W.F. Helfrich, Z. Naturforsh 28c, 693 (1973).

4. This term results in general from the balance of two energies; for example, a surface tension and a repulsive interaction between surfactant molecules.

5. S. Safran, L. A. Turkevich, P. Pincus, J. Physique Lett. 45, L-69 (1984).

6. D. Roux, C. Coulon, J. Physique 47, 1257 (1986).

7. J. Charvolin, J. Chem. Phys. 1, 80 (1983).

8. J. F. Sadoc and J. Charvolin, J. de Physique 47, 683 (1986).

9. D. M. Anderson, H. T. Davis, J. C. C. Nitsche, and L. E. Scriven, Phil. Mag. (to be published) and in *Physics of Amphiphilic Layers*, eds. J. Meunier, D. Langevin and N. Boccara (Springer Verlag, 1987).

10. P. G. de Gennes and C. Taupin, J. Phys. Chem. 86, 2294 (1982).

11. L. Peliti and S. Leibler, Phys. Rev. Lett. 54, 1690 (1985).

12. L. Peliti, this book; D. Nelson, this book.

13. P. G. de Gennes, *Scaling Concepts in Polymer Physics* (Cornell U Press, Ithaca, 1979).

14. M. B. Schneider, J. T. Jenkins, and W. W. Webb, J. Physique 45, 1457 (1984); I. Bivas, P. Hanusse, P. Bothorel, J. Lalanne and O. Aguerre-Charriol, J. Phys. (Paris) 46, 855 (1987).

15. E. A. Evans, Biophys. J. 14, 923 (1974).

16. A. G. Petrov and I. Bivas, Prog. Surf. Sci. 16, 389 (1984).

17. (a) R. Cantor, Macromolecules; (b) see Ref. 10; (c) S. Milner and T. Witten (to be published).

18. I. Szleifer, D. Kramer, A. Ben-Shaul, D. Roux and W. Gelbart Phys. Rev. Lett. 60, 1966 (1988).

19. J. N. Israelachvili, *Intermolecular and Surface Forces* (Academic Press, Orlando, 1985).

20. A. Parsegian, N. Fuller and R. P. Rand, Proc. Natl. Acad. Sci. 76, 2750 (1979).

21. D. Roux and C. Safinya, J. de Physique 49, 307 (1988).

22. S. Engstrom and H. Wennerstrom, J. Phys. Chem. 82, 2711 (1978).

23. R. P. Rand, Ann. Rev. Biophys. Bioeng. 10, 277 (1981).

24. W. Helfrich, Z. Naturforsch. 33a, 305 (1978).

25. W. Helfrich and R. M. Servuss, Il Nuovo Cimento 3, 137 (1984).

26. J. M. Dimeglio, M. Dvolaitsky, and C. Taupin, J. Phys. Chem. 89 871 (1985); J. M. Dimeglio, M. Dvolaitsky, L. Leger, and C. Taupin, Phys. Rev. Lett. 54, 1686 (1985); F. Larché, J. Appell, G. Porte, P. Bassereau, and J. Marignan, Phys. Rev. Lett. 56, 1700 (1986).

27. P. Bassereau, J. Marignan, G. Porte, J. Physique 48, 673 (1987).

28. C. R. Safinya, D. Roux, G. S. Smith, S. K. Sinha, P. Dimon, N. A. Clark and A. M. Bellocq, Phys. Rev. Lett. 57, 2718 (1986).

29. See Ref. 21.

30. C. R. Safinya, D. Roux, G. S. Smith and E. Sirota (to be published).

31. See Ref. 23.

32. See Ref. 19.

33. J. Marra and J. Israelachvili, Biochemistry 24, 4608 (1985).

34. P. G. de Gennes, *The Physics of Liquid Crystals* (Clarendon, Oxford, 1974).

35. P. Pieranski, Contemp. Phys. 24, 25 (1983); N. A. Clark et al., J. Phys. Colloq. C3, 43, 137 (1985).

36. A. Caillé, C. R. Acad. Scien. Ser. B 274, 891 (1972).

37. L. Gunther, Y. Imry and J. Lajzerowics Phys. Rev. A 22, 1733 (1980).

38. G. S. Smith, C. R. Safinya, D. Roux and N. Clark, Mol. Cryst. Liq. Cryst. 144, 235 (1987).

39. F. Nallet and J. Prost, Europhysics Lett. 4, 307 (1987).

40. P. Oswald, M. Allain, J. de Physique 46, 831 (1985).

41. P. C. Martin, O. Parodi, P. S. Pershan, Phys. Rev. A 6, 2401 (1972).

42. F. Brochard and P. G. de Gennes, Pramana 1, 23 (1975).

43. F. Nallet, D. Roux and J. Prost (to be published).

44. Y. Talmon and S. Prager, J. Chem. Phys. 69, 2984 (1978).

45. J. Jouffroy, P. Levinson and P. G. de Gennes J. de Phys. (Paris) 43, 1241 (1982).

46. B. Widom, J. Chem. Phys. 81, 1030 (1984).

47. S. A. Safran, D. Roux, M. Cates and D. Andelman, Phys. Rev. Lett. 57, 491 (1986); D. Andelman, M. Cates, D. Roux and S. A. Safran, J. Chem. Phys. 87, 7229 (1987).

48. W. Helfrich, J. Physique 46, 1263 (1985).

49. See Ref. 11.

50. M. Cates, D. Andelman, S. Safran and D. Roux, Langmuir (accepted).

51. S. Milner, S. Safran , M. Cates, D. Andelman and D. Roux, J. de Physique (to be published).

52. M. Cates, D. Roux, D. Andelman, S. Milner and S. Safran, Europhysics Letters 5, 733 (1988).

53. D. Gazeau, A. M. Bellocq, L. Auvray, T. Zemb and D. Roux (to be published).

54. R. Lipowsky and S. Leibler, Phys. Rev. Lett. 56, 2561 (1986).

INTRODUCTION TO CONVECTION

DAVID S. CANNELL and CHRISTOPHER W. MEYER

Department of Physics, University of California
Santa Barbara, CA 93106 USA

ABSTRACT. Rayleigh-Bénard convection, the flow between parallel horizontal surfaces maintained at sufficiently different temperatures, is presently the most extensively studied non-linear pattern forming system. This system, which can be very carefully controlled experimentally, exhibits phenomena such as defect motion, wavelength selection, pattern competition, and sensitivity to boundary conditions, which are believed to be of general importance in the behavior of non-linear pattern forming systems. By varying the lateral extent of the system the transition to turbulent flow may be studied in systems ranging from ones described well by a few coupled modes, to systems in which spatial degrees of freedom play a major role.

Rayleigh-Bénard convection is the flow which occurs in a fluid confined between parallel horizontal surfaces, each of which is isothermal, but which are maintained at sufficiently different temperatures as to cause convection. Experimental studies have focussed on measurements of the heat current carried by convection j^{conv}, the flow patterns which form, their relationship to j^{conv}, and the manner in which these patterns evolve, either in time, for a fixed temperature difference, or as the temperature difference is varied.

It is commonly believed (or at least hoped!) that the Oberbeck-Boussinesq approximation to the full equations of motion, contains most of the essential physics of this system. These equations follow from the continuity equation, the heat flow equation and the Navier-Stokes equation under the assumptions that the fluid is incompressible and that the only fluid property which depends on temperature is the density. They may be cast in dimensionless form by measuring length in units of the depth of the fluid layer d, and time in units of the vertical thermal diffusion time d^2/κ where κ is the thermal diffusivity. The fluid properties and applied temperature difference are characterized by two dimensionless ratios, the Rayleigh number R and the Prandtl number P. The results[1] are

$$\nabla \cdot \vec{v} = 0 \qquad (1a)$$

$$\frac{\partial \theta}{\partial t} + (\vec{v} \cdot \nabla)\theta = R\vec{v} \cdot \hat{z} + \nabla^2 \theta \qquad (1b)$$

$$\frac{1}{P}\left(\frac{\partial \vec{v}}{\partial t} + (\vec{v} \cdot \nabla)\vec{v}\right) = \theta\hat{z} - \nabla\pi + \nabla^2\vec{v} \quad . \qquad (1c)$$

with

$$R = \alpha g \Delta T d^3 / \kappa \nu \qquad (2)$$

and

$$P = \nu/\kappa \quad . \qquad (3)$$

Here $\vec{v}$ is the fluid velocity, θ is the deviation of the temperature from the linear purely conducting solution, π is the deviation of the pressure from the hydrostatic value corresponding to $\vec{v} = 0$, ν is the kinematic viscosity, α is the thermal expansion coefficient and ΔT is the temperature difference. Equations (1) govern the evolution of deviations from a state in which the fluid conducts but does not move.

Naively, one might expect convection to begin for any $\Delta T > 0$; after all the lower fluid layers are then less dense than the upper ones. This is not the case however, because the system must be able to release buoyant energy rapidly enough to overcome viscous losses, while transporting warm fluid upward and cold fluid downward in such a manner as to prevent thermal conduction from dissipating the buoyant force. Consequently convective flow begins only for $R \geq R_c = 1708$ (for the laterally infinite system). One discovers this theoretically by examining the stability of the conducting solution $(\theta = \vec{v} = \pi = 0)$ of Eqs. (1), with respect to infinitesimal perturbations. For $R < R_c$ all such perturbations decay in time, and the conducting solution is stable. For $R > R_c$ the conducting solution is unstable with respect to perturbations of finite wavevector over a range of wavevectors centered around the critical wavevector q_c, which corresponds to the only unstable perturbation at $R = R_c$. The resulting flow state consists of straight parallel rolls in a laterally unbounded system, as indicated schematically in Fig. (1).

The range of wavevectors over which the conducting state is susceptible to perturbation, as a function of the dimensionless stress parameter

$$\epsilon \equiv (R - R_c)/R_c \quad , \tag{4}$$

is indicated schematically by the dashed curve in Fig. 2. Such a perturbation grows exponentially in time within the linearized approximation to Eqs. (1), however the non-linear terms cause the flow to saturate at finite amplitude. The resulting roll flow is in turn susceptible to other perturbations which limit the range in both ϵ and q over which it may be observed. These limits, which are shown as solid lines in Fig. 2, may be calculated for straight rolls of uniform amplitude and fixed wavevector q, by retaining many modes and harmonics in the roll structure and examining its stability with respect to rather general perturbations[2] The results shown in Fig. 2 pertain to a Prandtl number of 7, which is typical of water. The entire stable volume (ϵ, q, P) for straight parallel rolls is known as the Busse balloon. These calculations are of great value, because despite the fact that most convective flows do not consist of straight parallel rolls, the same, or very similar, sorts of instabilities are generally observed in real flows.

Bounding the fluid laterally has a dramatic effect on the flow patterns which are actually observed. Only in long narrow channels or in a narrow annulus does one normally encounter straight rolls.[3] Instead one finds that complicated patterns, usually involving defects, are the rule rather than the exception.[4] For example, Fig. 3 illustrates some of the simpler stable patterns which may be observed in a circular container of moderate aspect ratio (radius/depth). In general the rolls tend to terminate normal to the walls, and they display resistance to bending or abrupt changes in wavevector. In large containers this results in very disordered patterns, which involve defects.

Defects offer mechanisms whereby pattern evolution and wavelength selection may occur. For example a grain boundary allows one set of rolls to adjust wavelength by changing the length of the rolls in the other set. Dislocations and disclinations can climb parallel to the roll axes to select one spacing over another,[5] while

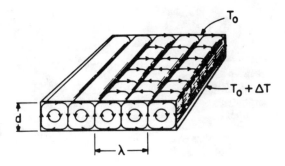

Fig. 1: Schematic diagram of straight parallel convective rolls. The wavelength of the flow is of order 2d.

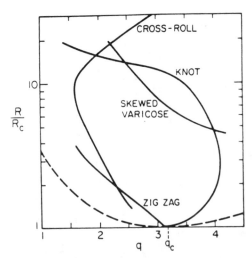

Fig. 2: Neutral stability curve (dashed line) and stability boundaries for straight roll flow in a fluid of Prandtl number 7, indicating the various perturbations to which the roll flow is unstable. Only within the innermost region bounded by the solid curves is flow in the form of straight rolls expected to be stable. For any given R/R_c, the conducting state however, is unstable with respect to perturbations corresponding to straight rolls, for the entire range of wavevectors bounded by the dashed curve.

focus singularities are observed to source or sink rolls,[4,6] probably because the flow is weaker in a region of strong curvature. The nature of these defects is indicated schematically in Fig. 4.

In a circular container of concentric rolls the pattern evolution with increasing ϵ, follows the trend toward longer wavelengths, first observed by Koschmeider[6] at high Prandtl number, but it does so in a very dramatic manner. As illustrated by the images of Fig. 5, the innermost roll moves off center as ϵ is increased before

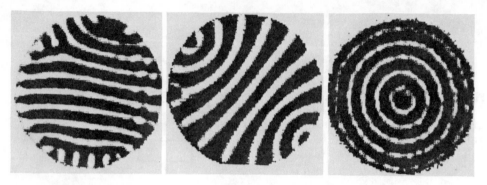

Fig. 3: Commonly observed convective patterns in cylindrical containers. The concentric pattern must be formed by steps or ramps in the temperature difference, at least for Prandtl numbers below 7 or so.

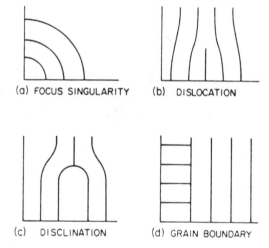

(a) FOCUS SINGULARITY   (b)   DISLOCATION

(c)   DISCLINATION   (d) GRAIN BOUNDARY

Fig. 4: Schematic of the structures of commonly observed defects in convection patterns.

disappearing to yield a concentric pattern of longer wavelength.[7] Surprisingly, the state shown in Fig. 5(b) is stable. This behavior is believed to be the result of large scale flows[8] induced by the roll curvature.

Probably the most drastic departure from the stable flow expected for straight rolls is the chaotic time dependence observed for small ϵ (ϵ of order 0.1) for $P \lesssim 7$, in large aspect ratio containers. Although first observed in heat transport measurements in helium,[9] room temperature experiments[10] reveal that this is the result of defect generation and motion. This motion may be periodic, with a period which is long compared even to the horizontal thermal diffusion time, or it may be chaotic, with defect nucleation occurring at random times. It is tempting to surmise that this behavior results from different mechanisms selecting different wavelengths, with the resulting competition producing a spread in wavevectors which exceeds the stability boundaries. This mechanism was originally proposed to explain effects observed in wavy Taylor vortex flow,[11] but it has also been examined for at least one convective flow.[12]

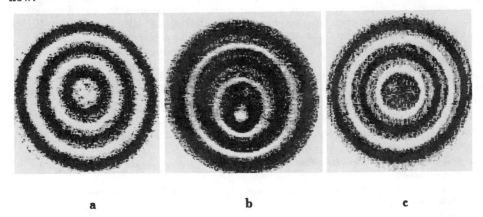

a   b   c

Fig. 5: Wavelength evolution with increasing ϵ for a pattern of concentric rolls in a circular container of aspect ratio 7.5. Image (a) corresponds to $\epsilon = 1.81$, image (b) to $\epsilon = 2.97$ and image (c) to $\epsilon = 3.17$

    In order to make progress with theoretical models of pattern behavior it is necessary to explore models which incorporate from the beginning the observed fact that convection does not consist of plane parallel rolls; instead the roll direction and amplitude vary, and defects represent severe local disturbances. Such an approach is implicit in the amplitude equation[13] framework. Where the amplitude and wavevector vary slowly, the flow may be described by a stream function ψ of the form

$$\psi = A(x, y, t)e^{iq_c x} + \text{c.c.} \tag{5}$$

In fact, for small ϵ, Eq. (5) can be derived from the Oberbeck-Boussinesq equations, and the equation governing the evolution of the complex amplitude A is

$$\tau_o \frac{\partial A}{\partial t} = \left\{ \epsilon + \xi_o^2 \left(\frac{\partial}{\partial x} - \frac{i}{2q_c} \frac{\partial^2}{\partial y^2} \right)^2 \right\} A - g \, |\, A\, |^2 \, A \quad , \tag{6}$$

where τ_o, ξ_o and g are real constants which set fundamental time, length and amplitude scales, respectively. The magnitude of A gives the local convective flow amplitude, while the gradient of its phase gives the deviation of the local wavevector from $\hat{x}q_c$. This deviation must however remain small; the roll axes are assumed to lie essentially in the $\hat{y}$ direction.

    Despite its limitations Eq. 7 does contain a great deal of the most obvious physics of the flow. For rolls of uniform amplitude, but of wavevector $q_c + \delta q$, the time independent solution is

$$A = g^{-1/2}(\epsilon - \delta q^2 \xi_o^2)^{1/2} e^{i\delta qx}, \tag{7}$$

which vanishes on the neutral stability curve $\epsilon_m = \delta q^2 \xi_o^2$ (remember here δq is the deviation of the wavevector from q_c). In addition, the stable solutions of Eq. (6) are themselves subject to the Eckhaus and zig-zag instabilities.

Coupled amplitude equations for rolls of different orientation can be used to study competition between squares, hexagons and straight rolls, and the amplitude equation can describe the temporal behavior[14] of the convective heat current following a transition from $\epsilon < 0$ to $\epsilon > 0$. It should be noted however that Eq. (6) corresponds to a system with a potential and thus exhibits purely relaxational dynamics. The steady state solutions are all time independent, and the equation can not describe the phase turbulence observed near threshold in low and moderate Prandtl number fluids.

This research is supported by Department of Energy Grant DOE 87ER 13738.

1. See, for instance, S. Chandrasekhar, *Hydrodynamic and Hydromagnetic Stability* (Oxford U.P., Oxford, 1961).

2. F. H. Busse, and R. M. Clever, J. Fluid Mech. 91, 319 (1979); R. M. Clever and F. H. Busse, J. Fluid Mech. 65, 625 (1974).

3. For a detailed study of straight rolls in a rectangular container see, P. Kolodner, R. W. Walden, A. Passner and C. M. Surko, J. Fluid Mech. 163, 195 (1986).

4. See, for example, V. Steinberg, G. Ahlers, and D.S. Cannell, Phys. Script. 32, 534 (1985) for results in circular containers; and J. P. Gollub, A. R. McCarriar and J. F. Steinman, J. Fluid Mech. 125, 259 (1982) for a study of the behavior in a large rectangular container.

5. A. Pocheau and V. Croquette, J. Physique (Paris) 45, 35 (1984).

6. E. L. Koschmeider, Adv. Chem. Phys. 26, 177 (1974).

7. V. Croquette and A. Pocheau, in *Cellular Structures and Instabilities*, ed. by J. Wesfreid and S. Zaleski, (Springer-Verlag, New York, 1984), p. 106; V. Steinberg, G. Ahlers and D. S. Cannell, Phys. Script. 32, 534 (1985).

8. A. Zippelius and E. D. Siggia, Phys. Rev. A 26, 178 (1982).

9. G. Ahlers and R.P. Behringer, Phys. Rev. Lett. 40, 712 (1978).

10. G. Ahlers, D. S. Cannell and V. Steinberg, Phys. Rev. Lett. 54, 1373 (1985); A. Pocheau, V. Croquette and P. Le Gal, Phys. Rev. Lett. 55, 1094 (1985).

11. G. Ahlers, D.S. Cannell and M. A. Dominguez-Lerma, Phys. Rev. A 27, 1225 (1983).

12. M. S. Heutmaker and J. P. Gollub, Phys. Rev. A 35, 242 (1987).

13. A. C. Newell and J. A. Whitehead, J. Fluid Mech. 38, 279 (1969); L. A. Segel, J. Fluid Mech. 38, 203 (1969).

14. G. Ahlers, M. C. Cross, P. C. Hohenberg and S. Safran, J. Fluid Mech. 110, 297 (1981).

YOSSI LEREAH AND GUY DEUTSCHER

ONSET OF CONVECTION

DAVID S. CANNELL, C. W. MEYER, and *GUENTER AHLERS*
Department of Physics, University of California
Santa Barbara, CA 93106 USA

ABSTRACT. The results of recent experiments probing the initial stages of pattern formation in Rayleigh-Bénard convection are presented. Unless extraordinary measures are taken, horizontal temperature gradients induced by the sidewalls result in flows which reflect the container geometry and penetrate inward from the walls. When such forcing is reduced sufficiently the initial stages are observed to involve random cellular flow emerging throughout the entire cell, and gradually healing to roll-like flow. Results for the convective heat current can be modeled using the amplitude equation with either a constant or a stochastic forcing term.

In this lecture the results of some recent experiments[1] exploring the very earliest stages of pattern formation in Rayleigh-Bénard convection will be presented. In these experiments the temperature difference across the fluid layer (water, 3.18 mm thick) was increased linearly with time, passing through the critical value ΔT_c at which convection began in steady state experiments, and continuing on upward. Both the patterns which formed when convection began, and the heat current carried by convection $j^{conv}(t)$ were measured for various ramp rates. The experiments were carried out in cylindrical cells of aspect ratio (radius/height) of 10.0. The basic issues which can be explored this way include the following:

(i) What sort of patterns emerge?

(ii) How are the patterns affected by the properties of the sidewalls which bound the fluid laterally?

(iii) How long is the onset of convection delayed after ΔT passes ΔT_c? This bears directly on the question of what is responsible for convection beginning in the first place.

(iv) Can simple models such as an amplitude equation or Lorenz model account for $j^{conv}(t)$?

The heart of the experimental apparatus is shown schematically in Fig. 1. The fluid was confined between an upper horizontal boundary of sapphire (3 mm thick) and a lower one of copper, about 6 mm thick. A resistive heater on the bottom of the copper plate, was used to pass heat through the fluid layer. Copper and sapphire were used because they are much better thermal conductors than the fluid, with sapphire being $\approx$ 80 times better and copper 650 times better. The upper surface of the sapphire was held at a fixed temperature by a stream of temperature controlled water which impinged on its surface from an array of jets.

The lower plate contained a temperature sensitive resistor (thermistor) which was able to determine its temperature with a resolution of $\approx 50\mu K$. Thus both the total heat current through the fluid, and the temperature difference across it were determined quite accurately, certainly to better than 0.1%. Of course, a certain amount of the heat input Q passed through the sidewalls and the insulation surrounding the cell. This loss is linear in the temperature difference, for static conditions, and it was determined by measuring ΔT and Q for $\Delta T < \Delta T_c$, where no convection occurred, and using the known thermal conductivity of water to determine the fraction of the total heat current which was flowing through the fluid, (typically $\gtrsim 50\%$). For $\Delta T > \Delta T_c$, j^{conv} was determined by subtracting the heat current through the walls and insulation from the total heat input.

Fig. 1: Schematic diagram of a convection cell used to impose well-defined boundary conditions on a horizontal layer of fluid.

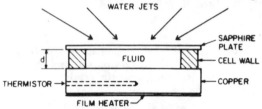

The determination of the convective heat current when the bottom plate temperature is time dependent is not nearly so straightforward as the procedure described above for the static case.[2] Physically the reason for this is that some of the heat input is being used to increase the bottom plate temperature, the mean temperature of the fluid, and that of the sidewalls. However, for the situation of a linear temperature ramp one can show that the appropriate relationship is given by

$$j^{conv}(t) = \left(\frac{Q(t)}{Q_c} - \frac{\Delta T(t)}{\Delta T_c} - \frac{C_1}{\Delta T_c} \frac{dT_b}{dt} \right) \left(1 + \frac{\lambda_o}{\lambda_w} \right) \qquad (1)$$

where the constant C_1 was determined by fitting in the non-convecting region, and λ_o/λ_w was determined directly from static measurements. Here time is measured in units of the vertical thermal diffusion time d^2/κ, where d is the thickness of the fluid layer, and κ is its thermal diffusivity.

The flow patterns were visualized using the shadowgraph technique. Light from a small lamp was focussed as well as possible onto a pinhole of diameter $\approx 0.5\,mm$, and the transmitted light was collected and collimated by a lens. A beam splitter deflected about 50% of this light down onto the cell which it passed through. It then reflected back up from the polished nickel plated copper bottom, passed through the beam splitter and was reflected by a mirror onto a translucent screen.[3] A video camera recorded images of this screen, under computer control, and these images were stored digitally by the computer. The images were formed of small areas called pixels, with a complete image being 256×256 pixels. The intensity falling on each pixel was recorded as an 8-bit binary number, giving a resolution of $\approx 0.4\%$ of full scale. The mechanism by which the image was formed is simply that the light rays were deviated toward the cold down-flowing regions which had a higher refractive index than the hot up-welling regions. This angular deviation resulted in an intensity modulation after the rays had travelled an appropriate distance. Frequently this modulation was sufficient to be seen by eye, but when the pattern was very weak it was still possible to resolve it by digital image processing techniques.

We used the procedure of taking a reference image under static conditions for $\Delta T \cong 0.8\Delta T_c$. Each subsequent image was divided by the reference image, pixel by pixel, and to print the resulting image ratio it was averaged over blocks of 3×3 pixels. In this way small deviations from unity could be enhanced and assigned a gray scale. The process of image division served to eliminate the effects of non-uniform illumination and non-uniform camera response, but did nothing to reduce noise. The effectiveness of this simple technique is illustrated by Fig. 2. Figure 2(a) is an image of a very weak convection pattern, which is undiscernable. Figure 2(b) is the ratio of 2(a) to the reference image, and one can barely make out that a pattern is visible. Considerably more improvement is possible however, by using a simple filter. Since we are only interested in structure on length scales of the order of a roll size (i.e., the cell height), we Fourier transform the image ratio (2c), and smoothly truncate all spatial Fourier components with wavevectors greater than a cutoff value. We used a $\cos^2$ cutoff to interpolate the filter function smoothly between unity inside the inner circle ($q = 1.3q_c \simeq 1.3\pi/d$), and zero outside the outer circle ($q = 2.6q_c$) shown in Fig. 2d, which i s the filtered Fourier transform. Figure 2(e) is the final image obtained by re-transforming the filtered Fourier transform.

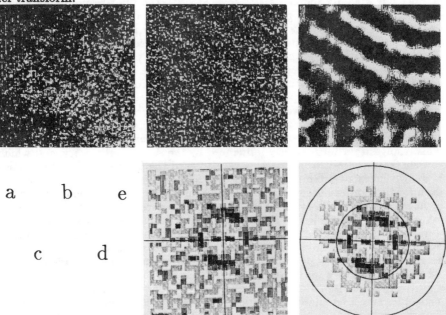

a   b   e

c   d

Fig. 2: Images of a weak convective flow, showing the results which may be obtained by image division and spatial filtering. Image (a) is the actual image; image (b) is the ratio of (a) to a reference image; image (c) is the Fourier transform of (b); image (d) is the filtered Fourier transform which has been cutoff using a $\cos^2$ function to interpolate between 1 and 0 at the two circles; and (e) is the result of retransforming (d).

The actual experiments involved ramping the heat current linearly in time while measuring the resulting time dependent bottom plate temperature and periodically taking images. After an initial transient, the bottom plate temperatures also increased linearly with time, and we defined $t = 0$ to be the time for which the temperature difference reached ΔT_c, the critical value for static conditions. It is convenient to consider both ramp rates and heat currents in dimensionless form, which we do by measuring ΔT in units of ΔT_c, and time in units of the vertical thermal diffusion time $t_v \equiv d^2/\kappa$. The relevant stress parameter is

$$\epsilon(t) \equiv \frac{R(t) - R_c}{R_c} = \frac{\Delta T(t)}{\Delta T_c} - 1 \quad , \tag{2}$$

where $R(t)$ is the time dependent Rayleigh number, and R_c is the critical Rayleigh number for the onset of convection under static conditions. Our ramps yielded

$$\epsilon = \beta t \tag{3}$$

after the initial transient had died. We were able to explore the range $0.01 \leq \beta \leq 0.30$.

As one steadily increases ΔT toward ΔT_c, the fluid responds by conducting thermally, but remains at rest, except for very small movements caused by thermal expansion. This generates temperature and pressure profiles which are presumably a solution of the equations of motion, even in the convecting regime $\Delta T > \Delta T_c$. This solution is however, not stable with respect to perturbations lying within the range of wavevectors corresponding to the neutral stability curve at $\epsilon = \epsilon(t)$. Thus as $\epsilon(t)$ increases from 0, the conducting solution becomes unstable with respect to an ever widening band of perturbations with wavevectors centered at q_c, the critical wavevector which is the wavevector of the rolls which form at $R = R_c$.

The above discussion raises the question of exactly what influences are responsible for initiating the flow. One very common form of influence is sidewall forcing.[2] This results in patterns clearly influenced by the sidewall geometry, and which tend to penetrate the cell from the sidewalls toward the interior. We believe this effect is due to horizontal temperature gradients near the sidewalls. If they are present even for time independent ΔT, we term them static. This results in the appearance of flow near the walls even for $\Delta T < \Delta T_c$. A more subtle form of forcing occurs whenever ΔT is time dependent, provided the fluid and sidewalls differ in thermal diffusivity. In this case changing ΔT with time results in lateral gradients, and again tends to force flows which mirror the sidewall geometry. For example one observes concentric rolls penetrating a circular container from the sidewalls following a rapid temperature jump from $\Delta T < \Delta T_c$ to $\Delta T > \Delta T_c$.

We have devised two methods of greatly reducing dynamic forcing. One consists of leaving a small inward projecting fin at the mid-height of the cell to prevent flow near the boundary. The other consists of bounding the fluid by sidewalls formed of a 5% polyacrylamide gel. The gel is 95% water, but damps any flow. It has a thermal diffusivity very nearly equal to that of the convecting fluid (water). Under these conditions one observes flow patterns at onset which consist of randomly arranged downward flowing cells surrounded by regions of upflow, as shown by the images of Fig. 3. Images a and b were obtained with $\beta = 0.01$, and image c corresponds to $\beta = 0.27$. Not only do these patterns appear random to the eye, they are not reproducible from one experimental run to the next. Any given arrangement of cells heals into a more roll-like pattern within a few vertical thermal diffusion times, as is noticeable in comparing images a ($t \cong 6.5$) and b ($t \cong 10$).

In addition to imaging the emerging flow, we have also measured $j^{conv}(t)$, as shown in Fig. 4. We have attempted to model $j^{conv}(t)$ using an amplitude equation[2,4] for uniform straight rolls,

$$\tau_o \dot{A} = \epsilon A - gA^3 + kA^5 + f \tag{4}$$

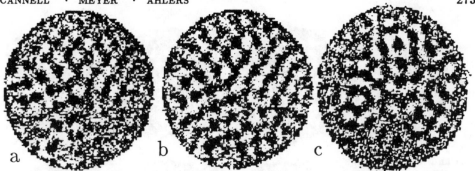

Fig. 3: Convective flow patterns formed by ramping at different rates, with $\beta = 0.01$ for images (a) and (b) and $\beta = 0.27$ for image (c). Image (a) was taken 6.5 vertical thermal diffusion times after crossing the static threshold, while (b) and (c) correspond to $t = 10$, and 1.5, respectively.

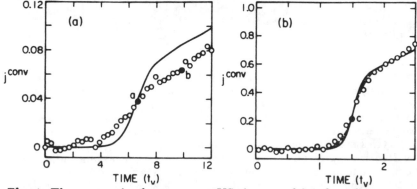

Fig. 4: The convective heat current VS time resulting from linear ramps, with (a) corresponding to $\beta = 0.01$ and (b) to $\beta = 0.27$. The solid curves result from the amplitude equation with a constant phenomenological forcing field, and the dashed curve in (b) is the result of using a Langevin noise source in the same equation. The solid circles show the times at which the images in Fig. 3 were taken.

where we identify $j^{\mathrm{conv}}(t)$ as $A^2(t)$, and the constants g and k are determined by fitting static measurements of j^{conv} VS ϵ. The term f is intended to model whatever forces are involved in initiating the flow. Some such term is necessary, since $A = 0$ remains a solution of Eq. 4 even for $\epsilon > 0$. We have used both constant values for f, and also replaced it by a Langevin noise term, hoping to distinguish between causal and stochastic forcing. The results are shown by the solid (causal) and dashed (stochastic) curves in Fig. 4, and neither seems preferable, although either accounts quite reasonably for the data (in the case of small ramp rates $j^{\mathrm{conv}}(t)$ is the difference of two large numbers and is probably not accurate enough to consider the deviation from the fit too seriously).

In addition to ramping ΔT linearly in time we have also modulated it sinusoidally with $\epsilon(t) = \epsilon_o + \delta \sin \Omega t$, for $\Omega = 1$. In essence, what we find is that memory of the previous pattern is lost whenever $\epsilon(t)$ reaches a sufficiently negative value during a cycle. Under these conditions the flow emerging during each cycle ap-

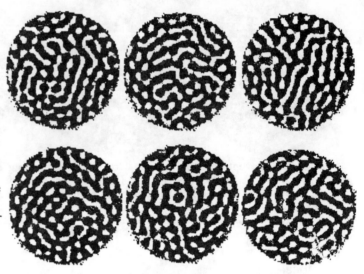

Fig. 5: Convection patterns emerging on consecutive cycles in a fluid subjected to sinusoidal modulation $\epsilon(t) = \epsilon_o + \delta \sin \Omega t$, for $\epsilon_o = 0.21$, $\delta = 0.51$ and $\Omega = 1$. Note that different patterns emerge on each cycle. These images were taken late enough in the cycle for the emerging cellular flow to have begun healing to a roll-like structure.

soidally with $\epsilon(t) = \epsilon_o + \delta \sin \Omega t$, for $\Omega = 1$. In essence, what we find is that memory of the previous pattern is lost whenever $\epsilon(t)$ reaches a sufficiently negative value during a cycle. Under these conditions the flow emerging during each cycle appears identical to the random cellular flow observed with ramps, and heals toward a roll-like pattern during the cycle. Under these conditions patterns emerging on succeeding cycles are different, as shown by the images of Fig. 5, which were taken for $\epsilon_o = 0.21, \delta = 0.51$. If, on the other hand, $\epsilon(t)$ does not become sufficiently negative, the pattern is preserved from cycle to cycle. Thus there is a curve $\epsilon_o^*(\delta)$ above which patterns repeat, and below which random flows emerge. We have measured this curve and it has been shown[5] to be quantitatively consistent with the results of our ramping experiments.

The amplitude of the Langevin noise term required to fit our results for $j^{\mathrm{conv}}(t)$ has been compared to what might be expected from thermal fluctuations,[6] and it was found to be about 7000 times larger in mean squared amplitude. Thinking that temperature fluctuations in the water jets playing on the upper surface of the sapphire might be implicated, we replaced the sapphire by a lucite-copper composite which would immensely reduce any fluctuations reaching the fluid. This resulted in no perceptible changes in our results, and it is not presently known what is actually generating the random flows we have observed.

This research is supported by Department of Energy Grant DOE 87ER 13738.

1. C. W. Meyer, G. Ahlers, and D. S. Cannell, Phys. Rev. Lett. **59**, 1577, 1987.

2. G. Ahlers, M. C. Cross, P. C. Hohenberg, and S. Safran, J. Fluid Mech. **110**, 297 1981.

3. A more complete description of the apparatus may be found in V. Steinberg, G. Ahlers, and D. S. Cannell, Phys. Script. **32**, 534 1985.

4. A. C. Newell and J. A. Whitehead, J. Fluid Mech. **38**, 279 1969; L. A. Segel, J. Fluid Mech. **38**, 203 1969.

5. J. Swift and P. C. Hohenberg, Phys. Rev. Lett. **60**, 75 1988.

6. H. van Beijeren and E. G. D. Cohen, Phys. Rev. Lett. **60**, 1208 1988.

WAVES AND PLUMES IN THERMAL CONVECTION

GIOVANNI ZOCCHI, STEVE GROSS and ALBERT LIBCHABER

The James Franck and Enrico Fermi Institutes
The University of Chicago
Chicago, IL 60637

The interface between two fluids is an interesting object because one can say simple things about its dynamics. For example, if the fluids have different densities the interface can sustain gravity-driven waves, like the surface of the sea. If there is a surface free energy this will give rise to the quite different phenomenon of capillary (surface tension driven) waves. These instabilities can be investigated easily in terms of a linear stability analysis. Then there are other, nonlinear objects which can propagate unchanged along an interface: an example is solitary waves in a channel.

Now suppose you were told that in a tank filled with water, where there is only one fluid and no interfaces, and where you have created a turbulent flow which is so complex that you can describe it only statistically, still you can observe these same phenomena. Where would be the place to look?

In any kind of flow, no matter how turbulent, there are regions that are special: these are the regions near solid boundaries. The flow inside a box can be quite messy, but the velocity has to go to zero at the walls, and so there is a whole layer along the walls where there is no turbulence because the velocity is too small. It is this boundary layer which has, as we will see, many of the properties of a real interface.

Our experimental setup is as follows: The flow is produced by convection: a cubic cell filled with water is heated from below (Rayleigh-Benard convection). With this geometry, the average flow takes the form of a big roll the size of the cell itself. Thus the situation near the top and bottom plates is that there is an average "wind" sweeping by parallel to the plates. Since the velocity is zero at the plates themselves, this is a shear flow. The average velocity of the main flow, let's call it u_0, increases as the dimensionless Rayleigh number

$$\text{Ra} = \frac{g\alpha\Delta L^3}{\nu\kappa} \tag{1}$$

is increased (g is the acceleration of gravity, α the thermal expansion coefficient, Δ the temperature difference between the top and bottom, L the size of the box, ν the kinematic viscosity and κ the thermal diffusivity). In fact by dimensional arguments

$$u_0 \sim \left(g\alpha\Delta L\right)^{1/2} \tag{2}$$

so

$$u_0 \sim \text{Ra}^{1/2} \tag{3}$$

The velocity given above is the "free fall velocity" over a distance L for a blob of fluid of temperature difference L with respect to the medium. This is not the

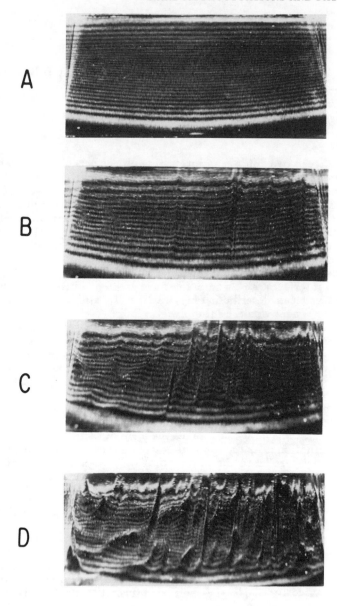

Fig. 1

only possible choice: another characteristic velocity is e.g., the viscosity limited terminal velocity of the blob. Also, there are corrections to the simple scalings given throughout here, for example a more refined model gives

$$u_0 \sim \mathrm{Ra}^{3/7} \tag{4}$$

See Ref. 3 for a detailed discussion.

The dimensions of our system are such ($L \sim 20$cm) that with a few degrees temperature difference across the cell we achieve $R_a \sim 10^8 - 10^9$. This is a regime in which the flow inside the cell is turbulent: the Reynolds number at the scale of the cell would be

$$\text{Re} = \frac{u_0 L}{\nu} \sim \text{Ra}^{1/2}\text{Pr}^{-1/2} \tag{5}$$

from the previous estimate for u_0 ($\text{Pr} = \nu/\kappa$) is the Prandtl number). The Prandtl number is of order one, so if Ra $\sim 10^8$ we get Re $\sim 10^4$. This is of course only an order of magnitude estimate: for instance, since there is a non-zero critical Rayleigh number Ra^{crit} for the onset of convection, the proper scaling will be in powers of (Ra - Ra^{crit}).

Let us now look at the situation of the temperature field. The temperature field is also "turbulent" because it is coupled to the velocity field through the convective term in the heat transport equation $(\mathbf{u}\vec{\nabla})T$; the relevant "Reynolds number" is here $u_0 L/\kappa$; since Pr ~ 1, this is also big. One can nonetheless ask for the average temperature (average in time) at any given point inside the cell. One will then find that the average temperature profile in the vertical direction is as follows: the entire temperature drop across the cell occurs in two thin layers close to the bottom and top plate; here the average temperature gradient is big, the instantaneous vertical component of the velocity is small, and the heat transport is diffusive. Everywhere else the average temperature gradient is almost zero, the instantaneous vertical component of the velocity is big, and the heat transport is convective. The cross-over between these two regions, that is the top of the thermal boundary layer, is our "interface".

To estimate the thickness δ of this thermal boundary layer, there is an argument[1] which says that the Rayleigh number *of the boundary layer*,

$$\text{Ra}^{\text{b.l.}} = \frac{g\alpha\frac{\Delta}{2}\delta^3}{\nu\kappa} \tag{6}$$

($\Delta/2$ is the temperature drop across *one* boundary layer), is constant, that is, is independent of the Rayleigh number of the box. In fact, we know that there is an upper bound to $\text{Ra}^{\text{b.l.}}$, because in the boundary layer the heat transport is diffusive, and so the boundary layer itself must be below the onset of convection, and the argument says that there is a regime where $\text{Ra}^{\text{b.l.}}$ goes about to the critical value and stays there.

Then, if $\text{Ra}^{\text{b.l.}}$ = const. we obtain

$$\delta \sim L\text{Ra}^{1/3}, \tag{7}$$

so if Ra is large we have indeed a very thin layer; in the experimental situation described above, $\delta \sim 1$mm, so $\delta/L \sim 10^{-2}$.

Let us summarize the relevant points: we have a thermal boundary layer which is close to onset of convection, and a shear flow; as Ra is increased, $\text{Ra}^{\text{b.l.}}$ stays constant or at most increases very slowly, but the Reynolds number of the boundary layer increases:

$$\frac{u_0\delta}{\nu} \sim \text{Ra}^{1/6} \tag{8}$$

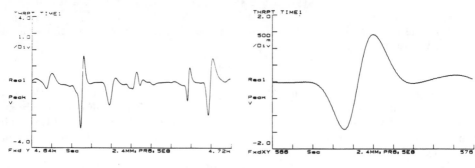

Fig. 2

(from the scaling for u_0 and δ). We will thus eventually reach a critical Ra at which instabilities of the boundary layer will set in, due to the shear flow.

In Fig. 1 we show pictures of these instabilities: the deformations of the pattern of straight lines which covers the bottom plate correspond to deformations of the average temperature field in the bottom boundary layer (the optical arrangement is described in Ref. 2). Figure 1A is below onset (Ra $< 10^8$), Fig. 1B above (Ra $= 8.6 \cdot 10^8$); here we see that the instability takes the form of single, solitary wave like objects with a plane front which travels in the direction of the main flow (from left to right in the pictures). Figures C and D show the evolution for increasing Ra ($1.9 \cdot 10^9$ and $3.8 \cdot 10^9$ respectively); the last picture, with its breaking waves, resembles the sea in a storm!

We can measure the temperature signal associated with these waves, by means of bolometers suspended inside the cell. Figure 2 shows a time recording of the temperature difference between two probes placed close to each other at the same height; the waves have a characteristic "down-up" signature, reproduced enlarged in the second half of the picture. Preliminary results show that these are dispersive waves, with the velocity increasing with the wavelength, but also dependent on the amplitude and in general the shape of the wave.

These are not the only kind of instabilities which occur in the boundary layer. As with a real interface, which can sustain long wavelength gravity waves and short wavelength capillary waves, here too we can excite short wavelength ripples, which travel much faster than the big waves shown in the pictures, and indeed much faster than the velocity of the main flow, and quite independently of its direction. We obtain this by shaking the top of the boundary layer with a vibrating wire. The resemblance of this phenomenon to that of capillary waves at an interface is so striking that one is led to ask if it may be possible to associate a surface energy to the thermal boundary layer and treat it in the same way.

We observe yet another kind of instability, which is thermal plumes emitted from the boundary layer into the bulk. These are important because they change the statistics of the temperature field in the bulk,[3] but we won't go into this effect here.

1. W. V. R. Malkus, Proc. Roy. Soc. (London) A225, 185-212 (1954).

2. S. Gross, G. Zocchi and A. Libcaber, Comptes Rendus de l'Academie des Sciences (submitted).

3. B. Castaign et al, J. Fluid Mech. (submitted).

AN INTRODUCTION TO MULTIFRACTAL DISTRIBUTION FUNCTIONS

B. B. MANDELBROT

Physics Department, IBM T.J. Watson Research Center
Yorktown Heights NY 10598, USA
    and

Mathematics Department, Yale University
New Haven CT 06520

ABSTRACT. This text (an abridged version of a forthcoming detailed paper) is addressed to both the beginner in multifractals and the seasonal professional. An alternative presentation of this subject is put forward, extending and streamlining the original approach in Mandelbrot 1974. Of central concern is the study of low-dimensional cuts of high-dimensional multifractals. The generalization from fractal *sets* to multifractal *measures* involves the passage from geometric objects characterized primarily by one number, namely a fractal dimension, to geometric objects characterized primarily by a function, which may be $\rho(\alpha)$, namely a limit probability distribution plotted suitably, on double logarithmic scales. In terms of the function $f(\alpha)$ used in the approach of Frisch-Parisi and of Halsey et al., one has $\rho(\alpha) = f(\alpha) - \max f(\alpha)$, and $f(\alpha)$ need not be positive.

1. Introduction and Summary

This text hopes to serve both the reader who is not yet fully familiar with multifractal *measures* (they are *not* sets!), and the reader who is already familiar with them via the approach of Frisch and Parisi 1985, as adopted by Halsey et al. 1986. The alternative approach used here is a streamlining and development of our early work of 1968-1976, in which the multifractals were first studied and applied, and more particularly of Mandelbrot 1974. Our goal is not to teach manipulations, but to present the reader who is new to multifractals with what we believe is the most understandable and the simplest form of their theory, and to provide the skilled reader with the surprisingly simple explanation of the formal manipulations with which he is familiar. In one phrase, the generalization from fractal sets to multifractal measures involves the passage from geometric objects characterized primarily by one number, to geometric objects characterized primarily by a function, for example by a "probability" distribution plotted suitably.

1.1. SUMMARY FOR THE READER NEW TO MULTIFRACTALS

The work is best summarized as follows. The notion of self-similarity extends readily from fractal sets to measures, a measure being simply a way of specifying a method of spreading mass, or probability, or other "stuff," over a supporting set. The distribution may be spread on a Euclidean "support," like an interval or a square, or it may be restricted to a fractal support like a Cantor set.

In order to specify a self-similar method of spreading stuff around, one standard first characteristic is a sequence of moments, or the cumulant generating function, which is denoted in this context by $\tau(q)$. Another equivalent characteristic is a limit distribution function. This limit is like—but not exactly like—an ordinary probability distribution function. Also, the renormalization must follow a very unusual and ill-known path: one needs a *multiplicative* renormalization that is substantially different from the *additive* renormalization that is familiar to every student of the central limit theorem. Because of this renormalization, the limit distribution function is best plotted with doubly logarithmic coordinates, as a function we shall denote by $\rho(\alpha)$. The alternative expression $f(\alpha) = \rho(\alpha) + E$, where E is the dimension of the measure's support, has become entrenched in the literature to specify the limit distribution.

To go from $\tau(q)$ to $f(\alpha)$, one takes a direct or inverse Legendre transform. This property follows immediately from the Lagrange multipliers approach to thermodynamics and to the Gibbs distribution, which has long been familiar to every physicist. Later on, a full mathematical justification of the formalism, valid in a broader context in which $f(\alpha)$ can very well be negative, is provided by reference to existing (but little-known) limit theorems of probability due to Harald Cramèr, and concerned with "large deviations."

On a plot of $f(\alpha)$, the quantities $-\tau(q)$ and D_q are the ordinates of the intercepts of that tangent to $f(\alpha)$, whose slope is q, with the vertical axis and with the main bisector of the coordinate axes.

Mandelbrot 1974 considers two distinct kinds of random self-similar multifractals, respectively called *conservative* (or *microcanonical*) and *canonical*. This distinction is crucial to the most sophisticated part of the theory of multifractals, which is our study of low-dimensional cuts of multifractals embedded in a high dimensional space. It is also necessary to understand the very peculiar standing of the lognormal distribution, which is widely mentioned, when applying multifractals to turbulence but is widely misunderstood, and to understand why the "principle of lognormality" claimed in 1962 by Kolmogorov is logically untenable. Observe that Frisch and Parisi 1985 have noted that while their approach stemmed from Mandelbrot 1974, it was less general because it did not accommodate the lognormal.

1.2. SUMMARY FOR THE READER FAMILIAR WITH THE MULTIFRACTAL FORMALISM, AS DERIVED BY FRISCH AND PARISI

This work is best summarized as follows. The function $\tau(q)$ with which you are familiar is a standard probabilistic tool to represent measures, called cummulant generating function.

The quantity α with which you are familiar is a standard notion, called Holder exponent. It is *not* a dimension nor a "pointwise dimension." Many confusions are avoided if the term "dimension" (in all its multiple and still multiplying forms!) is reserved to *sets* and never used to apply to *measures*.

The function $f(\alpha)$ with which you are familiar *is not*, again, a new concept to be labeled at will (for example as the "spectrum of singularity"). It is simply equal to $\rho(\alpha) + E$, where $\rho(\alpha)$ is a limit probability distribution function plotted on doubly logarithmic coordinates, and E is the dimension of the set that supports the measure.

The quantities $D_q = \tau(q)/(q - 1)$ are the "critical dimensions." Then we introduced in Mandelbrot 1974, through a theorem asserting that along a cut of dimension D through a measure, one has $\langle \mu^q(dx) \rangle = \infty$ when $q > q_{crit}(D)$. The function inverse of the function $q_{crit}(D)$ is D_q.

The attractiveness of multifractals may to some extent be due to mystery. All mystery is eliminated, however, when one understands the nature of the formal manipulations. But does it *really* matter that $\tau(q)$, α and $f(\alpha)$ are properly labeled? To respond, let us note that our approach obtains the Legendre transform between $\tau(q)$ and $f(\alpha)$ via Lagrange multipliers, as is usual in first courses of thermodynamics. In thermodynamics, one way of making the use of these multipliers rigorous after the fact is to follow the Darwin-Fowler method of steepest descents, which is unfortunately no longer familiar to every physicist. Thus, in effect, the Frisch-Parisi method skips Lagrange multipliers, and proceeds to the Darwin-Fowler method immediately. (Would a teacher of thermodynamics dare proceed in this way in a first course?)

The preceding remarks help explain that, in our experience, the Frisch-Parisi approach to multifractals has generated confusion. No one has an intuition of what a "spectrum of singularities" can or cannot be like. Proper foundations flush out this confusion. As a foremost example, the Frisch-Parisi intrepretation of $f(\alpha)$ as a fractal dimension has led to paradoxes linked to negative $f(\alpha)$. These paradoxes had begun as surprising anecdotal evidence, and have of course ended by being explained away by suitable special developments of the theory. They do not even arise in our properly probabilistic approach.

2. Spatial Variability Beyond Fractal Homogeneity

The notions of set and of measure are sharply distinct. However, given a classical set, e.g., an interval or a square, one can associate with it a uniform measure and the set and its uniform measure are mathematically equivalent. Next, consider a fractal set constructed recursively, e.g., a Sierpinski gasket; the most natural measure, again called "uniform," gives equal weight to each of its thirds. Again, the set and its uniform measure are equivalent. More generally, the mass distribution has been called "fractally homogenous," in my book, *FGN*, when the measure is the Hausdorff measure. However, this measure is very special, and in many cases very unrealistic. This was stressed in *FGN*, p. 375 ss, in a section titled "nonlacunar fractals," a clumsy term soon replaced by the Frisch-Parisi coinage *multifractals*. Let us quote the from the subsection on *Relative Intermittency*.

"The phenomena to which [multi]fractals are addressed are scattered throughout this Essay, in the sense that many of my case studies of natural fractals negate some unquestionable knowledge about Nature.

"We forget in Chapter 8 that the noise that causes fractal errors weakens between errors but does not desist.

"We neglect in Chapter 9 our knowledge of the existence of interstallar matter. Its distribution is doubtless *at least* as irregular as that of the stars. In fact, the notion that it is impossible to define a density is stronger and more widely accepted for interstellar than stellar matter. To quote deVaucouleurs, 'it seems difficult to believe that, whereas visible matter is conspicuously clumpy and clustered on all scales, the invisible intergalactic gas is uniform and homogenous ... [its] distribution must be closely related to ... the distribution of galaxies ...'

"And in Chapter 10 the pastrylike sheets of turbulent dissipation are an obviously over simplified view of reality.

"The end of Chapter 9 mentions very briefly the fractal view of the distribution of minerals. Here, the use of closed fractals implies that, between the regions where copper can be mined, the concentration of copper vanishes. In fact, it is very small in most places, but cannot be assumed to vanish everywhere.

"In each case, [portions of space] of less immediate interest were artificially emptied to make it possible to use *closed* fractal sets, but eventually these areas must be filled. This can be done using a fresh hybrid [now called *multifractals*. A multifractal] mass distribution in the cosmos will be such that no portion of space is empty, but, [given two] small thresholds θ and λ, a proportion of mass at least $1 - \lambda$ is concentrated on a portion of space of relative volume at most θ."

3. The Principal Ideas Underlying Multifractal Measures

3.1. AN OLD BUT GOOD ILLUSTRATION, AND THE MEANING OF "SINGULAR" MEASURES

Multifractals are old, insofar as we had first developed the basis of this technique in the years 1968 to 1976, in order to study different aspects of the intermittency of turbulence. Since "to see is to believe," let us begin by examining Fig. 1. It reproduces the earliest illustration of a multifractal, and first appeared in our earliest full paper on this topic, Mandelbrot 1972.

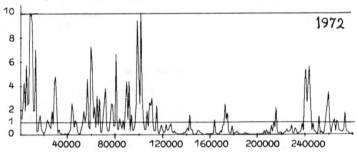

Fig. 1: The earliest simulation of a sample from a multifractal measure, namely the limit lognormal measure (or "Mandelbrot's 1972 measure").

The horizontal axis shows "time" divided into small boxes of width Δt, and the vertical axis shows the sequence of the "measures" of these boxes, i.e, of the stuff within these boxes. If the total integral measure over the time span $[0, 1]$ is set to 1, one can think of *the measure in a box* as *the probability of hitting this box*. The measure of the interval $[0, t]$, call it $\mu([0, t])$, is of course an increasing function of t, and what is plotted here in the sequence of its incremetns. The increments' values are joined to form a curve, nevertheless this curve is *not* what it seems to be. It is not the graph of a function, and it cannot readily be transformed into one.

For contrast and familiar background, let us draw an analogous diagram for a measure having a continous density. By dividing by Δt, we would obtain an approximation to the graph of the function representing this density, and a first characterization of our measure would be provided by the distribution of this approximate density along the horizontal.

In the present instance, however, the situation is extremely different. By design, the measure is approximately self-similar, in the sense discussed in Section 2.3. It follows that this measure grossly fails to have a density, nor is it discrete. A measure that is continuous but has no density is called "singular" (a technical term due to Lebesque). For example, if the Δt is halved, the sharing of the measure in an original Δt between the two halves is usually very unequal. Therefore, there is no way of getting the wiggly line on Fig. 1 to approximate a function, and there is no notion of asymptotic "distribution" for the values of this measure. Fortunately, there is a very useful substitute.

3.2. THE NOTIONS OF LIMIT PROBABILITY DISTRIBUTION $\rho(\alpha)$, AND OF $f(\alpha)$

Take different values of Δt, and, for each value of Δt, plot the corresponding measure distributions on doubly logarithmic coordinates. The measures we want to call multifractal have the following property. When both logarithmic coordinates of the plots drawn for different Δt's are reduced by the same factor $\log \Delta t$, the reduced plots of the distribution converge to a limit as $\Delta t \to 0$. This property can be turned around, and used to *define* the notion of multifractal (but one must realize that the convergence to the limit may be slow).

The reduced horizontal logarithmic coordinate is denoted by α, and will be seen to be a quantity called Holder exponent. The reduced vertical logarithmic coordinate corresponding to the limit will be denoted by $\rho(\alpha)$. It will be seen that it is negative for all α, except the value when $\rho(\alpha)$ reaches its maximum.

As has been first pointed out by Frisch and Parisi 1985, it is convenient to also introduce a quantity denoted by $f(\alpha) = \rho(\alpha) + 1$. The only case Frisch and Parisi have considered is the case $f(\alpha) \geq 0$. If so, one can interpret $f(\alpha)$ as being the fractal dimension of a suitable set. The replacement of $\rho(\alpha)$ by $f(\alpha)$ has virtues in some cases, but our feeling is that, fundamentally, it hides the nature of the multifractals.

Until the advance preprint of Frisch and Parisi was distributed in 1983, multifractals continued to develop only in the sense that the mathematical background was very much extended (e.g.. Kahane and Peyrière 1976). But they did not receive new applications, nor were they mentioned in *Physical Review Letters*. Their spread is a recent phenomenon, and most readers who have heard of them are likely to know presentations that follow the approach common to Frisch and Parisi and to Halsey et al. Unfortunately, the algebra of these presentations is needlessly complicated, artificial, and of limited applicability and the terminology of Halsey et al. hides the extremely simple and almost familiar nature of the underlying structure. We shall, therefore, adopt the notation of Halsey et al., but follow our original approach in the form into which it has lately developed.

3.3. SELF-SIMILAR MEASURES AND BEYOND

We need some definitions concerning measures $\mu(S)$. These $\mu(S)$ will be positive; therefore, again, those not familiar with measure can think of μ as being the probability of hitting the set S. The multifractal measures obtained as a result of multiplicative cascades are the closest analog among measures of the exact self-similar fractal sets. Recall that a fractal set is exactly self-similar, if it can be decomposed into parts, each of which is obtained from the whole by transformation variously called isotropic contraction or contracting similitude, C. Such a set if fully determined by a collection of contractions. For example, each third of a basic fractal called Sierpinski gasket is obtained from the whole by a contracting similitude of ratio $r = 1/2$. Starting with any triangle in a "prefractal collection of triangles," the interpolation of the shape itself continues without regard to the "past" construction steps.

Now suppose that a (positive) measure $\mu(P)$ is defined for each third of the gasket, for each third of a third etc.. When the part P' is obtained from the part P by the contracting transformation C, so that $P' = C(P)$, the conditional measure of P' in P is defined, just like a conditional probability, that is, by the ratio $\mu(P')/\mu(P)$ of the measure $\mu(P')$ to the conditioning measure $\mu(P)$. Now the idea of self-similarity for a measure is that the interpolation of the measure carried by a triangle is a prefract collection of triangles also continues without regard to the "past" steps. That is, as the parts contract, the measures they carry contract proportionately. To express this idea formally, take a second contracting transformation $\mathcal{L}$, and compare $\mu(P')/\mu(P)$ with $\mu[\mathcal{L}(P')]/\mu[\mathcal{L}(P)]$. If these conditional measures are identical, the measure μ will be called a *strictly self-similar multifractal*.

A random measure is called *statistically* self-similar if, given one or a finite collection of non overlapping parts $P_\gamma = C_\gamma(P)$, the distribution or the joint distribution of the quantities $\mu(P_\gamma)/\mu(P)$ depends only on the contractions C_γ.

Side remark. In a more general mathematical fractal set, the parts are obtained from the whole by transformations that are *non linear*. Examples where the contractions are in some sense *near linear* include the Julia sets of polynomial maps. The corresponding multifractals include the harmonic measures on these sets. Other examples of multifractal measures concern the limit sets of groups based upon inversions in circles (*FGN*, Chapters 18 and 20). The limit set itself a straight line, as in the example examined by Gutzwiller and Mandelbrot 1988. Finally, the "fat fractals" (new term for the fractals in *FGN*, Chapter 15) and the *Mandelbrot set* involve essentially non-linear transformations. As yet, there is no general agreement about which transformations are acceptable in defining the terms "fractal" and "multifractal."

3.4. THE BASIC NON RANDOM SELF-SIMILAR MULTIFRACTALS

3.4.1. *Basic background: the binomial multifractal measure.* To construct this measure, given m_0 satisfying $1/2 < m_0 < 1$ and $m_1 = 1 - m_0$, we spread mass over the halves of every dyadic interval, with the relative proportions m_0 and m_1. If $t = 0 \cdot \eta_1 \eta_2 \ldots \eta_k$ is the development of t in the binary base 2, and φ_0 are φ_1 the relative frequencies of 0's and 1's in the binary development of t, the binomial measure assigns to the dyadic interval $[dt] = [t, t + 2^{-k}]$ of length $dt = 2^{-k}$ the mass

$$\mu(dt) = m_0^{k\varphi_0} m_1^{k\varphi_1}.$$

Adapting the classical notion of Hölder exponent to apply to the interval $[dt]$, we write

$$\frac{\alpha = \log[\mu(dt)]}{\log(dt)} = -\varphi_0 \log_2 m_0 - \varphi_1 \log_2 m_1,$$

and $0 < \alpha_{\min} = -\log_2 m_0 \leq \alpha \leq \alpha_{\max} = -\log_2 m_1 < \infty$. (As already mentioned, the Hölder exponent has been given many new names. For example, it has been relabeled "dimension" by Hentschel and Procaccia 1983, or as "pointwise dimension," but the use of the term "dimension" *must* be reserved to sets.)

The number of intervals leading to φ_0 and φ_1 is $(k\varphi_0)!(k\varphi_1)!/k!$, giving the box fractal dimension

$$\delta = \log[(k\varphi_0)!(k\varphi_1)!/k!]/\log(dt).$$

For large k, the Stirling approximation yields

$$\delta = -\varphi_0 \log_2 \varphi_0 - \varphi_1 \log_2 \varphi_1.$$

Thus, α determines φ_0, hence $\delta = f(\alpha)$.

A theorem by Eggleston relates δ to the Hausdorff dimension (see Billingsley 1967).

3.4.2. *The multinomial measure.* To construct a multinomial measure of base $b > 2$, we require b masses m_β $(0 \le \beta \le b - 1)$. The b-adic intervals characterized by the frequencies φ_β of the digits β in the base-b development $0 \cdot \eta_1 \eta_2 \ldots \eta_k$ yield

$$\alpha = -\sum \varphi_\beta \log_b m_\beta \quad \text{and} \quad \delta = -\sum \varphi_\beta \log_b \varphi_\beta.$$

Now, the points (α, δ) cover a domain shown in black on Fig. 2.

Fig. 2: Rough idea of the domain of (α, δ) for a multinomial multifractal with $b = 4$. The upper boundary defines the function $f(\alpha)$. Here, all the m_β are different, $\alpha_{min} = \min(-\log_b m_\beta) > 0$, and $\alpha_{max} = \max(-\log_b m_\beta) < \infty$.

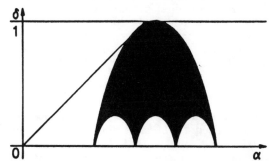

3.4.3. *The Lagrange multipliers argument, and the Legendre and inverse Legendre relations of the Gibbs theory.* The collections of b number φ_β's yielding the same α are dominated by the highest dimension term. This term maximizes $-\sum \varphi_\beta \log_b \varphi_\beta$, given $-\sum \varphi_\beta \log_b m_\beta = \alpha$, and $\sum \varphi_\beta = 1$. The classical method of Lagrange multipliers introduces a multiplier q, with $-\infty < q < \infty$, and yields

$$\varphi_\beta = \frac{b^{q \log_b m_\beta}}{\sum b^{q \log_b m_\beta}} = \frac{m_\beta^q}{\sum m_\beta^q}.$$

Define the quantity $\tau(q) = -\log_b \sum m_\beta^q$, which the mathematical statisticians call "cumulant generating function." In terms of $\tau(q)$, the Lagrange multipliers determine q and $f(\alpha)$ from α by

$$\alpha = -\sum \varphi_\beta \log_b m_\beta = -\frac{\partial}{\partial q} \log_b \sum m_\beta^q;$$

$$\max \delta = f(\alpha) = -\frac{\sum (q \log_b m_\beta - \log_b \sum m_\beta^q) m_\beta^q}{\sum m_\beta^q}.$$

That is,

$$\alpha = \frac{\partial \tau(q)}{\partial q} \quad \text{and} \quad f(\alpha) = q \frac{\partial \tau}{\partial q} - \tau = q\alpha - \tau.$$

Figure 2 is now replaced by its upper boundary, which is the graph of a function $f(\alpha)$. Clearly, the α and δ of the multinomial can satisfy $\alpha > 0$ and $\delta \ge 0$, hence

$f(\alpha) \geq 0$, $\alpha_{\min} > 0$, $f(\alpha_{\min}) \geq 0$ and $f'(\alpha_{\min}) = \infty$, and $\alpha_{\max} < \infty$, $f(\alpha_{\max}) \geq 0$ and $f'(\alpha_{\max}) = -\infty$. Multifractals that are not multinomial, yet possess these properties we call "pseudo-multinomial."

Formally, q = inverse temperature, τ = Gibbs free energy, and f = entropy.

3.4.4. *The term "multifractal formalism" and the question of actual computation.*
The equations $\alpha = \tau'$ and $f = q\alpha - \tau$ are the "multifractal formalism." Frisch and Parisi 1985 and Halsey et al. 1986 obtain the same result via a steepest-descent argument, which experts will recognize as identical to the Darwin-Fowler justification of the Lagrange multipliers procedure. This is *not* the right way to proceed, just as *no one* will think of teaching thermodynamics by describing Darwin Fowler without first presenting the Lagrange multipliers. Therefore, Sec. 3.4.3 has taken the path towards the same formalism that involves the least effort and the fullest understanding. Section 3.5 describes the next simplest generalization.

A considerable literature has developed around ways of using this formalism. This literature is, obviously, unaffected by a change in the foundations.

3.5. THE RANDOM 1974 MULTIFRACTAL MEASURES

3.5.1. *Generalization of the multiplicative measures.*
Now we proceed to the exactly renormalized "1974 multiplicative multifractals" introduced in Mandelbrot 1974. First, let us observe that the $f(\alpha)$ of a multinomial measure is unchanged if the indexes of the masses m_β are shuffled at random before each stage of the cascade that distributes the mass within each of the kth level intervals into masses within the $(k+1)^{\text{st}}$ level intervals. (Note that one must not shuffle mass between *all* the kth level intervals!) Next, let us generalize from multifractals in one dimensional t to multifractals in E-dimension $\mathbf{X}$. To do so, suppose that $b = B^E$, with positive integers B and E. With no change in the algebra, the multinomial measure with random weight assignment can be interpreted as spread on cells in a E-dimensional cube of base B. The weights in the B cells along a $1-d$ cut through this measure are a random selection of B weights among the possible weights m_β. To achieve an idea of these cuts, one may imagine them on the two-dimensional sample measures shown in Meakin; 1988, Figs. 4.4 and 4.5. Now we get to a centrally important point. When $b/B \gg 1$, the weights in space still add to 1, but these constraints hardly matter. The B weights along a cut are for all practical purposes statistically independent. This fact motivates our "canonical" measure, which assumes them to be independent. Independence has many vary important effects. Now the masses are values of a random multiplier M that can take either of b renormalized values $m'_\beta = m_\beta B^{E-1}$, with the probability $1/b$ for each value.

Now we are finished with motivating preliminaries. The processes we are really interested in results from a cost-free generalization, wherein the mases are random variables that satisfy only $M \geq 0$ and $\langle M \rangle = 1/B$.

We seek the mass $\mu(dx)$ in the B-adic interval of length B^{-k} starting at $x = 0 \cdot \eta_1 \eta_2 \ldots \eta_k$. In the multinomial case, this was a product of k "low frequency" multipliers, one for each cascade stage of frequency $< B^k$. The same is true for a general multiplicative process in space. Because of exact renormalizability, the measure in a space cell of side B^{-k} only depends upon the k first cascade stages. Along a cut, however, the measure is *not* exactly renormalizable. We also need an essential additional "high frequency" factor $\Omega(\eta_1, \ldots \eta_k)$, which is due to stages of frequency $> B^k$. Mandelbrot 1974 has shown that Ω is the mass in $[0,1]$ after a canonical cascade has continued over infinitely many stages. Thus,

$$\mu(dx) = \Omega(\eta_1, \ldots \eta_k) M(\eta_1) M(\eta_1, \eta_2) \ldots M(\eta_1, \ldots \eta_k) \ldots.$$

Here, the successive M are identically distributed and independent, and so are the random factors Ω $(\eta_1 \ldots \eta_k)$. Writing

$$\alpha_L = -(1/k) [\log_B M(\eta_1) + \log_B M(\eta_1, \eta_2) \ldots]; \qquad \alpha_H = -(1/k) \log_B \Omega,$$

yields $\alpha = \alpha_L + \alpha_H$. We first examine α_L, then α_H.

The term α_H is negligible for $k \to \infty$, and the term α_L is the average of k independent random variables.

Scientists tend to believe that they know enough about the behavior of such sums, and that "loose" physical arguments suffice. Here, however, what they know is not relevant and "physical" arguments must be replaced by careful probability theory; see Sects. 3.5.3 and 3.5.4. In order to understand better the general results described below in Sect. 3.5.3., let us begin by explicit special examples.

3.5.2. *Examples of sums of independent random variables, for which the probability distribution is known in analytically closed form for both the individual addends and their sum.* There are several such examples. In each case, we denote the density by $p(x)$ for each addend, by $p_k(x)$ for the sum k addends, by $p_k(k\alpha))$ for the average of k addends. We also consider the quantity $[p_k(k\alpha)]^{1/k}$ or $(1/k) \log p_k(k\alpha)$, which amounts to an alternative unfamiliar renormalization of the distribution of the average α.

(A) The Gaussian density $p(x) = (\pi\sigma)^{-1/2} \exp(-x^2/2\sigma^2)$. Here $[p_k(k\alpha)]^{1/k} = (\pi\sigma^2 k^2)^{1/2k} \exp(-\alpha^2/2\sigma^2)$. Thus, $(1/k) \log p_k(k\alpha) \to \exp(-\alpha^2/2\sigma^2) = \rho(\alpha)$.

(B) The Poisson distribution of mean δ, that is, $p(x) = e^{-b}\delta^x/x!$ Here $p_k(x) = e^{-k\delta}(k\delta)^x/x!$ By the Stirling approximation, $(1/k) \log p_k(k\alpha) \to -\delta + \alpha(\log \delta a/\alpha) = \rho(\alpha)$.

(C) The Gamma density of parameter δ, that is $p(x) = x^{\delta-1} e^{-x}/\Gamma(\delta)$. The sum of two Gammas of respective parameters δ' and δ'' is a Gamma of parameter $\delta' + \delta''$. There, $p_k(x) = x^{k\delta-1} e^{-k}/\Gamma(k\delta)$. By the Stirling approximation, $(1/k) \log p_k(k\alpha) \to \delta \log(\alpha/\delta) - \alpha + \delta = \rho(\alpha)$.

(D) The binomial distribution. The argument in Section 3.4.1 can be reinterpreted as showing that for the average of k binomial quantities, $(1/k) \log p_k(k\alpha) \to \rho(\alpha) = f(\alpha) - 1$, where $f(\alpha)$ is the "entropy" function described in 3.4.1.

Observations about the examples. In each example, $\rho(\alpha) \leq 0$, with equality for one value of α. However, before the asymptotic range is reached, each of the above quantities $(1/k) \log p_k(k\alpha)$ is > 0 over a range of α's near the maximum of $\rho(\alpha)$. That is, direct estimates from data collected for a finite k overestimate $\rho(\alpha)$ near its maximum. In the binomial case, the overestimate remains visible evern when $k = 56$.

Conclusion from the examples. To form the quantity $(1/k) \log p_k(k\alpha)$ is to renormalize the average in a "new" way that greatly accentuates the low probability "tails," and de-emphasizes the central "bell." In the examples open to explicit calculation, we have found that the newly renormalized average still converges to a limit, and also that all the various limits differ from one another; e.g., the limits for the non-Gaussian cases differ from the limit relative to the Gaussian case. In the following Sect. 3.5.4., this surprising result will be shown to hold very generally. It is the second most important probabilistic fact underlying the multifractals (the most important being the behavior of moments along low dimensional cuts; Sects. 3.5.6 and 3.5.8).

3.5.3. Generalization of the multifractal formalism by application of the Cramèr limit theorem.

The above examples are very special cases of the "large deviations theorems" of H. Cramèr, which happen to be available "off-the-shelf," a pleasant surprise; see Book 1984, Chernoff 1952, Daniels 1954, 1987. They prove that, as $k \to \infty$,

$(1/k)\log_B$ (probability density of α_L) converges to a limit, to be denoted as $\rho(\alpha)$.

The quantities $(1/k)\log_B$ (probability of $\alpha_L > \alpha > \langle\alpha\rangle$) (resp., of $\alpha_L < \alpha < \langle\alpha\rangle$) converge to the same limit. It is easily shown that

$$f(\alpha) = \rho(\alpha) + \max f(\alpha) = \rho(\alpha) + \text{dimension of the measure's support.}$$

It is a noteworthy fact that in the generalized Gibbs formalism resulting from the Cramèr theorem, different M's yield different $f(\alpha)$'s, and conversely.

Obviously, Cramèr-type theorems extend to the case when the factors M are weakly dependent or weakly non-identical.

The special "pseudo-multinomial" situation requires $\Pr\{M_{\max}\} \geq B^{-1}$ and $\Pr\{M_{\min}\} \geq B^{-1}$. But a first feature of our 1974 multiplicative measures is that they allow $\alpha_{\min} = 0$ and $\alpha_{\max} = \infty$ and $f(\alpha_{\min}) = \log_B[\Pr\{M_{\max}\}] + 1 < 0$ and $f(\alpha_{\max}) < 0$.

3.5.4. Comments concerning lognormality.

Section 3.5.3 ought to surprise those many readers who know the literature to the effect that $\log_b M(\eta_1) + \log_b M(eta_1, \eta_2) + \ldots$ is asymptotically Gaussian, so that α is asymptotically lognormal. These assertions result from the application of a different renormalization, one that leads to the classical central limit theorem. It is indeed correct that the central limit theorem yields information about the multiplicative multifractals. Also, this information is universal, but it only implies that $\rho(\alpha)$ and $f(\alpha)$ are parabolic in the central bell near their maximum. Away from the maximum, the behavior of $\rho(\alpha)$ and $f(\alpha)$ is not universal.

There is a seeming paradox here. On the one hand, the probability outside of the central bell tends to 0 as $k \to \infty$, meaning that the tails become thoroughly insignificant. In the limit $k \to \infty$, the most probable value, the expectation and the other usual parameter of location all converge to each other. On the other hand, those "negligibly" few values in the tails are so huge that their contributions to all the moments of order $q \neq 0$, and to $\tau(q)$, are predominant. Moreover, the moments and $\tau(q)$ depend on the exact $f(\alpha)$, that is, are non-universal.

In any event, the functions $(p)\alpha$ and $f(\alpha)$ are not like those from the lognormal M, except if the M are lognormal. Furthermore, lognormal M's require very special precautions.

3.5.5. *The meaning of negative values for the dimensions* $f(\alpha)$. When $f(\alpha)$ is viewed as a fractal dimension, $f(\alpha) < 0$ is impossible. However, numerous authors have shown that restricting multifractals to $f(\alpha) \geq 0$ leads to self contradiction (Cates and Deutsch 1987), or is otherwise not acceptable (Fourcade et al. 1987, Fourcade and Tremblay 1987).

From the B^k data in a single sample of the cut measure, one can estimate $f(\alpha)$ if the probability of α is $> B^{-k}$, which means $f(\alpha) \geq 0$. The range of the α's is not expanded by increasing B, because the population from which we sample also depends upon B. This invalidates the "intuitive ergodic" belief that the sample value necessarily converges to the population expectation as the sample increases. The α's with $f(\alpha) \geq 0$ will be called *manifest*, while the α's with $f(\alpha) < 0$ will be called *latent* (= present but hidden). For negative latent dimension in a different context, see B. Mandelbrot 1984. It is quite obvious (after the fact) than "generically" one should expect the overall original measure in $E - d$ to include cells of higher or lower density, "hotter" or "colder," than the hottest or coldest cells along the 1-dimensional cut.

To measure for $f(\alpha)$ for latent α's one may combine data from many independent samples of size B^k. A far more enlightening procedure is to examine a higher-dimensional cut from the same measures.

Note that the "full" $\tau(q)$ and $D_q = \tau(q)/(q-1)$ are analytic functions, but the "manifest" $\tau(q)$ or D_q evaluated on the basis of manifest α's are incorrect (truncated and biased) for large $|q|$'s.

3.5.6. *The prefactors* $\langle \Omega^q \rangle$ *and the critical* q_{crit}. Now let us focus on Ω. It can be shown that $\sum M_\beta \Omega_\beta \equiv \Omega$ ($\equiv$ means "identical in distribution"), so that Ω is the fixed point of operation of randomly weighted averaging using as weights the random quantities M. We have already noted taht, on the logarithmic scale of the Hölder α, the factor $\alpha_H = -(1/k)\log_B \Omega$ is asymptotically negligible as $k \to \infty$.

But in the scale of μ, the inclusion of Ω yields $\langle \mu^q(dx) \rangle = \langle \Omega^q \rangle (dx)^{\tau(q)+1}$, therefore brings in new moment prefactors $\langle \Omega^q \rangle$. These prefactors, reflecting high frequency effects from scales $< B^{-k}$, not only are non-trival, but have far reaching effects, because *they need not be finite*. Mandelbrot 1974 shows (and Kahane and Peyrière 1976 then proves rigorously) that $\langle \Omega^q \rangle < \infty$ holds if $q < q_{\text{crit}}$, where q_{crit} is the solution of the equation $\tau(q) = D_q = 0$. When $q > q_{\text{crit}}$, then $\tau(q) < 0$ and $D_q < 0$. Without the divergent $\langle \Omega^q \rangle$, the inequality $\tau(q) < 0$ would have meant that $\langle \mu^h(dx) \rangle$ actually *decreases* as dx increases, an extreme anomaly, but one that vanishes if one recalls that $\langle \Omega^q \rangle = \infty$.

Furthermore, since the model allows $M_{\text{max}} > 1$, one may have $\alpha_{L\text{min}} < 0$ (with $f(\alpha_{L\text{min}}) < 0$), with the implication that a subinterval B^{-k-1} can include more mass than its embedding interval B^{-k}. The paradox of these α's, to be called *virtual*, can be shown to vanish if one takes account of the moments of Ω.

Mandelbrot 1974 conjectures (J. Guivarc'h has proven this fact in a private communication) that $\Pr(\Omega > \omega) \sim \omega^{-q_{\text{crit}}}$ and the same behavior holds for $\mu(dx)$ if dx is well above the inner cutoff (if there is a cutoff > 0). This prediction of the model is extremely important to data analysis, but it only holds when the cascade proceeds down to scales well below dx. In the case of turbulence, this requires an extremely high Reynolds number R, and indeed there are reports of ill-behaved sample moments of high q. But for the lower R reached in the laboratory, Ω is effectively tuncated at a level increasing with R, which avoids divergent moments. Experimentalists are urged to perform this comparison.

3.5.7. *Sequences of cuts with different B's but the same multipliers.* To renormalize once for all B's, it suffices to consider a cascade that multiplies the densities (instead of the masses) by a fixed *weight* $W = bM$, satisfying $\langle W \rangle = 1$. A 1974 measure in E-dimensions, with weight W and base b, if intersected by a fractal of dimension $D = \log B / \log b < E$, yields a measure with the same W and base B. In fact, by using random cuts, one may view B as continuously variable.

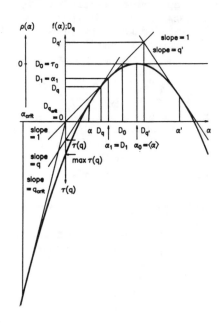

Fig. 3: Enhanced multifractal diagram. One ordinate scale shows $\rho(\alpha) = \lim_{k \to \infty}(1/k)\log_B$ (probability) versus the Hölder $\alpha = -(1/k)\log_b$ (measure). A second ordinate scale shows the function $f(\alpha)$. It is well-known that $q = f'(\alpha)$ and $-\tau(q)$ is the ordinate of the intercept of the tangent of slope q by the vertical axis. Let us add the observation that D_q is the ordinate of the intercept of this tangent by the main bisector of the axes.

Now consider for different B's or D's the enhanced graphs showing $f(\alpha)$, $\tau(q)$ and D_q (as in Fig. 3), and translate the origins to the points of coordinates $\alpha = \alpha_0 = \langle \alpha \rangle$ and $f(\alpha) = D_0$. One can show that the translated graphs *coincide* if one includes the manifest, the latent and the virtual α's. As a corollary, increasing B eventually changes latent α's into manifest α's; negative $f(\alpha)$'s into positive $f(\alpha)$'s one can estimate through the B^k data from a single sample. Also, increasing B eventually changes virtual α's into latent, then into manifest ones. Decreasing B eventually reaches $D = D_0$, beyond which the measure becomes degenerate (Mandelbrot 1974, Kahane and Peyrère 1976).

As corollary, the manifest portion of $f(\alpha)$ fails to represent a 1974 multifractal in an intrinsic fashion. But a full description of the effects of the same W in different D_0's is provided in the context of Cramèr's limit theorem by the intrinsic quantities $\alpha - \langle \alpha \rangle$, $\rho(\alpha)$, and $D_q - D_0$.

3.5.8. *The D_q as critical dimensions.* The last task is to explain why, for 1974 measures, D_q is a *critical* dimension for the moments of order q. Mandelbrot 1974. Later, the D_q were written down, without any motivation, in Hentschel and Procaccia 1983 (to call $D - q$ a dimension can make no geometric sense when $D_q > E$ for low enough q's.) The key to our motivation for D_q, again, resides in the study of low dimensional cuts of higher dimensional multifractals with the same $f(\alpha)$.

The discussion is simplest when $D_{\max} < \infty$. The main step is to embed the cascade in a space of dimension $E > D_{\max}$. We know that the equation $D_q = 0$ defines the q_{crit} relative to the property that $\langle \mu^q(dx) \rangle < \infty$ and that the variability of the Ω's (which are independent in a sample) overwhelms the variability of the low frequency component of $\mu(dx)$. Therefore, the law of large numbers implies that along the cut with $D > D_q$, the qth sample moment is the sum of many contributing terms, each of which is negligible in relative value. Along a cut with $D < D_q$, to the contrary, $\tau(q) < 0$, which can be shown to imply that, while $\langle \mu^q \rangle = \infty$, all the percentiles of the sample q^{th} moment tend to 0 with dt. Hence, the sample moment is very small with a very high probability, and instances when it is *not* very small occur because of the exceptional contribution of a single huge sample addend.

Thus, the set of points that contribute to the sample q^{th} moment of μ is hit by spaces of dimension $> D_q$, and missed by spaces of dimension $< D_q$.

Many thanks to Jens Feder and Tamas Vicsek for pointing out rough and/or obscure spots in earlier versions of this text.

1. Billingsley, P., *Ergodic Theory and Information* (J. Wiley, New York, 1967), p. 139.
2. Book, S. A., in *Encyclopedia of Statistical Sciences* 4, 476 (1984).
3. Cates, M. E. and Deutsch, J. M., *Phys. Rev. A* 35, 4907 (1987).
4. Chernoff, H., *Ann. Math. Stat.* 23, 493 (1952).
5. Daniels, H. E., *Ann. Math. Stat.* 25, 631 (1954).
6. Daniels, H. E., *International Statistical Review* 55, 37 (1987).
7. Fourcade, B., Breton, P. and Tremblay, A.-M.S., *Phys. Rev. B* 36, 8925 (1987).
8. Fourcade, B. and Tremblay, A.-M.S., *Phys. Rev. A* 36, 2352 (1987).
9. Frisch, U. and Parisi, G., in *Turbulence and Predictability in Geophysical Fluid Dynamics and Climate Dynamics*, International School of Physics "Enrico Fermi," Course 88, edited by M. Ghil et. al. (North-Holland, Amsterdam, 1985), p. 84.
10. Grassberger, P., *Phys. Lett.* 97A, 227 (1983).
11. Gutzwiller, M. C. and Mandelbrot, B. B., *Phys. Rev. Lett.* 60, 673 (1988).
12. Halsey, T. C., Jensen, M. H., Kadanoff, L. P., Procaccia, I. and Shraiman, B. I., *Phys. Rev.* A33, 1141 (1986).
13. Hentschel, H. G. E. and Procaccia, I., *Physica (Utrecht)* 8D, 435 (1983).
14. Mandelbrot, B. B., in *Statistical Models and Turbulence* (Lecture Notes in Physics, Vol. 12), Proc. Symp., La Jolla, Calif., M. Rosenblatt and C. Van Atta, eds. (Springer-Verlag, New York, 1972), p. 333.
15. Mandelbrot, B. B., *J. Fluid Mech.* 62, 331 (1974); also *Comptes Rendus* 278A, 289, 355.
16. Mandelbrot, B. B., *The Fractal Geometry of Nature* (New York, W. H. Freeman, 1982).
17. Mandelbrot, B. B., *J. Stat. Phys.* 34, 895 (1984).
18. Mandelbrot, B. B., *Fractals and Multifractals: Noise, Turbulence and Galaxies*, Selecta, Vol. 1 (New York, Springer, in press).
19. Meakin, P., in *Phase Transitions and Critical Phenomena* 12, eds. C. Domb and J. L. Lebowitz (Academic Press, London, 1988), p. 335.
20. Meneveau, C. and Sreenivasan, K. R., *Phys. Rev. Lett.* 59, 1424 (1987).
21. Prasad, R. R., Meneveau, C. and Sreenivasan, K. R., *Phys. Rev. Lett.* 61, 74 (1988).

MULTIFRACTALS IN CONVECTION AND AGGREGATION

MOGENS H. JENSEN
NORDITA
DK-2100 Copenhagen, Denmark

1. Introduction

Patterns found in convection and aggregation are often multifractals and can be characterized by a continuum of exponents. This differs from critical phenomena where typically a finite number of relevant exponents is needed. The set of exponents for a multifractal is conveniently presented as an $f(\alpha)$ spectrum, i.e., a spectrum of pointwise dimensions. We shall illustrate the formalism by a Rayleigh-Benard convection experiment developed by Libchaber and co-workers. The convective state exhibits an unstable mode. This mode is coupled to an external oscillation and it is possible to drive the system towards the onset of chaos via quasiperiodicity, by tuning the ratio between the two frequencies to an irrational number. At the critical point an attractor is extracted from a stroboscopic temperature signal and the corresponding $f(\alpha)$ spectrum is computed. A similar scenario is found in the theory of dynamical systems with two characteristic frequencies. The simplest such systems are discrete mappings expressed in terms of angular variables, circle maps. For circle maps one can tune the ratio of the frequencies to the same irrational value as in the convection experiment. At the critical point where chaos sets in, an attractor is extracted. The associated $f(\alpha)$ spectrum is calculated and a renormalization group analysis establishes it to be universal. This is a theory of no adjustable parameters which can be compared to the experiment. Good agreement is found providing a strong evidence that the convection experiment and circle maps belong to the same universality class.

Patterns found in aggregation can be modeled by Julia sets of conformal mappings. From the dynamics on the Julia sets, the $f(\alpha)$ can easily be computed. We find that the tip structure is stable and is basically "universal" for all types of growth patterns. The fjords between the arms of the aggregate are on the other hand largely screened form the diffusive field, leaving the possibility of a complete screening of some fjords in probabilistic diffusion. This causes a straight part in the $f(\alpha)$ spectrum and can be interpreted as a phase transition in the underlying thermodynamic formalism.

2. Mode-Locking and Quasiperiodicity

The behavior of a system exhibiting two coupled frequencies is very rich. We consider a situation where the system itself exhibits an internal frequency ω_1 (for example an unstable mode), and a second oscillation of frequency ω_2 is introduced from outside. When the two frequencies couple non-linearly the results will manifest themselves as resonances every time a harmonic of the one frequency is sufficiently close to a harmonic of the other frequency

$$n\omega_1 \simeq m\omega_2 \qquad or \qquad \frac{\omega_1}{\omega_2} \simeq \frac{m}{n}. \tag{1}$$

These resonances remain stable even if some of the parameters (as ω_2 or its amplitude A_2) are slightly changed. A typical phase diagram is shown in Fig. 1. The K-parameter is related to the strength of the coupling between the frequencies and the Ω-parameter is proportional to the external frequency ω_2. Any resonance, with ω_1/ω_2 rational, fills up a finite area and forms a tongue which widens as K is increased. The figure is called an "Arnold-tongue diagram" for systems with two frequencies.[2] The diagram in Fig. 1 has been constructed from very simple dynamical systems, circle maps,[1-5] which are mappings in terms of an angular variable θ_n on the form

$$\theta_{n+1} = \theta_n + \Omega - Kg(\theta_n) \tag{2}$$

where K and Ω are constants and g is a periodic function. This mapping can be found from the dynamics on a cut through the two-torus generated by the two frequencies ω_1, ω_2.[5] The cut of a torus results in a curve which is equivalent to a circle, therefore the term "circle-map." As the periodic function we shall use $g(\theta_n) = 1/2\pi \sin(2\pi\theta_n)$. From the cut on the torus it is easy to see that the ratio between the frequencies, ω_1/ω_2, is equal to the winding number[2,3]

$$W = \lim_{n \to \infty} \frac{\theta_n - \theta_0}{n}. \tag{3}$$

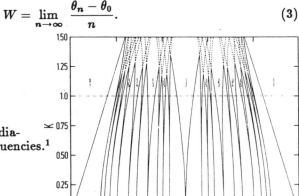

Fig. 1: An "Arnold-tongue" diagram for a system with two frequencies.[1]

In Fig. 1 we have shown the parameter space, K as a function of Ω, for the circle map Eq. (2). The lines are the borders of the Arnold-tongues where W is rational. Any rational number will appear as a tongue in this diagram. As K increases beyond 1 the tongues will start to overlap, which give rise to hysteresis and chaotic behavior. Thus $K=1$ is the critical line for which chaos sets in. The corresponding diagram for the experiment, described in detail in the next chapter, is shown in Fig. 2. Note the tongues and the similarity with Fig. 1.

Until now we have discussed the behavior of rational winding numbers. What happens if the winding number is irrational? In that case the frequencies do not resonate and the motion is quasiperiodic (or incommensurate). The parameter values in the $K - \Omega$ diagram for which W is a specific irrational is a line. As K increases, these lines are squeezed more and more between the tongues and finally, at $K=1$, the quasiperiodic motion is no longer possible and becomes chaotic. This outlines a route to chaos via quasiperiodicity. In this way the $K=1$ line is the "reverse" to the real axis: On the real axis the irrationals have full measure and the

Fig. 2: The phase-diagram, A_2 versus ω_2, for the convection experiment on mercury. Note the Arnold tongues and the similarity to Fig. 1. The inserts are magnifications around two irrationals, the golden mean and the silver mean.[6,7]

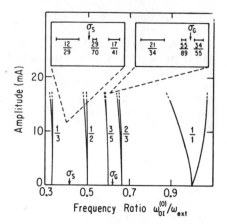

rationals zero measure. On the $K=1$ line the rationals have broadened to the full measure and the irrationals are of zero measure. This is indicated in Fig. 3 where W is shown as a function of Ω at $K=1$, giving a staircase-structure with a step for every rational number. The interesting point is that the staircase is seen again on any small scale and thus exhibits fractal properties.[1] Actually, it was specifically found that measure of the rational steps sums to one showing that the irrationals live on a Cantor set (of zero measure). This set has a universal dimension of $D=0.8700$.[1] Within a few percent, this value has been confirmed in many experiments, including the experimental results shown in Fig. 2.[6] In the following we shall focus on a particular irrational number, the golden mean $w^\star = \sqrt{5} - 1/2$. The reason for choosing this point is that the number has a particular nice continued fraction expansion (consisting of all 1's) so it is possible to do a nice renormalization group calculation.[5]

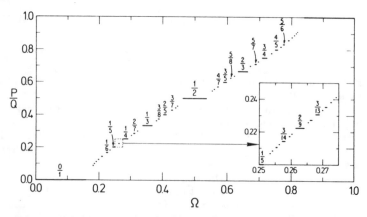

Fig. 3: The winding number, W, versus Ω for the circle map Eq. (2) for $K=1$. There is a step for each rational number.[1]

3. The Experimental Set-Up

The Rayleigh-Benard experiment we study has been developed by Stavans, Heslot, and Libchaber.[6,7] The experimental cell is shown in Fig. 4a and the general set-up in Fig. 4b. The cell has a height of 0.7 cm and a width of 1.4 cm such that precisely two convective rolls are present (Fig. 4a). The cell is filled with fluid mercury which has a very low Prandl number.

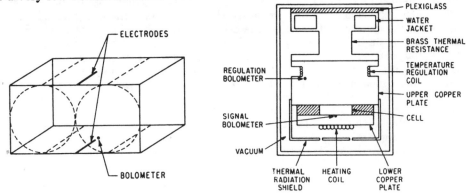

Fig. 4: (a): A schematic diagram of the Rayleigh-Benard cell. The height of the cell is 0.7 cm. The ac current is passed through the electrodes and the experimental signal is taken from the bottom bolometer. (b) The experimental set-up containing the cell in Fig. 4a. The set-up is discussed in details in Refs. 6 and 7.

In the experiment, only the temperature of the upper plate is regulated. The temperature is controlled by the regulation bolometer near the heating coil (Fig. 4b). The temperature stability is better than 10^{-5} °C. A temperature signal of the flow is measured by the bolometer at the bottom plate.

The convection starts when the temperature difference between the top and the bottom is about 3.0°C; two convective rolls appear. When the temperature difference is increased further, the convective rolls become unstable against an oscillatory instability. The wavy instability travels along the rolls and creates an oscillatory temperature variation at the bottom bolometer. Thus it is easy to extract the internal frequency $\omega_{OI} = \omega_1$ of this oscillation. The typical period of the oscillation is a few seconds.

The second frequency is introduced into the system from outside. Two thin line electrodes are attached to the top and bottom plates. The external frequency is generated by rectangular ac current pulses whose width is about 1/10 of the period of the oscillatory instability. A typical value of the pulse amplitude is 20 mA. The whole system is embedded in a magnetic field of 200 G parallel to the convective rolls. As mercury is conductive, a Lorentz force acts on the fluid horizontally and transverse to the rolls axes. The action of the Lorentz force is to bend the convective rolls, thereby creating an ac vertical vorticity component in the flow. By varying the amplitude and the frequency of the injected current, Stavans et al[6,7] mapped out a number of Arnold tongues, as shown in Fig. 2. As expected the width of the tongue decreases rapidly as the denominator of the rational number increases.

Still, it was possible to observe tongues with denominators up to 200. This is of importance because in the following we shall in particular focus at the behavior at golden mean winding number which is indicated in the magnified region in Fig. 2.

To approach the golden mean the experimentalists follow a series of rational tongues, with winding numbers w_n, converging in the limit to $w^\star$. These rationals are written in terms of Fibonacci numbers F_n, $w_n = F_{n-1}/F_n$, where $F_{n+1} = F_n + F_{n-1}$, $F_0=0$, $F_1=1$.[4,5] In this way it is possible to approach $w^\star$ to an accuracy of 10^{-5} in the experiment. Simultaneously, the amplitude of the ac signal is increased. We want the experiment to be at the critical point where chaos sets in, this point is located by finding the place where broad band noise in the Fourier spectrum of the signal, sets in.[6] For a specific value of the external frequency and amplitude, the experiment is now at criticality with golden mean winding number. From the temperature measured in units of the external period by the bottom bolometer, we obtain a strobed signal T_n. We need to use the external period as the sampling time if we wish to compare the results with results from circle maps. Fig. 5 shows the attractor with T_{n+1} plotted against T_n. We observe an interesting structure with a large variation in density. Our goal is to study the scaling properties of this attractor in detail, i.e. its multifractal properties which is done in Sec. 5.

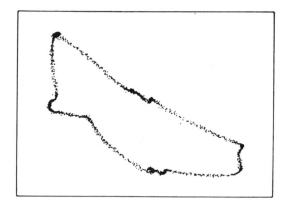

Fig. 5: The experimental attractor with golden mean winding number. Plotted is a phase space in the strobed temperature signal, T_{n+1} versus T_n. Note the variation in density along the attractor.

Similar to the experiment, we construct an attractor for the circle map model. The winding number should be $w^\star$ at the point where chaos sets in. We already know that $K=1$ for criticality and we have to calculate Ω such the W, Eq. (3), is equal to $w^\star$. Again we approach the golden mean with Fibonacci tongues[4] and finally find the value $\Omega_{gm}=0.6066....$ The corresponding attractor with the angular variable, θ_n, plotted on the circle is shown in Fig. 6. Again there is a large variation in density, a variation we wish to quantify through the multifractal formalism and compare the results with the results obtained from analyzing the experimental attractor, Fig. 5.

4. Multifractal Formalism

A multifractal is composed of infinitely many interwoven "single" fractals. The nature of such behavior was first discussed by Mandelbrot[8] and has recently been intensively studied by Parisi and co-workers[9] Halsey et al[10] and many others.[11-15]

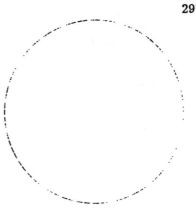

Fig. 6: The "theoretical" attractor of the circle map, Eq. (2), with golden mean winding number and $K=1$. Note also here the variation in density along the attractor.

One dimension is not sufficient for characterization of a multifractal. More information is given by a continuum of dimensions, the Renyi dimensions D_q,[16] which were introduced in the context of dynamical systems by Hentschel and Procaccia.[17] A similar characterization is obtained by the spectrum scaling indices, the $f(\alpha)$ spectrum, defined by Halsey et al.[10] The D_q-function and the $f(\alpha)$ spectrum are related by a Legendre transformation. We shall outline this formalism and apply it on the attractors shown in the previous Chapter.

A starting point is a probability density, or a fractal measure, on a fractal set. In the case of the attractors in Figs.5,6 the measure is given by the density of the points. The set is partitioned into N pieces of sizes ℓ_i and the probability of the i'th piece is denoted p_i. Then a partition sum is constructed[10]

$$\Gamma(q,\tau) = \sum_{i=1}^{N} \frac{p_i^q}{\ell_i^\tau}. \tag{4}$$

As $l = \max_i \ell_i \to 0$, three possibilities may occur. If $\tau > \tau(q)$ then $\Gamma(q,\tau)$ diverges to infinity; if $\tau < \tau(q)$ then $\Gamma(q,\tau)$ converges to zero. At $\tau = \tau(q)$ the partition sum will tend to a constant thus defining a functional relation between τ and q. $\tau(q)$ is related to the generalized dimensions by $\tau(q) = (q-1)D_q$. Next we assume that the density in the i'th piece scales like a power law in the $l \to 0$ limit

$$p_i \sim l^{\alpha_i}. \tag{5}$$

Eq. (5) defines the pointwise dimension, α_i, for any point of the set. Let us say that the pointwise dimension of a specific point on the set is equal to α_1. This value may be found also in many other points. Actually, α_1 is found on a subset of the whole set of a dimension $f(\alpha_1)$.[10] Therefore, this set with dimension $f(\alpha_1)$ is one the of the interwoven subfractals. Similarly, another value of the scaling exponent, α_2 say, is found on a subset of dimension $f(\alpha_2)$ and so on. In general α assumes values over an interval and a continuous $f(\alpha)$ spectrum is defined on this interval. We now transform the partition sum Eq. (4) into an integral in α. As a typical length scale we use $l = \max_i \ell_i$. The number of times α assumes a value in the interval $[\alpha', \alpha' + d\alpha']$ is then

$$d\alpha'\rho(\alpha')l^{-f(\alpha')}. \tag{6}$$

Inserting this and the scaling ansatz Eq. (5) into Eq. (4) we obtain

$$\Gamma(q, \tau) = l^{-\tau} \int d\alpha' \rho(\alpha') l^{q\alpha' - f(\alpha')}. \tag{7}$$

In the limit $l \to 0$, the minimal value of the exponent to l under the integral will dominate so we perform a saddle-point approximation

$$\frac{d}{d\alpha'}[q\alpha' - f(\alpha')]\,|_{\alpha'=\alpha(q)} = 0 . \tag{8}$$

This leads to the following Legendre transformation which is used to calculate the $f(\alpha)$ spectrum[10]:

$$\frac{d\tau}{dq} = \alpha , \quad \tau(q) = \alpha q - f \tag{9}$$

$$\frac{df}{d\alpha} = q , \quad \frac{d^2 f}{d\alpha^2} < 0.$$

We observe that the $f(\alpha)$ spectrum is convex and that the slope at each point is q. As $q \to \infty$ the largest p_i (i.e. the most concentrated part of the multifractal) dominate the partition sum Eq. (4). This corresponds to the place where the $f(\alpha)$ curve vanishes with infinite slope (if there is no phase transition, see §7) which is at its left-most part for the minimum α value. As $q \to -\infty$ the smallest p_i dominate (i.e. the least concentrated part) and at the corresponding α-value the right-most part of the $f(\alpha)$ curve vanishes with (negative) infinite slope.

The simplest multifractal one can imagine is a two-scale Cantor set, shown in Fig. 7a. On the second level the partition function for this set is very simple

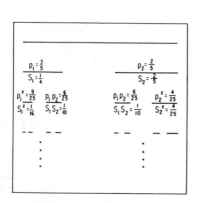
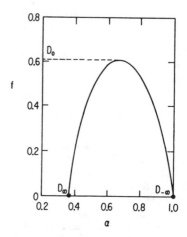

Fig. 7: (a) A two-scale Cantor set constructed from two length scales s_1, s_2 and two values of the measure, p_1, p_2 (note that in the formalism, $s_i \sim \ell_i$). (b) The corresponding $f(\alpha)$ spectrum (Ref. 10).

$$\Gamma = \left(\frac{p_1^q}{\ell_1^\tau} + \frac{p_2^q}{\ell_2^\tau} \right), \tag{10}$$

and will on higher levels follow a binomial expansion.[10] We calculate $\tau(q)$ and obtain the $f(\alpha)$ spectrum shown in Fig. 7b. The spectrum has the expected convex form, is smooth and continuous and its maximum point is the dimension D_0. The minimum value of α is $\alpha_{min} = \log p_1 / \log \ell_1$ and the maximum value is $\alpha_{max} = \log p_2 / \log \ell_2$.

5. Universality of $f(\alpha)$ Spectra: Theory and Experiment

The first non-trivial set we analyze is the attractor of the circle map with golden mean winding number, Fig. 6. The attractor should be partitioned into small boxes of sizes ℓ_i and the measures p_i in every box must be calculated. In practice we assign the same probability to each box, $p_i = \frac{1}{N}$ where N is the number of points on the attractor. The length scales are then just the distances from a point to its nearest neighbor. From the partition sum, Eq. (4), we obtain the function $\tau(q)$ and via the Legendre transform, Eq. (9), the $f(\alpha)$ spectrum. This spectrum is shown as the curve in Fig. 8. The maximum and minimum values of α are related to special point on the attractor, see Refs. 4 and 10. This spectrum quantifies the global scaling structure of the attractor. But it is just "some curve" and we can ask so what ? The interesting point about this curve is that it is universal for the class of critical circle maps with golden mean winding number at the onset of chaos. This has been shown in a renormalization group calculation by Kadanoff.[18] Therefore we rush and compare it to experiments at the same point of criticality, i.e. at the onset of chaos and golden mean winding number. We already discussed a set with these properties, the attractor from the Rayleigh-Benard experiment shown in Fig. 5. We partition the set into small boxes and calculate the density of points in each box. In practice one embeds the experimental signal in 3-d (to avoid accumulations caused by 2-d projections).[19] Then the density in each box is estimated by the return times of the dynamics. The corresponding $f(\alpha)$ values is shown as the dots in Fig. 8. The error bars are due to finite amount of data points ($\sim$5000) and a long time drift in the experiment. The error bars are largest for high α values which correspond to the place on the attractor with low statistics of points. The error here is around 10%. At the rest of the curve the typical error is 5%. We see that within the error bars the theory and the experiment agree.[19]

What are the conclusions that can be drawn from Fig. 8? The results indicate that the forced Rayleigh-Benard experiment and the critical circle map are in the same universality class. Although the theoretical attractor, Fig. 6, and the experimental attractor, Fig. 5., look quite different, they belong from a scaling point of view to the same class of multifractals. Thus the $f(\alpha)$ formalism is a convenient tool to quantitatively characterize fractal sets and compare them to each other. In a way the $f(\alpha)$ spectrum is analogous to a set of critical exponents such as those defined for second order phase transitions, for example. But for critical phenomena one usually finds a discrete set of independent exponents thus being fundamentally different from multifractals who exhibit a continuum of independent exponents. The universality of the $f(\alpha)$ spectrum with golden mean winding number has been confirmed in a completely different experiment, a driven photo conductor.[20] The system is driven so hard that it is in a regime of negative differential resistance. This gives rise to a spatial modulation of charge in the photo conductor which

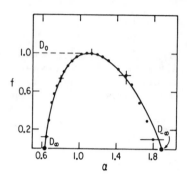

Fig. 8: The $f(\alpha)$ spectrum for the "theoretical" attractor, Fig. 6, is shown as in curve. The $f(\alpha)$-values of the experimental attractor, Fig. 5, is shown as the dots. The error bars are due to a small long term drift in the experiment and a finite number of data points.[19]

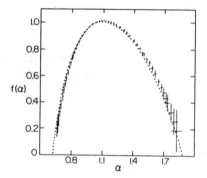

Fig. 9: The $f(\alpha)$ spectrum for a driven photo conductor with the frequency ratio at the golden. Here the dots represents the theoretical curve and the lines are the experimental results with error bars.[20] Compare with the results from the Rayleigh-Benard experiment in Fig. 8.

results in a very stable internal oscillation (ω_1). The conductor is driven by an external ac current of frequency ω_2. The two frequencies couple and we have the situation described earlier. The system is driven to chaos at the ratio of the frequencies equal to the golden mean. A strobed attractor (of $\sim$ 100000 points!) is obtained in the current signal and Fig. 9 shows the resulting $f(\alpha)$ spectrum. The agreement between theory and experiment is striking showing that also this experiment belongs to the universality class described by the critical circle map. The Rayleigh-Benard experiment has further been studied in the subcritical regime with the corresponding $f(\alpha)$ spectra calculated and compared to theoretical results.[21]

Another universality class of multifractal sets, which has been exploited both theoretically and experimentally, is found in the period doubling scenario to chaos.[22] At the onset of chaos the attractor is a multifractal set. The $f(\alpha)$ spectrum is universal in this case as well.[10] The Rayleigh-Benard experiment will for a variety of choices of the parameters go through period doubling sequences ending up in chaos. Such sequences are found inside the Arnold tongues.[23] Again we calculate the $f(\alpha)$ curve for the experimental attractor and compare it to the theory. Fig. 10

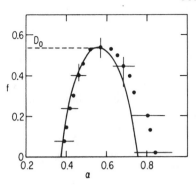

Fig. 10: The $f(\alpha)$ spectrum for an attractor at the accumulation point of period doublings. The experimental results are represented by the dots.[23]

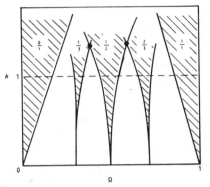

Fig. 11: We consider strange attractors at a crossing point between two Arnold tongues, specifically at the crossing of the 1/2 tongue with the 2/3 tongue.

shows the results. We find a fairly good agreement between theory and experiment, confirming the universality class for period doubling attractors.[23] The error bars here are slightly larger than in the quasiperiodic experiment because of fewer data points.

6. Multifractal Scaling Within the Chaotic Regime

The $f(\alpha)$ spectra discussed in the previous sections are all obtained at a critical point where chaos sets in. What happens when we move into the chaotic regime, i.e. beyond the $K=1$ line in Fig. 1 ? As seen in the figure, the structure is infinitely complicated, with tongues overlapping here and there. To be able to make any prediction relevant for an experiment we have to choose a specific point in the diagram. A natural choice is a crossing point between two tongues, say the 1/2 with the 2/3 tongue, as indicated in Fig. 11. Above the critical line the chaotic attractor is wrinkled[24] and a one-dimensional model is no longer adequate. As a natural extension of the 1-d circle map, Eq. (2), we therefore allow for variations in the radius of the attractor and write a mapping on the form[24,25]

$$\theta_{n+1} = \theta_n + \Omega - Kg(\theta_n) + bh(r_n) \qquad (11)$$

$$r_{n+1} = bh(r_n) - Kg(\theta_n)$$

We choose g to be a sin-function as in the previous cases. The function $h(r)$ can in principle be anything. The parameter b is a measure for the dissipation in the system, in the infinite dissipation limit $b \to 0$ we recover the 1-d map Eq. (2). As $b \to 1$ the dissipation becomes less and allows for a wrinkling of the torus. In general, it is difficult to calculate an $f(\alpha)$ spectrum for an object in 2-d just from the definition of the partition function, Eq. (4). Basically, we should find a generating partition for the set, i.e. a partition that maximizes the partition function. A covering of the set with square boxes will not be a generating partition and will consequently lead to poor convergence of the $f(\alpha)$ spectrum. For strange attractors, however, one can construct a generating partition around the unstable periodic orbits on the attractor.[25,26] Thus we have to find and encode the periodic orbits. For a two-frequency problem with a phase diagram similar to the one in Fig. 1, determined by a 1-d circle map, this task is fairly simple: A theorem states that at a crossing between two tongues, m/n and m_1/n_1, any rational number between these two will be represented by an unstable periodic orbit on the attractor.[2] This is not necessarily so in 2-d. Fig. 12a shows a plot of the "presence" of an orbit with a given rational number P/Q as a function of the parameter b, for the case $h(r) = r$. We observe that any periodic orbit will eventually disappear as b increases towards 1. If orbits indeed have disappeared, a theory is difficult. However Fig. 12a indicates, that the disappearances seems to occur from the side and moves inwards. Therefore we conjecture that there is a critical value of b, b_c, below which all orbits are present.[25] This appears indeed to be the case. It is even reasonable that an experiment will be described by a dissipation below b_c as the numerical value $b_c \sim 0.25$ is fairly large. We choose a parameter value below b_c, $b=0.2$, and find all orbits up to a certain length.[25] We do the partition around these orbits and obtain the $f(\alpha)$ spectrum this way. Fig. 12b is the result and we find excellent convergence within this approach. Notice that this spectrum is quite different from the spectrum at the onset on chaos. Now all α-values are larger than 1, i.e. the attractor in non-singular. Also all f-values are larger than 1.

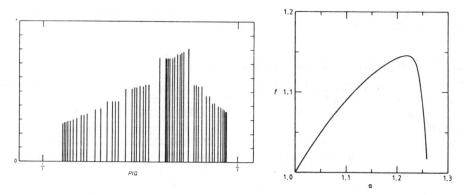

Fig. 12: (a) Value of b where orbits of winding number P/Q exist in the model Eq. (11). Note that any orbits will eventually disappear at some b-value. (b) The associated $f(\alpha)$ spectrum for $b=0.20$ at the crossing point between $1/2$ and $2/3$.[25]

The important question is: is the spectrum in Fig. 12b universal ? The answer is yes and no. The "left" part of the spectrum going down to $\alpha=1$ is universal for the class of systems located at a crossing point between two tongues. The "right" part of the spectrum is on the other hand not universal. We find though that it is "structurally stable" in the sense that it only changes very little from system to system (see Ref. 25 for a discussion of other systems). Moving away from the critical point thus means that we lose full universality of the multifractal spectra. This is very much like in critical phenomena: the renormalization group is only valid at the critical point and universality is not present away from it.

7. Thermodynamic Formalism, Inversion of $f(\alpha)$ Spectra and Phase Transitions

The multifractal description is a generalization of the "thermodynamic formalism" as developed by Bowen,[27] Sinai[28] and Ruelle.[29] The analogy to thermodynamics can be seen from the partition sum Eq. (4). Keeping all the p_i's equal ($=a^{-n}$, n is an index of the refinement) and using the substitutions[30] $q \to F$, $\ell_i = e^{\log \ell_i} \to e^{E_i}$, $\tau \to (1/kT)$ we obtain the standard partition sum for a "free energy" $F(T)$

$$e^{n \log a \, F} = \sum_i e^{-E_i/kT}. \tag{12}$$

The thermodynamic formalism is very useful. For refining sets, as a period doubling attractor or the orbit with golden mean rotation (Fig. 6), one can write the formalism in terms of a transfer matrix from one level to the next in the refinement.[30] The elements in the transfer matrix are the elements in Feigenbaums scaling function. Using this technique one can "invert" an $f(\alpha)$ spectrum and learn about the underlying dynamics that gave rise to the spectrum. Lets say we from the Rayleigh-Benard experiment are given two spectra: the "dots" in Fig. 8 and the "dots" in Fig. 10. Can we from this information say something about the underlying dynamics. The surprising answer is: Yes! We shall not go in detail here but refer the reader to Refs. 30 and 31. From the inversion the winding number is obtained thus allowing us to distinguish an attractor with golden mean rotation from an attractor at the period doubling critical point. Also one gets out element of the transfer matrix which will tell us about the critical behavior of the underlying dynamical system. So this is like "retrieving whole potatoes from mashed potatoes" (an authentic quotation from Ref. 31)!

The thermodynamic function $F(T)$ (i.e $q(\tau)$) is usually smooth and all its derivatives are smooth. In some cases, however, one can find a discontinuity in the n'th order derivative indicating a n'th order phase transition.[32] Here we shall study examples of first order transitions. This type of transition usually yields $f(\alpha)$ spectra with straight sections. Fig. 13 is the $f(\alpha)$ spectrum of the Henon strange attractor which consists of a smooth section with $\alpha > 1$ (i.e. a non-singular part) and then a phase transition down to the point $(0.72,0)$.[33] We have shown that this transition takes place between the smooth non-singular part and the turnbacks of the attractor where the measure is singular.[34] One can argue that $\alpha_{min}=0.76$ and also that $f(\alpha_{min}) \sim 0.26$ because the turnbacks live on a set roughly equal to the cross section of the Henon attractor. We shall not go in detail and refer to.[34] It is also possible to introduce an order parameter for such transitions. Fig. 14 shows an order parameter for a first order phase transition in the thermodynamics of a 1-d dynamical system at the point where chaos is fully developed.[35] We see that the order parameter becomes increasingly sharp as the set is refined. We can thus

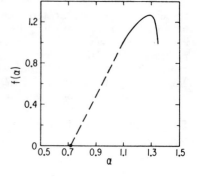

Fig. 13: The $f(\alpha)$ spectrum for the Henon attractor calculated from unstable periodic orbits.[33] Note the straight part of the spectrum indicating a phase transition in the underlying thermodynamic formalism.

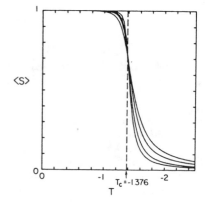

Fig. 14: An order parameter for a first order transition in a one-dimensional dynamical system. The order parameter corresponds to a thermal average of the magnetization. The various curves describe different system sizes and we see clear finite size effects.[35]

use finite size scaling for first order transitions and find a different universality class than in usual transitions.[36]

8. Tips and Fjords on Aggregates

In the remaining part of these notes we turn to study the multifractal properties of fractal aggregates, like diffusion limited aggregates or viscous fingers in a porous medium. We want to quantify the "branchiness" of such structures and develop a quantitative description of their growth. Several calculations of $f(\alpha)$ spectra for aggregates have been performed, see Fig. (15). These calculations usually relay on various box-counting methods using the partition sum, Eq. (4), over the boxes. It is typical for these methods that the convergence for the tips (i.e. the $\alpha's < 1$) is very good but the convergence for the fjords (i.e. the $\alpha's > 1$) is extremely poor. In Fig. (15) we see that some calculations obtain very large α values in the fjord regime and altogether there are big discrepancies between the various calculations.

We shall argue it is not at all surprising when these calculations disagree in the fjord regime. This is due to the fact that in probabilistic diffusion or growth there can be fjords which are highly screened from the growth field. The typical behavior is that the more screened a fjord is, the larger the corresponding α value. Actually,

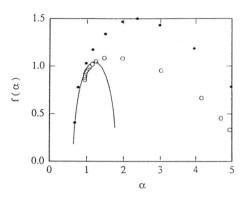

Fig. 15: Multifractal $f(\alpha)$ spectra calculated from an ensemble of DLA clusters (black dots),[39] an experiment on
non-newtonian fluids (circles)[40] and
an experiment on viscous fingers (curve).[41]

it is even possible that some fjords are completely screened, causing $\alpha_{max} \to \infty$:
this appears as a phase transition in the corresponding $f(\alpha)$ spectrum. On the other
hand, the tip regime is very stable and insensitive to changes of the model and we
expect to find a more or less universal growth-spectrum for typical aggregates. This
is consistent with previous results shown in Fig. 15: their $f(\alpha)$ spectra agree up to
$\alpha=1$, i.e., in the tip regime. We base our intuition on calculations with fractal sets
found in the theory of iterations of conformal mappings, using them as a probe for
fractal aggregates.[37,38] There we can calculate the $f(\alpha)$ spectra with accuracy, not
attainable by a box counting method.

9. Julia Sets as Model Aggregates

The use of conformal mappings and their Julia sets as models for diffusion aggregates is based on an analogy between a diffusion field and electrostatics. To be
more specific: a DLA cluster, for instance, is generated by diffusing particles and
the diffusion is described by the Laplace equation. Thus we have a field which is
a solution to the Laplace equation with the boundary condition that the field is
zero on the aggregate. The probability of growth is then the gradient of the field
at the boundary. This is equivalent to finding the electrostatic potential around
the aggregate with the boundary condition that it is zero at the aggregate. The
gradient of the potential is the electric field, thus the electric field is the growth
measure, or the harmonic measure.[42]

As a simple prototype conformal map we consider

$$f(z) = z^2 + c, \tag{13}$$

with complex parameter c.[43,44] For a specific value of c, the corresponding Julia set
is the boundary of the set

$$\{z| \lim_{n\to\infty} |f^n(z)| < \infty\}. \tag{14}$$

For beautiful pictures of such sets, see the book by Peitgen and Richter.[44] We
are interested in cases where the Julia set has resemblance to fractal aggregates.
Thus, first of all the set should have branches on top of branches and be "skinny"

(i.e. have small or zero measure). Examples of such sets are obtained for c values corresponding to the Misiurewicz points.[45] Figure 16 shows an example of such set, it is chosen to have five "arms" in order to stress the resemblance with probabilistic diffusion aggregates (by varying c one can obtain any number of arms). Our goal is to solve the Laplace equation around this set, i.e. to find the potential lines around it. Douady and Hubbard[46] found that the potential at the point z is given by

$$U(z) = \lim_{n \to \infty} \frac{1}{2^n} \log |f^n(z)|. \tag{15}$$

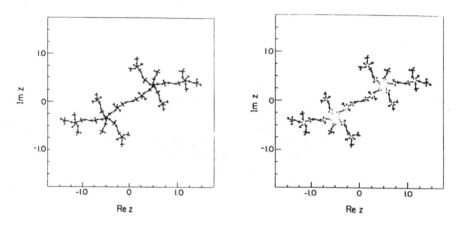

Fig. 16: (a) A Julia set with five arms for the map, Eq. (13), with $c = -0.636754 - 0.685031i$. Note its visual resemblance with a fractal aggregate. (b) The same Julia set found from inverse iterates of the map, Eq. (13). The density of the inverse iterates is proportional to the electric field and is thus equivalent to the harmonic measure on a "real" aggregate.[37]

Notice that $U = 0$ on the Julia set, i.e. it is grounded. The pointwise dimension α_i at any point z_i on the Julia set, is

$$U(z_i + \epsilon) \sim \epsilon^{\alpha_i}, \qquad \epsilon \to 0. \tag{16}$$

The minimum value $\alpha_{\min}$ is found at one of the fixed points, $z_0 = (1 + \sqrt{1 - 4c})/2$, of Eq. (13)[37]

$$\alpha_{\min} = \frac{\log 2}{\log f'(z_0)} \tag{17}$$

$\alpha_{\min} < 1$ and this point is thus located at the tip with the strongest divergence of the electric field ($\vec{E} = \nabla U$).

In practice, we do not have to do the cumbersome job and calculate the α value in each point on the set from Eq. (16). Starting at the tip z_0 and iterating the map

Eq. (13) backwards we obtain all other tips on the set, and the corresponding α's are found from the derivative of the map in a given point.[37] Similarly, we obtain all other fjords from backwards iterates of one fjord. Fig. 16b shows a plot of all these backwards iterates. Calculating the $f(\alpha)$ spectrum for these type of sets is thus very easy; we use backward iterates and we obtain the spectrum shown in Fig. 17a. The analogy to the diffusion aggregates is eve n stronger: the density of inverse iterates on Fig. 16b is equal to the electric field on the surface and is thus exactly equivalent to the harmonic measure of a growing aggregate!

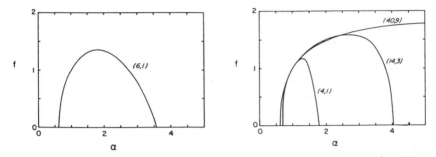

Fig. 17: (a) The $f(\alpha)$ spectrum for the Julia set in Fig. 16. (b) The $f(\alpha)$ spectra for a series of Julia sets with increasing density of tiny branches. Note that α_{max} becomes increasingly large. For the parameter values, see Ref. 37.

We have obtained an accurate $f(\alpha)$ spectrum of the set in Fig. 16a, the spectrum has a tip regime ($\alpha < 1$) and a fjord regime ($\alpha > 1$) and now what? To obtain a comparison with a real aggregate we should add many more tiny branches to the Julia set in Fig. 16. This can be done by following a series of Misiurewicz points corresponding to higher order unstable cycles, thus adding more branches.[45] The corresponding $f(\alpha)$ curves are shown in Fig. 17b.[37] The results are interesting! We see that the tip regime is basically unchanged in the series. The fjord regime however varies wildly: α_{max} eventually diverges to infinity. This shows that some fjords are completely screened. By analogy, in real aggregate some fjords are expected to be completely inactive and not grow at all. $\alpha_{max} \to \infty$ can be interpreted as a phase transition in the $f(\alpha)$ spectrum at the Hausdorff dimension. Of course this observation explains the experimental findings in Fig. 15. There is no reason that the calculations should agree for $\alpha > 1$ as some fjords are screened and the calculation here is meaningless. Note, however, that the tip regimes computed from the Julia sets are in agreement with the tip regimes of the experiments in Fig. 15. The tip structure is responsible for the growth and does not care about the fjords that are left behind, more or less screened. The possibility even exists that some fjords will have exponentially small field giving $\alpha_{max} = \infty$ (for instance, a rectangular channel has exponentially small field from solving Laplace equation). Also, one can show that the curve has to go through $(\alpha, f) = (1, 1)$[37] which again is in agreement with both numerics and experiments, Figs. 15, 17.

Acknowledgements: This review is based on very exciting and fruitful collaborations with P. Bak, D. Bensimon, T. Bohr, P. Cvitanović, M. J. Feigenbaum, J. A.

Glazier, G. H. Gunaratne, T. C. Halsey, L. P. Kadanoff, A. Libchaber, I. Procaccia, B. Shraiman and J. Stavans. I thank them sincerely. Also I am grateful to P. Cvitanović for a careful reading of the manuscript.

1. M. H. Jensen, P. Bak, and T. Bohr, Phys. Rev. Lett. 50, 1637 (1983); Phys. Rev. A 30, 1960 (1984).

2. V. I. Arnold, Am. Math. Soc. Trans. , Ser. 2 46, 213 (1965).

3. M. R. Herman, in Geometry and Topology, ed. J. Palis, Lecture Notes in Mathematics 597, 271 (Springer, Berlin, 1977).

4. Scott J. Shenker, Physica 5D, 405 (1982)

5. M. J. Feigenbaum, L. P. Kadanoff, and Scott J. Shenker, Physica 5D, 370 (1982); S. Ostlund, D. Rand, J. P. Sethna, and E. D. Siggia, Physica 8D, 303 (1983).

6. J. Stavans, F. Heslot, and A. Libchaber, Phys. Rev. Lett. 55, 596 (1985).

7. J. Stavans, Phys. Rev. A 35, 4314 (1987).

8. B. B. Mandelbrot, J. Fluid. Mech. 62, 331 (1974).

9. U. Frisch and G. Parisi, Varanna School LXXXXVIII, M. Ghil, R. Benzi, and G. Parisi, eds. , North-Holland, New York (1985), p. 84; R. Benzi, G. Paladin, G. Parisi, and A. Vulpiani, J. Phys. A 17, 352 (1984).

10. T. C. Halsey, M. H. Jensen, L. P. Kadanoff, I. Procaccia, and B. I. Shraiman, Phys. Rev. A 33, 1141 (1986).

11. P. Grassberger, Phys. Lett. 107A, 101 (1985).

12. R. Badii and A. Politi, Phys. Rev. Lett. 52, 1661 (1984).

13. T. Bohr and D. Rand, Physica 25D, 387 (1987).

14. L. de Arcangelis, S. Redner, and A. Coniglio, Phys. Rev. B 31, 4725 (1985).

15. R. Blumenfeld, Y. Meir, A. B. Harris, and A. Aharony, J. Phys. A 19, L791 (1986).

16. A. Renyi, Probability Theory (North-Holland, Amsterdam, 1970).

17. H. G. E. Hentschel and I. Procaccia, Physica 8D, 435 (1983).

18. L. P. Kadanoff, J. Stat. Phys. 43, 395 (1986); D. Bensimon, M. H. Jensen, and L. P. Kadanoff, Phys. Rev. A 33, 3622 (1986).

19. M. H. Jensen, L. P. Kadanoff, A. Libchaber, I. Procaccia, and J. Stavans, Phys. Rev. Lett. 55, 2798 (1985).

20. E. G. Gwinn and R. M. Westervelt, Phys. Rev. Lett. 59, 157 (1987).

21. J. A. Glazier, G. H. Gunaratne, and A. Libchàber, Phys. Rev. A 37, 523 (1988).

22. M. J. Feigenbaum, J. Stat. Phys. 19, 25 (1978); J. Stat. Phys. 21, 669 (1979).

23. J. A. Glazier, M. H. Jensen, A. Libchaber, and J. Stavans, Phys. Rev. A 34, 1621 (1986).

24. T. Bohr, P. Bak, and M. H. Jensen, Phys. Rev. A 30, 1970 (1984).

25. G. H. Gunaratne, M. H. Jensen, and I. Procaccia, Nonlinearity 1, 157 (1988).

26. P. Cvitanović, G. H. Gunaratne, and I. Procaccia, Phys. Rev. A (1988).

27. R. Bowen, in "Equilibrium States and the Ergodic Theory of Anisov Diffeomorphisms," Lecture Notes in Mathematics 470 (Springer, Berlin, 1975).

28. E. B. Vul, Ya. G. Sinai, and K. M. Khanin, Usp. Mat. Nauk 39, 3 (1984) [Russ. Mat. Surveys 39, 1 (1984)].

29. D. Ruelle, Statistical Mechanics, Thermodynamic Formalism (Addison-Wesley, Reading, 1978).

30. M. J. Feigenbaum, J. Stat. Phys. 46, 919 (1987); 46 925 (1987).

31. M. J. Feigenbaum, M. H. Jensen, and I. Procaccia, Phys. Rev. Lett. 56, 1503 (1986).

32. P. Cvitanović, in proceedings of XIV Colloqium on Group Theoretical Methods in Physics, ed. R. Gilmore (World Scientific, Singapore 1987); in Non-Linear Evolution and Chaotic Phenomena, eds. P. Zweifel, G. Gallavotti and M. Anile (Plenum, New York, 1988).

33. G. H. Gunaratne and I. Procaccia, Phys. Rev. Lett. 59, 1377 (1987).

34. M. H. Jensen, Phys. Rev. Lett. 60, 1680 (1988).

35. T. Bohr and M. H. Jensen, Phys. Rev. A **36**, 4904 (1987).

36. Y. Imry, Phys. Rev. B **21**, 2042 (1980).

37. T. Bohr, P. Cvitanović, and M. H. Jensen, Europhys. Lett. (1988).

38. I. Procaccia and R. Zeitak, preprint.

39. C. Amitrano, A. Coniglio, and F. di Liberto, Phys. Rev. Lett. **57**, 1098 (1986).

40. J. Nittmann, H. E. Stanley, E. Touboul, and G. Daccord, Phys. Rev. Lett. **58**, 619 (1987).

41. K. J. Måløy, F. Boger, J. Feder, and T. Jøssang, in Time-Dependent Effects in Disordered Materials, Eds. R. Pynn and T. Riste, p. 111 (Plenum, New York, 1987).

42. N. G. Makarov, Proc. London Math. Soc. **51**, 369 (1985); T. C. Halsey, P. Meakin, and I. Procaccia, Phys. Rev. Lett. **56**, 854 (1986).

43. B. B. Mandelbrot, Fractal Geometry of Nature, (Freeman, San Francisco, 1982).

44. H.-O. Peitgen and P. H. Richter, The Beauty of Fractals (Springer, Berlin, 1986).

45. J. Myrheim and P. Cvitanović, Phys. Lett. **94A**, 329 (1983); Comm. Math. Phys. (to appear).

46. A. Douady and J. H. Hubbard, CRAS (Paris) **294**, 123 (1982).

Luca Peliti

"HANDS ON" EXPERIMENT ON THE BEACH
(STRENGTH TEST OF RE-INFORCED SANDPILE)
BY ARMIN BUNDE, MURIELLE DEUTSCHER AND AVRAHAM SIMIEVIC

MULTIFRACTAL ANALYSIS OF SEDIMENTARY ROCKS

JAN PETTER HANSEN, J. L. MCCAULEY, JIRI MULLER
and A. T. SKJELTORP

Institute for Energy Technology
Box 40, 2007 Kjeller, Norway

Geometry may be the most important physical property that distinguishes sedimentary rocks from all other porous media. Fractal analysis is becoming a common tool for analyzing geometry of objects occurring in nature. There have been several attempts to characterize sedimentary rocks using the "Hausdorff-Besicovitch" fractal dimension. Recent developments in the fractal theories have suggested that a full treatment of geometric shapes should come through its multifractal description.[1]

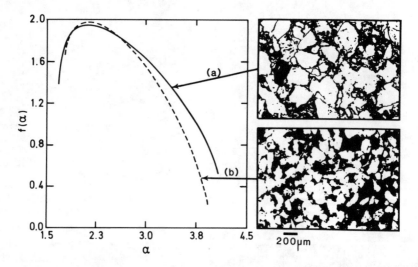

Fig. 1

In the right hand side of Fig. 1 we display digitized representations of optical micrographs of thin sections cut from core samples which were collected from two different oil field reservoirs (denoted as "a" and "b"). It is clear that pore distribution is quite different in both cases. This is indeed reflected in the multifractal analysis (represented by the $f(\alpha)$ curves in Fig. 1) which we performed on these two samples. Such analysis should give us better understanding of several issues[1,3] such as flow property of rocks, relationship between porosity and permeability, diagenesis, wetting, fingering etc. Core samples, thin sections and a "multifractal kit" will be demonstrated during the "hands on" session.

1. Articles in *Time-Dependent Effect in Disordered Materials* edited by R. Pynn and T. Riste (Plenum, New York, 1987).
2. J. P. Hansen and A. T. Skjeltorp, *Phys Rev. B* (in press).
3. J. P. Hansen, J. L. McCauley, J. Muller and A. T. Skjeltorp, to be published.

PHASE TRANSITION ON DLA

JYSOO LEE, PREBEN ALSTRØM, and H. EUGENE STANLEY
Center for Polymer Studies and Department of Physics
Boston University, Boston, MA 02215 USA

We have carried out the first exact enumeration approach to multifractal spectra of DLA, using methods similar to Nagatani renormalization group. Specifically, we form the partition function $Z(\beta, L) = \sum_\alpha C_\alpha \sum_i p_{i,\alpha}^\beta$, where C_α is the weight of configuration α, $p_{i,\alpha}$ is growth probability of site i of configuration α ,and L is the length scale of clusters. The "free energy" of the system is calculated from $L^{-F(\beta,L)} = Z(\beta, L)$, the "energy" from $E(\beta, L) = \partial F(\beta, L)/\partial \beta$, and the "specific heat" from $C(\beta, L) = \partial E(\beta, L)/\partial \beta$,
    We report three results:

(i) The specific heat develops a singularity near $\beta_c \sim -1$

(ii) Near $\beta_c \sim -1$ there are large fluctuations of the energy, and

(iii) The minimum growth probability (corresponding to the maximum energy) does not scale as a power law in L, but rather falls off as $p_{min} \sim \exp(-aL^2)$.

Based on these results, we argue that there is probably a well defined critical value, $\beta = \beta_c$ exists in DLA. Associated with this critical point seem to be singularities entirely analogous to those characterizing critical points in thermal systems.

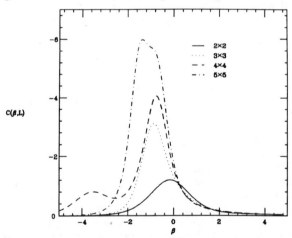

Fig. 1: Temperature dependence of specific heat of DLA.

DISORDERED PATTERNS IN DETERMINISTIC GROWTH

TAMAS VICSEK
Budapest P. O. Box 76
H-1325, Hungary

ABSTRACT. Methods developed for the investigation of dynamical systems are used to characterize the degree of randomness of spatially disordered chains of atoms and growing unstable interfaces. Our calculations indicate that the experimentally observed geometry of viscous fingering patterns can not be described in terms of low-dimensional chaos.

1. Introduction

The growth of unstable interfaces usually leads to structures which look random even if the given process is described by deterministic equations. Examples for such phenomena are discussed for example in the proceedings of the 1985 Cargese school on Growth and Form, edited by Stanley and Ostrowsky. When one is interested in the stochastic nature of the geometrical properties of an interfacial pattern the following questions arise naturally:

(i) How random are these structures? and
(ii) What is the origin of their randomness?

These are far from trivial questions since two disorderly-looking patterns may possess very different degrees of randomness, depending on their complexity. The problem of characterizing the complexity of a spatial configuration is the subject of considerable current interest (Grassberger 1986). There seems to be three basic cases: *ordered, complex (disordered), and completely random (analogous to white noise)*. As far as concerning the second question about the origin of randomness, the simpler situation is realized when the development of an unstable spatial pattern is dominated by imposed random fluctuations. Naturally, in this case the geometry of the interface is expected to fall into the third category (noise). However, much less is known about the situation when the fluctuations can be neglected and it is the *initial condition* only which determines the final configuration through *deterministic equations* having relatively simple structure.

In order to address the above questions we shall use the following approach. We chose two specific systems (which are known to represent a large number of relevant physical phenomena). Next, we shall look for *analogies* between chaotic behavior in dynamical systems and spatial properties of these systems. During the past decade powerful methods have been elaborated for the treatment of chaotic signals observed as a function of time in processes which can be described using systems of non-linear ordinary differential equations (see e.g. Berge et al 1986). Application of these methods to the characterization of spatial patterns should provide additional insight into their structural properties.

Our final goal is to investigate questions (i) and (ii) in the context of interfacial growth under far-from-equilibrium conditions. In particular, we would like to characterize the geometry of disordered patterns evolving in two-phase fluid flows. However, let us first review the simpler case of an equilibrium system, the so called

Frenkel-Kontorova model. This model represents a suitable basis for demonstrating the usefulness of methods and techniques worked out for systems exhibiting temporal chaos in the studies of spatially disordered configurations.

2. Spatial Chaos in Equilibrium Models

Let us consider the following simple model originally due to Frenkel and Kontorova (1938). The system they introduced consists of an array of atoms which are connected with harmonic springs and interact with a periodic potential of period b (Fig. 1). For example, this model can be used to describe interacting gas atoms absorbed on a crystalline substrate. The energy can be written in the form

$$F = \sum_i (x_{i+1} - x_i - a_0)^2 + V \left[1 - \cos\left(\frac{2\pi}{b}\right) x_i\right], \qquad (1)$$

where x_i is the position of the ith atom and a_0 is the lattice constant which would be realized in the absence of the periodic potential (i.e., for $V = 0$). Several other equilibrium models can be shown to have analogous Hamiltonians (Bak 1982).

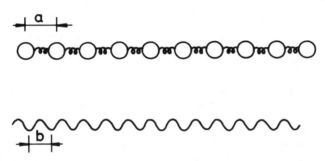

Fig. 1: Schematic representation of the Frenkel-Kontorova model.

As it was discussed by Aubry (1979) and Bak (1982) in details, depending on the strength of the potential three basic types of configurations are possible:

(a) In the incommensurable state (small V) the configuration is periodic, but its period is not commensurable with b.

(b) For strong enough potential the structure relaxes into a state with a period which is a simple rational factor of b.

(c) We are interested in the third possibility, when the system has stable or metastable configurations with no periodic structure.

The configurations minimizing the energy are the solutions of the equations

$$\frac{\partial F}{\partial x_i} = (x_{i+1} - x_i) - (x_i - x_{i-1}) + \frac{2\pi V}{b} \cos\left(\frac{2\pi}{b} x_i\right) = 0. \qquad (2)$$

The above set of equations can be written in the form of the following recurrence relations

$$x_{i+1} = x_i + w_i + \frac{2\pi V}{b} \cos\left(\frac{2\pi}{b} x_i\right), \qquad (3)$$

$$w_{i+1} = x_{i+1} - x_i.$$

Thus, the problem of determining the positions of the atoms is reduced to iterating (3) starting with some x_0 and x_1. Remarkably, (3) is equivalent to the area-preserving standard map (Chirikov 1979) well known in the theory of dynamical systems. It can be shown numerically that there exists a region of the (x_0, x_1) points such that the "trajectories" (configurations) starting from this region are chaotic. From the numerical point of view this means that the iterates fill a small part of the phase space.

In this way, completely deterministic equations may lead to random spatial configurations. The behavior is chaotic because the positions of the atoms are practically stochastic (unpredictable), and an arbitrarily small change in the initial point inevitably leads to a finite deviation in the whole structure.

3. Disordered Patterns in Fluid Flows

3.1. EQUATIONS

Fluid flows are known to exhibit several kinds of instabilities. Most of these is described by the Navier-Stokes equation

$$\nabla p = \mu \nabla^2 \vec{v} + \rho \partial \vec{v} / \partial t + \rho (\vec{v} \nabla) \vec{v}, \tag{4}$$

where the gravitational effects have been neglected. In the above equation p denotes the pressure, ρ and μ are respectively the density and the viscosity of the fluid and $\vec{v}$ is the local velocity. Equation (4) is non-linear and contains both space and time partial derivatives.

In general, the behavior of hydrodynamical systems can be complex both in *space and time* depending on the actual physical conditions (e.g., geometry, viscosity, velocity). Temporal chaos can be indicated using a detector at a fixed point and recording the velocity dependent signal as a function of time. Above a critical value of the Reynolds number the signal becomes stochastic indicating the onset of chaos. One way to understand this phenomenon is to study the Lorentz model (Lorentz 1963) which is a set of ordinary differential equations (with no spatial derivative).

From the experimental point of view, the complexity of the temporal behavior is examined using the measured time series to calculate the Fourier transform of the signal. Another alternative is to construct various plots from the time delayed data. In the latter case one obtains a set of points in the d_m-dimensional space whose correlation dimension is an important characteristics of the dynamics (these techniques will be discussed later).

Let us next develop a similar framework for the investigation of *viscous fingering patterns* which are also determined by the Navier-Stokes equations with the appropriate boundary conditions. The phenomenon of viscous fingering takes place in two-phase flows, when a less viscous fluid is injected into a more viscous one under circumstances leading to a fingered interface. Here we shall discuss experiments, where the conditions are such that most of the terms in the Navier-Stokes equation can be neglected and the resulting mathematical problem corresponds to Laplacian growth.

At the end of the last century Hele-Shaw introduced a simple system to study the flow of water around various objects for low Reynolds numbers. The cell he designed consists of two transparent plates of linear size w separated by a relatively small distance b (typical sizes are in the region $w \sim 30$ cm and $b \sim 1$ mm). The viscous fluids are placed between the plates and pressure can be applied for example at the centre of the upper plate (radial version) of the cell.

The relation of viscous fingering to Laplacian growth can be shown by assuming that the plates are horizontal, and the flow in the x, y plane has a velocity profile $v(z) = [v_x^2(z) + v_y^2(z)]^{1/2}$ which is approximately parabolic in the direction z perpendicular to the plates

$$v(z) = a\left(\frac{b^2}{4 - z^2}\right).$$ (5)

Furthermore, we assume that $v_z = 0$ and $\partial v_x/\partial x = \partial v_y/\partial y = 0$. For the average velocity one has

$$\bar{v}(z) = \frac{1}{b}\int_{-b/2}^{b/2} v(z) = \frac{ab^2}{6}.$$ (6)

For small b the first term of the right-hand side in (4) dominates, because it is proportional to $1/b^2$. Inserting (5) into (4) (where the second and the third term of the right-hand side is neglected) and using (6) we get

$$\vec{v} = -\frac{b^2}{12\mu}\nabla p,$$ (7)

The above equation represents the so called Darcy's law expressing the fact that for small b the average velocity is proportional to the local force. Assuming that the fluids are incompressible one arrives at the Laplace equation

$$\nabla^2 p = 0,$$ (8)

for the pressure distribution p, from the condition that the divergence of the velocity vanishes. The first boundary condition to the above equation is essentially equivalent to (8) evaluated at the interface.

When writing down the boundary condition for the pressure jump Δp at the interface, one has to take into account the specific geometry of a Hele-Shaw cell. Because of the quasi-three-dimensional nature of the cell, an accurate expression for Δp contains at least three terms (Bensimon et al 1986). Keeping only the most relevant term in the dimensionless version of the second boundary condition it has the form

$$\Delta p \simeq -d_0\kappa,$$ (9)

where d_0 is the so-called capillary length and κ denotes the local curvature of the interface in the $x - y$ plane. Finally, it is assumed that the pressure has a constant value far from the interface. The traditional experiment is carried out using two immiscible, Newtonian fluids with a high viscosity ratio. For example, air can be used to displace glycerin or oil.

Note that Eq. (8) derived from (4) does not contain time dependence explicitly [only through the boundary condition corresponding to (7)]. In the radial Hele-Shaw cell *no steady-state fingers* can develop in the cell, because of the Mullins-Sekerka instability (Mullins and Sekerka 1963) which leads to the growth of disordered interfaces shown in Fig. 2. even for low Reynolds numbers. Interestingly, the question whether such viscous fingering patterns become fractals (Mandelbrot 1982)

Fig. 2: Interface of a representative viscous fingering pattern obtained in the radial Hele-Shaw cell containing glycerin, for relatively large pressure of the injected air.

in the large size limit has not been satisfactorily answered yet. (In a cell with imposed randomness the growing interface has a fractal geometry (Maloy et al 1985, Chen and Wilkinson 1985, Lenormand 1986). Non-Newtonian, miscible fluids also give rise to fractal patterns (Daccord et al. 1986).)

Stochastic dynamics and disordered interfaces represent only two of the possible behaviors related to the Navier-Stokes equation. Turbulence, which is a phenomenon corresponding to a simultaneous chaotic behavior in space and time is perhaps the most complex case described by (4). However, here we shall concentrate on the geometrical properties of the interfaces in fluid flows, not treating time dependence. To characterize the patterns we determine $\kappa(s)$ which is the *local curvature* of the interface as a function of the *arc length* s (distance measured along the surface). In the following $\kappa(s)$ will be used as an *analogue of the time series* measured in the experiments on dynamical systems.

The most typical regimes are schematically shown in Fig. 3. In each case the starting configuration is a simple, but irregular interface.

(i) If the less viscous fluid is pumped out from the cell, the interface is stable and the resulting pattern is a circle. Let us now construct a plot of $\kappa(s + \Delta s)$ against $\kappa(s)$ in a manner similar to that used in the theory of chaos to find attractors from a single time series. For a circle we obtain a single point in analogy with the fix points defined for dynamical systems.

(ii) Injecting the less viscous fluid into a system with anisotropy the situation changes qualitatively (anisotropy can be introduced into experiments on viscous fingering using anisotropic plates or fluids (Ben-Jacob et al 1985, Buka et al 1986). In this case nearly periodic structures are observed in the vicinity of the tips stabilized by the anisotropy. This behavior in a crude approximation corresponds to a limit cycle in the $\kappa(s + \Delta s)$ versus $\kappa(s)$ plot.

(iii) Injecting the less viscous fluid into a cell with no anisotropy one obtains patterns without any apparent symmetry. Accordingly, the data seem to be randomly scattered in the related plot of κ values.

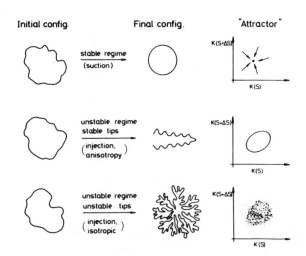

Fig. 3: These schematic plots show the three main regimes which are observed in viscous fingering experiments under various conditions.

3.2. DEGREE OF RANDOMNESS OF VISCOUS FINGERING PATTERNS

(V. K. Horváth and T. Vicsek, to be published)

We have seen in the previous section that disordered patterns growing in an experiment on viscous fingering can be described by a simple set of deterministic equations. Of course, this statement is valid only if the flow is not dominated by the small irregularities of the plates of the Hele-Shaw cell. We carried out the experiments using high quality, carefully cleaned glass plates. Gaseous nitrogen was injected into glycerin and the resulting interfaces were stored in a PC using a digital image processor. A typical pattern is shown in Fig. 2.

In order to extract quantitative information from the experimental patterns the local curvature, $\kappa(s)$, is determined as a function of the arc length at each pixel point of the digitized image. Then this set of approximately 15,000 data is treated as a "time series" and its Fourier spectrum is calculated using a fast Fourier transformation algorithm. Figure 4a shows part of the long series of data for $\kappa(s)$ together with the Fourier spectrum obtained for the complete data set (Fig. 4b). The tips of the fingers and the relatively sharp corners at the inner ends of fjords appear as peaks distributed in an apparently random manner. Accordingly, the Fourier spectrum has no well defined, singular peaks which would indicate periodic behavior.

In search for an underlying structure behind the seemingly stochastic behavior one can apply a technique analogous to the construction of "time delayed" plots. In our case this means of plotting the data as points in a d_m-dimensional "phase" space using as coordinates the curvature values obtained for the arc lengths $s, s + \Delta s, \ldots, s+(d_m-1)\Delta s$. This method has been found to be useful for the visualization of low-dimensional chaos in a number of dynamical systems (e.g., Turner et al 1981). According to our results for $d_m = 2$ and 3, the data points $\vec{\kappa}_s = \{\kappa(s), \ldots, \kappa(s + (d_m-1)\Delta s)\}$ are scattered randomly in such length shifted plots. The distribution of points shows no signs of the presence of an attractor corresponding to a low-

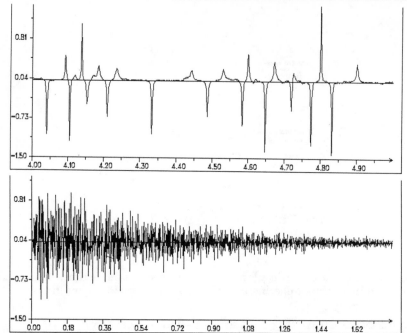

Fig. 4: (a) Part of the long series of curvature values (1000 out of 16200) determined for the pattern displayed in Fig. 2 as a function of the arc length. (b) The Fourier spectrum calculated for the curvature data.

dimensional spatial chaos.

The degree of randomness can also be examined by estimating the correlation dimension D_2 associated with the length shifted set of $\vec{\kappa}_s$ data. This method is especially useful for $d_m > 3$. It has been shown that $D_2 \leq D$, where D is the fractal dimension of the $\{\vec{\kappa}_s\}$ set. It follows from the embedding theorem that with growing d_m the correlation dimension converges to a finite value if the behavior of the system can be described in terms of a finite-dimensional chaos. The dimension at which this convergence is completed is called embedding dimension. For a completely random set of points (i.e., for points produced by a random number generator) $D_2 = d_m$ for all d_m, while, in turn, $D_2 < d_m$ indicates an underlying structure in the data. To determine D_2 one calculates the correlation function $c(\epsilon)$ from the expression

$$c(\epsilon) = \frac{1}{n^2}[\text{number of pairs of points } (s, s') \text{ with } |\vec{\kappa}_s - \vec{\kappa}_{s'}| < \epsilon], \qquad (10)$$

and uses the relation

$$c(\epsilon) \sim \epsilon^{D_2}, \qquad (11)$$

where ϵ is supposed to be small.

Using the curvature data the values $D_2 \simeq 1.85$, 2.7, 3.4, and 3.7 have been obtained for the correlation dimension for $d_m = 2$, 4, 6, and 10, respectively. These results suggest that the correlation dimension does not saturate for small d_m. On the other hand D_2 is found to be definitely smaller than d_m. At the present stage our data are not appropriate for deciding what happens in the $d_m \to \infty$ limit.

4. Conclusions

Although the development of the interface in a two-phase fluid flow is prescribed by deterministic equations, the observed viscous fingers do not seem to have regular structure. In addition to the experiments, this statement was supported numerically by Sander et al (1985) who solved the corresponding equations in a computer. Their results demonstrate that the randomness of patterns is not likely to be due to the spatial irregularities of the plates used in the Hele-Shaw cells.

The reason for the stochastic behavior is the instability of the interface: perturbations beyond a surface tension dependent wave length grow indefinitely. These perturbations are present in the initial condition, i.e., the starting shape of the interface is never perfectly symmetric. Consequently, if we wrote up a series for the initial interface it would contain infinitely many non-zero terms. In the absence of stabilizing effects (e.g., anisotropy), there is no mechanism which would drive the system into any of the regular shapes. Note that in systems with low-dimensional temporal chaos the onset of chaotic behavior usually takes place as one of the parameters of the system is changed. In our case the role of this parameter is possibly played by the anisotropy.

Finally, applying the techniques widely used for the characterization of temporal chaos, it was not possible to detect the existence of an attractor associated with the spatial behavior of the interface of viscous fingers. This fact suggests that in spite of their deterministic origin, growing Laplacian interfaces should be described in terms of turbulent behavior in space rather than by low-dimensional chaos.

I thank P. Bak, J. Kertész, P. Szépfalusy and T. Tél for useful discussions.

1. S. Aubry, in *Solitons and Condensed Matter Physics* edited by A. R. Bishop and T. Schneider (Springer, New York, 1979), p. 264.
2. P. Bak, *Rep. Prog. Phys.* 45, 587 (1982).
3. E. Ben-Jacob, Y. Godbey, N. D. Goldenfeld, J. Koplik, H. Levine, T. Mueller, and L. M. Sander, *Phys. Rev. Lett.* 55, 1315 (1985).
4. A. Buka, J. Kertész, and T. Vicsek, *Nature*, 323, 424 (1986).
5. D. Bensimon, L. P. Kadanoff, S. Liang, B. I. Shraiman, and L. Tang, *Rev. Mod. Phys.* 58, 977 (1986).
6. P. Berge, Y. Pomeau, and C. Vidal, *Order Within Chaos* (Wiley, New York, 1986).
7. J. D. Chen and D. Wilkinson, *Phys. Rev. Lett.* 55, 1982 (1985).
8. B. V. Chirikov, *Phys. Rep.* 52, 263 (1979).
9. G. Daccord, J. Nittmann, and H. E. Stanley, 1986 *Phys.Rev. Lett.* 56, 336 (1986).
10. Y. I. Frenkel and T. Kontorova, *Zh. Eksp. Teor. Fiz.* 8, 1340 (1938).
11. P. Grassberger and I. Procaccia, 1983 *Phys. Rev. Lett.* 50, 346 (1983).
12. P. Grassberger, *Physica* 140A, 319 (1986).
13. J. S. S. Hele-Shaw, *Nature* 58, 34 (1898).
14. V. K. Horváth and T. Vicsek (to be published).
15. R. Lenormand, *Physica* 140A, 114 (1986).
16. E. N. Lorentz, *J. Atmos. Sci.* 20, 130 (1963).
17. K. J. Måløy, J. Feder and J. Jossang, *Phys. Rev. Lett.* 55, 2681 (1985).
18. B. B. Mandelbrot, *The Fractal Geometry of Nature* (Freeman, San Francisco, 1982).
19. W. W. Mullins and R. F. Sekerka, *J. Appl. Phys.* 34, 323 (1963).
20. L. M. Sander, P. Ramanlal and E. Ben-Jacob, *Phys. Rev.* A32, 3160 (1985).
21. H. E. Stanley and N. Ostrowsky (eds), *On Growth and Form* (Martinus Nijhoff, Dordrecht, 1985).
22. J. S. Turner, J. C. Roux, W. D. McCormick and H. L. Swinney, *Phys. Lett.* 85A, 9 (1981).

$1/f$ VERSUS $1/f^\alpha$ NOISE

MICHAEL F. SHLESINGER[1] and BRUCE J. WEST[2]

[1] *Physics Division*
Office of Naval Research
800 North Quincy Street
Arlington, VA 22217 USA

[2] *Institute for Nonlinear Science, R-002*
University of California at San Diego
La Jolla, CA 92093 USA

and

La Jolla Institute
La Jolla, CA 92037 USA

ABSTRACT. A generic mechanism for generating the ubiquitous phenomenon of $1/f$ noise is reviewed. In this approach, $1/f$ noise is tied directly to a probability limit distribution for a product of random variables which can explain its widescale occurrence. A scaling hypothesis is invoked which provides the basis for a crossover from $1/f$ to $1/f^\alpha$ noise. Examples of these ideas are given in small particle statistics, the solar electron velocity distribution, income distributions, scientific productivity, bronchial structure, and cardiac activity.

1. Introduction: $1/f$ Noise

A random process $I(t) = \langle I \rangle + \delta I(t)$, with a correlation function $C(t) = \langle \delta I(t)\delta I(0)\rangle$ has a power spectrum $S(f)$ defined here as

$$S(f) = \mathrm{Re} \int_0^\infty \exp(2\pi i f t)C(t)dt. \qquad (1)$$

If $C(t) = \exp(-t/\tau)$ then

$$S(f) = \frac{\tau}{1 + \omega^2\tau^2}, \qquad\qquad \omega = 2\pi f. \qquad (2)$$

This noise only involves a single time scale. For $\omega\tau \ll 1$, $S(f)$ is flat and is called white noise.

It has been known since the 1920's that voltage fluctuations in carbon resistors have a power spectrum with the low frequency behavior of $S(f) \sim 1/f$. This is called "one over f" noise. White noise, which is well understood, and related to thermal equilibrium, has been considered normal, and $1/f$ noise has been considered as an anomaly. However, the examples of $1/f$ noise have become so widespread that it is clearly a ubiquitous phenomenon and must be considered natural and perhaps even inevitable.

One approach to understanding $1/f$ noise is to consider a distribution of time scales $D(\tau)$, so

$$S(f) = \text{Re} \int_0^\infty e^{i\omega t} e^{-t/\tau} D(\tau) d\tau, \tag{3a}$$

with

$$\int_0^\infty D(\tau) d\tau = 1. \tag{3b}$$

Van der Ziel[4] in 1950 emphasized that a scale invariant $D(\tau) d\tau = d\tau/\tau$ produces $1/f$ noise, i.e., from Eqs. (2) and (3).

$$\int_0^\infty \frac{d\tau}{1 + \omega^2 \tau^2} = \left(\frac{\pi}{2}\right)\left(\frac{1}{\omega}\right).$$

A generic mechanism generating this $D(\tau)$ was lacking, and in any event $D(\tau) = 1/\tau$ is not normalizable so high and low cutoffs are necessary. We show in the next section that the lognormal distribution is a natural candidate for $D(\tau)$. It is normalizable, produces $1/f$ noise, and it is a probability limit distribution.[8]

2. The Lognormal Distribution: From Kolmogorov's Rocks to the Solar Wind

Although the lognormal distribution was introduced in 1879, let us follow Kolmogorov's 1941 derivation which involved investigating the distribution of crushed ore sizes. Let X_0 be the initial size of a rock and suppose that it undergoes a succession of random breaks to attain sizes $X_1, X_2, \ldots X_N$. The difference $X_{n-1} - X_n$ is a random portion λ_n of X_{n-1}, i.e.,

$$X_{n-1} - X_n = \lambda_n X_{n-1} \qquad (0 \le n \le N), \tag{4}$$

where λ_n is uniformly distributed between 0 and 1. Dividing (4) by X_{n-1} we form the sum (and go to the continuum limit)

$$\sum_{n=1}^N \left[\frac{X_{n-1} - X_n}{X_n}\right] \rightarrow \int_{X_0}^{X_n} \frac{dx}{x} = \lambda_1 + \cdots + \lambda_N. \tag{5}$$

The rhs has a normal distribution so $\ln(X_N/X_0)$ is normal and the probability density $f(x)$ of X_N/X_0 is lognormal, i.e.,

$$\begin{aligned} f(x)dx &= \frac{1}{(2\pi\sigma^2)^{1/2}} \exp\left(-[\ln x/\langle x \rangle]^2/2\sigma^2\right) d\log x, \\ &= \frac{1}{x}\frac{1}{(2\pi\sigma^2)^{1/2}} \exp\left(-[\ln x/\langle x \rangle]^2/2\sigma^2\right) dx, \end{aligned} \tag{6}$$

where $\langle x \rangle$ and σ^2 are the mean and variance of the distribution, and these depend on the moments of the λ_i. The larger the σ, the more extended will be the $f(x) = 1/x$ behavior. We can rewrite (4) as

$$X_N = \prod_{n=1}^N (1 - \lambda_n) X_0, \tag{7}$$

i.e., if X_N can be expressed as a product of random variables then under mild conditions it will have a lognormal distribution. We now give a few illustrative examples.

Shockley[5] argued that several factors enter into the publication of a paper and applied these ideas to explain his discovery that the number of publications for a group of scientists has a lognormal distribution. For a manager of science this could be used to argue that bonuses should be proportional to the log(productivity) rather than to the productivity directly, i.e., let each be rewarded according to the logarithm of his worth.

In a similar manner, one can explain the lognormal distribution of electron velocities v in the solar wind.[6] A high energy distribution of solar electrons loses energy to a thermal low energy electron distribution because the scattering cross-section varies as $1/v^4$. Basically, high energy electrons only see low energy ones. Each collision reduces the velocity magnitude of a high energy electron in the same manner that Kolmogorov's rocks were reduced in size.

Many other processes have been found to be lognormal.[7] For example, rainfall amounts depend multiplicatively on several factors and have been found to be lognormal. Thus the input into rivers is lognormal so it is not surprising that river height fluctuations are $1/f$. The detailed behavior of $S(f)$ for $D(\tau)$ being lognormal, including the range of $1/f$ noise has been calculated.[8]

3. A Richness of Scales: $1/f^\alpha$ Noise

It is not hard to describe a product of factors which determines an earned income. In fact, the U.S. income distribution is lognormal from about 5% to 95% of the population. However, there is a crossover and the last few percent follow a power law distribution. It was suggested that if $g(x)$ represents the lognormal (with mean $\langle x \rangle$), then some fraction of the population (proportional to γ) receives additional income from investments and this group follows a law $(1/N)g(x/N)$ which has mean value $N\langle x \rangle$. Now some fraction γ of this group can invest investment profits and follow a distribution $(1/N^2)g(x/N^2)$. The total normalized income distribution $G(x)$ would then be

$$G(x) = (1-\gamma)\left\{ g(x) + \frac{\gamma}{N}g\left(\frac{x}{N}\right) + \frac{\gamma^2}{N^2}g\left(\frac{x}{N^2}\right) + \cdots \right\}. \tag{8}$$

$G(x)$ satisfies the scaling equation

$$G(x) = \frac{\gamma}{N}G\left(\frac{x}{N}\right) + (1-\gamma)g(x). \tag{9}$$

Although the solution to (9) is complicated, it contains the solution to the homogenous equation $G_h(x) = (\gamma/N)G_h(x/N)$, which can be verified by substitution that

$$G_h(x) = \frac{A(x)}{x^{1+\mu}}, \tag{10}$$

where $\mu = \ln(1/\gamma)/\ln N$ and $A(x) = A(x/N)$, which implies that $A(x)$ is periodic in $\ln x$ with period $\ln N$. In the income distribution example, $G(x)$ would look lognormal for a fraction $(1-\gamma)$ of the population and then turn over smoothly to a power law. Although Eq. (9) was introduced independently in the present context,[8] it was previously used by Novikov[13] in a model of the generation of intermittency in a turbulent flow.

Let us now apply these scaling ideas to the structure of biological systems.[9-11] In our first use of Eq. (8) the succeeding terms represented a branching out of income sources. In this next example, we consider the actual branching of the bronchial tree. The human lung has irregularities and a richness of structure along with organization. If one starts counting from the trachea, the branches emanating down can be counted and labeled by a generation number z (over 20 generations have been counted). Let $d(z)$ be the average diameter of an airway in the zth generation.

Classical scaling,[12]

$$d(z) = \gamma d(z-1), \qquad \gamma < 1, \tag{11}$$

seems intuitively obvious, i.e., the diameter of the present generation is some fraction of the previous one. However, this only allows for a single scale, $z_0 = \ln(1/\gamma)$, to appear; the solution to (11) is

$$d(z) = d(0)e^{-z/z_0}, \tag{12}$$

which does not fit experiment. After the first few generations, it appears that scaling applies in the form

$$d(z) = \frac{\gamma}{N} d\left(\frac{z}{N}\right) + \text{analytic function } (z).$$

The later generations become self-similar with a distribution of diameters. The solution, as before, is

$$d(z) = \frac{A(z)}{z^{1+\mu}}, \qquad \mu = \frac{\ln(1/\gamma)}{\ln N}$$

with $A(z)$ a logarithmically oscillating function of z. This oscillating power law does provide an excellent fit to data for the average diameter $d(z)$. It is the first example we know of where the oscillating factor in (10) is important. The solution for $d(z)$ incorporates a richness of scales which is a more appropriate candidate to describe a complex system than the single scale classical scaling.

As a last example consider the His-Purkinjee system, a branched structure which innervates the heart.[10] This structure takes a single nerve pulse and branches it out so that it reaches the cardiac muscle in a distribution of sites, each pathway with its own time scale. The ensuing spatial-temporal richness which this system generates from a single pulse is the most likely source of $1/f$ noise found in healthy EKG's which is superimposed on sharp peak representing the steady mean interbeat frequency. The $1/f$ noise is a signature of chaotic jitter in the heartbeat which prevents the mode-locking into a perfect oscillation. A mode-locked state would be insensitive to feedback and would imply a loss of communication between the heart and the rest of the body. The $1/f$ noise provides a flexibility (a rich frequency content) in which the cardiac oscillator can respond to changes. The appearance of mode-lock spectral peaks has been observed in the EKG's of patients undergoing heart attacks, lending credulence to the idea that mode locked oscillations without jitter are unhealthy. Also, EKG's of older patients show a diminished range of $1/f$ noise lending credulence to the idea that chaos is healthy.

In any event, a nonbranched single time scale system innervating the heart would have the Lorentzian power spectra $S(\omega) = \tau/[1 + \omega^2 \tau^2]$. The His-Purkinjee branched network is assumed to provide a distribution $D(\tau)$ of time scales following the behavior of (10), $D(\tau) \sim \tau^{-1-\alpha}$, which, using Eq. (3), yields

$$S(f) \sim 1/f^{1-\alpha}. \tag{13}$$

The exponent α can be positive or negative so long as $D(\tau)$ (including cutoffs) is normalized. The exponent α may also depend on other variables such as temperature, pressure, etc., depending on the system.

4. Summary

We have shown how a product of random variables generically leads to the lognormal distribution and to $1/f$ noise. Our scaling equation for branched processes was shown to provide a turnover to a power law distribution which provided for an infinity of scales with none of them dominating. This produced $1/f^\alpha$ noise. In our view, millions of years of evolution has chosen branched structures, and noisy chaotic dynamics for higher organisms. The human body abounds in such complex systems, the heart, the chordia tendinae, the biliary network, the vascular tree, and the urinary collecting system to name a few. Biological fractal structures cover much space, with little mass are fault tolerant in structure and function, and provide array of real systems whose nature and health are of immediate and immense importance to mankind. We predict that the ideas on fractals and dynamics wil find its most useful applications in the biological sciences.

1. P. Dutta and P. M. Horn, Rev. Mod. Phys. 53, 497 (1981).
2. M. Weismann, Rev. Mod. Phys. 60, 537 (1988).
3. *Ninth Int. Symposium on Noise in Physical Systems*, ed. C. M. Van Vliet (World Science Pub., Singapore, 1987), and previous volumes.
4. A. Van der Ziel, Physica 16, 359 (1950).
5. W. Shockley, Proc. IRE 45, 279 (1957).
6. M. F. Shlesinger and M. Coplan, J. Stat. Phys. 52, 1423 (1988).
7. J. Atchison and J. C. Brown, *The Lognormal Distribution* (Cambridge U Press, New York, 1963).
8. E. W. Montroll and M. F. Shlesinger, Proc. Nat. Acad. Sci. (USA) 79, 338 (1982); J. Stat. Phys. 32, 209 (1983).
9. B. J. West, V. Bhargava, and A. L. Goldberger, J. Appl. Phys. 60, 189 (1986).
10. A. L. Goldberger, V. Bhargava, B. J. West, and A. J. Mandell, Biophys. J. 48, 525 (1985).
11. B. J. West and A. L. Goldberger, Am. Sci. 75, 354 (1987).
12. E. R. Weibel and D. M. Gomes, Science 137, 577 (1962).
13. E. A. Novikov, Sov. Phys. Dokl. 11, 497 (1966).

A SIMPLE MODEL OF MOLECULAR EVOLUTION

L. PELITI

Dipartimento di Scienze Fisiche and Unitá GNSM-CISM,
Universitá di Napoli, Mostra d'Oltremare, Pad. 19,
I-80125 NAPOLI (Italy)

Although we have now been accustomed to see growth in aggregates, crystals, or fractures, it is true that we normally associate growth with life. And in some sense we feel that life as we know it originated from some fluctuation which took place in the early Earth: this is why I feel that a talk on models of the origin of life can fit in the subject-matter of this School without too much stretching.

I believe that most of the work to be done to understand the origin of life is of biochemical nature: to clarify the detailed molecular mechanisms which might have emerged, leading from the "primeval soup" to the first cell. Unfortunately I am not at all qualified to work in this fascinating area. There is however a problem against which a few theoretical physicists have tried their forces in recent years: it goes under the slightly misleading name of the "origin of biological information." I quote[1]: "We would like to understand in principle how, given a collection of related molecules (for example, guanine and cytosine) and random events, biologically useful information can spontaneously arise in the form of molecular sequences along macromolecular strings."

We know that the macromolecules of life (nucleic acids and proteins) are, in the words of Schrödinger, "aperiodic crystals": they are neither fully repetitive nor completely random. How did such structures arise? Anderson[2] pointed out that the key point is to understand the concomitant features of stability of these structures from generation to generation and of their diversity as the essential ingredients for evolution to start acting. On the other hand, Eigen,[3] on identifying the "hypercycle" as the unit of molecular reactions which could insure the emergence of biological information, stressed the need of the coexistence of several individually reproducible molecular species to overcome the "information barrier"—the fact that a system capable of insuring its own replication must possess a minimal amount of complexity. A physical system exhibiting the coexistence of stability and diversity is the spin glass: at low temperatures, the thermodynamical states are close to energy minima (stability); but frustration in spin-spin interactions leads to a great deal of essentially equivalent, yet different, minima (diversity). This prompted Anderson and collaborators[1,2] to introduce a model of protopolynucleotide chains (the precursors of the nucleic acids) which grow and reproduce by a pairing mechanism which corresponds to the template replication mechanism of present-day DNA, and which are selected with a probability defined in terms of a fitness function H having the structure of a spin glass Hamiltonian.

The work I have done in collaboration with Cettina Amitrano and Mohammed Saber[4] at the University of Naples addresses the question whether the class of models considered by Anderson and collaborators could lead to minimal complexity systems such as those postulated by Eigen. The answer is negative, but a great deal is learned along the way.

We take a very simple model of protopolynucleotide chains of *fixed* length N. Each monomer can be of one of two species, A or B, which can be aptly represented by an Ising spin $\sigma_i^\alpha = \pm 1$, where α labels the chain, and i labels the monomer. We also assume that some replication mechanism exists with a low error rate, so that at each generation slightly different variants of the previously present chains appear in the soup. We work for convenience with a fixed number M of chains. We adopt a mutation and selection procedure defined as follows: (i) *mutation*: one chooses a small fraction τ among the $N \cdot M$ Ising spins present in the soup, and changes their state as follows: $\sigma_i^\alpha \rightarrow -\sigma_i^\alpha$; (ii) *selection*: for each chain a one computes the fitness function $H(\underline{\sigma}^\alpha = H(\sigma_1^\alpha, \ldots, \sigma_N^\alpha)$, and then removes the chain with a death probability p given by

$$p = \{1 + \exp[\beta(H(\underline{\sigma}) - H_0)]\}^{-1}.$$ (1)

The coefficient β is a sharpness parameter, and H_0 is a threshold. This form of the death probability was suggested by Anderson.[2] Remark that the convention in biological literature is the opposite of what is natural for physicists, since *higher* values of H are considered as more favorable. After selection has taken place, one fills in the gaps in the population by making a suitable number of copies of randomly chosen chains among the surviving ones.

We have simulated the model with different choices of the fitness function $H(\underline{\sigma})$. One choice was a spin glass Hamiltonian, as originally suggested by Anderson[2]:

$$H_A(\underline{\sigma}) = \Sigma J_{ij}\sigma_i\sigma_j.$$ (2)

The sum runs over all pairs of spins belonging to the chain: for each pair, J_{ij} is an independent random variable (we took it to be equal to ± 1 with equal probability). Actually it is much simpler to consider the limit case in which $H(\underline{\sigma})$ is a completely uncorrelated random function. The importance of correlation in this context has been emphasized by Kauffman.[5] In practice the correlation one considers is that between the values of $H(\underline{\sigma})$ for different configurations σ and σ', as a function of their Hamming distance d_H, i.e., of the number of spins which are different in the two configurations:

$$d_H(\underline{\sigma}, \underline{\sigma}') = \frac{\Sigma(1 - \sigma_i\sigma_i')}{2}.$$ (3)

It is possible to define a whole spectrum of more or less correlated functions. The most correlated (after the constant!) is of the form

$$H(\underline{\sigma}) = \Sigma h_i\sigma_i,$$ (4)

since changing the state of one spin produces a change of a fraction $0(1/N)$ of the total value. The spin glass function [Eq. (2)] is still quite correlated, yielding a variation $0(1/\sqrt{N})$. A fully uncorrelated (or *rugged*) fitness function is one in which to each of the 2^N possible configurations corresponds an independent random variable. In the spin glass literature this is known as the Random Energy Model.[6] It may be conceived as a spin model with many-spin interactions. We take this form of the fitness function, and we shall see later how introducing correlations modifies our conclusions. We take therefore $H(\underline{\sigma})$ to be an independent random variable for each configuration $\underline{\sigma}$, say, uniformly distributed between -1 and +1.

Let us now take what may appear as a trivial limit, namely $\beta = 0$. Then the death probability is equal to 1/2, independently of the configuration. This situation has been investigated in detail by Zhang et al.[7] Let us first consider the case of a varying population size M. It is easy to show that if we have at the beginning M individuals, and each is equally likely to die or to produce an offspring at each generation, after a number of generations proportional to M the probability that none at all has survived is essentially equal to 1. Let us now start with M chains, which mutate and are selected just as we said. As some of them die and are replaced, the population organizes into families, descendent of the original ones. But the number of families, just as it happens with human families in closed valleys, gets smaller and smaller with time. Assume there are F families at a given time. Each has on average M/F representatives, therefore has the chance of disappearing after M/F generations. Hence the number of families decreases by one over this number of generations:

$$dF \propto - \left(\frac{F}{M} \right) dt. \qquad (5)$$

By integrating this equation, we see that after a number of generations of the order of M (up to logarithms) all original families, except one, have disappeared. Now, if the mutation rate τ is very small, the Hamming distance $\tau N M$ between any two surviving chains will not be very large. We will have therefore a population of very similar chains, which is called a "quasi-species" in the trade.[3] The mutation speed of the quasi-species is τN mutations per generation, independently of the size M of the population. To see this, consider that if we look back M generations in time we identify the common ancestor of all surviving chains. Hence if we look at the evolution of the population as a function of time, we see a "tube" of Hamming width $\tau N M$ around this common ancestor. The width of the tube depends on M, but its location is just the location of one single chain, which evolves at its fixed rate independently of M. These results remain qualitatively the same, provided that there is a finite death probability per each generation.

Let us now look at the opposite limit, $\beta \to \infty$. Here the situation is simple: either $H(\underline{\sigma}) > H_0$, and the chain survives; or $H(\underline{\sigma}) < H_0$, and the chain dies. But since $H(\underline{\sigma})$ is not correlated, the forbidden configurations with $H(\underline{\sigma}) < H_0$ are distributed at random among all possible ones. Each chain has an equal probability of stepping onto one of them as any other. Therefore we have a uniform, finite value of the death probability, and the results we just mentioned apply. With one exception: that the common ancestor is constrained to walk only over the permitted configurations. Now, if H_0 is small, only few points are forbidden: the center of the distribution wanders slightly more slowly, but as free as before. But as H_0 increases, and more and more points become forbidden, the remaining ones become less and less connected, until, at a certain threshold H_c, they split into finite, mutually isolated clusters. This is nothing else as the percolation transition, seen here on the hypercube formed by the 2^N possible configurations, instead of in Euclidean space as we are accustomed. It has been recently characterized by Flesselles and collaborators.[8] The critical value H_c separates a percolating regime from a trapping regime. In the percolating regime the largest cluster contains points with mutual Hamming distance of order N. Above this critical value, the Hamming width of any connected cluster is finite. The transition is sharp in the $N \to \infty$ limit. In our problem, if the population wanders on the percolating cluster, it will eventually drift far away from its starting point: i.e., each of its units will eventually mutate. But if we are in the trapped regime, the population is not able to mutate very far away, and will keep forever the memory of its initial configuration. Let us remark that in either case there is no tendency towards adaptation, i.e., the average values of H do not appear to increase as time goes on. We have thus defined what is usually called a neutralist model[9] of evolution.

What changes are introduced in this picture by correlation? The transition between the two regimes still takes place: it is just the prototype ergodicity breaking transition, seen here in the microcanonical rather than the canonical ensemble. Only, because allowed and forbidden configurations are not so intimately intertwined, the allowed clusters are much more compact and the average number of dying chains (which try to cross the frontier) is much smaller. As a consequence the population is much more dispersed and, in the trapped regime, may even end up more or less filling in one of the allowed clusters. But coexistence of stability and diversity is never observed.

What about adaptation? I think that the only way of getting it in this model is to take β large (but not infinity) and H_0 also large. The surviving probability gets very small, and depends exponentially on H: hence each small improvement can become very important. Simulationwise this means that one must go to large populations. Presumably the population will stick to a local maximum of H, but from time to time, when a mutant hits by chance a better optimum, it will rapidly switch to this new location. This "hopping regime" is reminiscent of the "punctuated equilibria" suggested by Eldredge and Gould[10] on the basis of paleontological data.

A way to obtain a sufficiently diverse and stable chain population is to introduce chain interactions in the picture, thus giving up the notion of a God-given fitness function H. Several attempts are being made in this direction, and I hope that some understanding will soon be reached.

1. D. S. Rokhsar, P. W. Anderson, and D. L. Stein, J. Mol. Evol. 23, 119 (1986).
2. P. W. Anderson, Proc. Natl. Acad. Sci. USA 80, 3386 (1983).
3. M. Eigen, Adv. Chem. Phys. 38, 221 (1980).
4. C. Amitrano, L. Peliti, and M. Saber, C. R. Acad. Sci. Paris, III (submitted).
5. S. Kauffman, *Origins of Order: Self-Organization and Selection in Evolution* (in press).
6. B. Derrida, Phys. Rev. B 24, 2613 (1981).
7. Y. C. Zhang, M. Serva, and M. Polikarpov, Phys. Rev. Lett. (submitted).
8. J. M. Flesselles, J. A. Campbell, R. Jullien, and R. Botet, J. Phys. A: Math. Gen. (in press).
9. M. Kimura, *The Neutral Theory of Molecular Evolution* (New York: Cambridge University Press, 1983).
10. N. Eldredge and S. J. Gould, in *Models in Paleobiology* ed. T. M. Schopf (San Francisco: Freeman, 1972), p. 82.

DAVID NELSON

SCALE INVARIANT SPATIAL AND TEMPORAL FLUCTUATIONS IN COMPLEX SYSTEMS

PER BAK, CHAO TANG, and KURT WIESENFELD
Brookhaven National Laboratory
Upton, New York 11973 USA

ABSTRACT. Dynamical systems with many degrees of freedom can evolve (or be driven) into a stationary state exhibiting scale invariant spatial and/or temporal fluctuations. General arguments are supported by numerical simulations of simple models. Scaling exponents are defined and calculated. We suggest that the concept presented here has wide potential applications, including fractal patterns, $1/f$ noise, randomly pinned surfaces or interfaces, glasses, and earthquakes. Some of the applications are discussed in detail.

1. Introduction

Because of the pedagogical effort by Mandelbrot[1] and others we have learned that many objects in nature, ranging from mountain landscapes to electric breakdown and turbulence, have a self-similar fractal spatial structure. This is by no means a trivial observation since this implies that the systems are correlated over large distances. Much effort has been put into computer simulation and characterization of these objects. However, the empirical geometrical observation does not serve as an explanation.[2] It seems obvious that in order to understand the origin of self-similar structures one must understand the nature of the dynamical processes that created them: The temporal and spatial properties must necessarily be completely interwoven.

Scaling self-similar structures are well known from systems undergoing second order phase transitions. If a parameter, such as the temperature is varied carefully, it is possible to bring the system to a critical point, where the spatial and temporal correlation functions are invariant under rescaling: the system has no intrinsic macroscopic length and time scales. We shall argue that extended dynamical systems by themselves may evolve into a critical point,[3] with no help from an experimentalist turning the knobs. The spatial structures that we observe can be visualized as snapshots of a "slow" quasi-stationary process. The critical state is a global attractor of the dynamics.

Because of the ubiquity of the phenomena that we wish to explain, no solution can be both complete and general. We shall limit ourselves to investigating a few simple models constructed from physical intuition derived from observing some specific phenomena in nature. We would like to think of these models as "Ising models" for extended dynamical systems, which despite their extreme simplicity may encompass the relevant physics. Maybe we shall even observe "universality" in the sense that exponents, scaling relations, etc. are the same as for real systems. Our hope is that the readers imagination will allow him to see the general philosophy behind the models, and maybe discover an application of his own.

Our idea can be easily visualized in a "toy" example. Consider building a sandpile from scratch by adding particles randomly and very slowly (Fig. 1). At the beginning, the pile is quite flat. The addition of a single particle causes (at most) a small local rearrangement. The correlation length is very short. As we continue, the pile gets steeper and steeper, and some of the "avalanches" become larger and larger. Finally the pile will reach a statistically stationary state, where its slope will grow no more and the rearrangements following each addition of particles take place on any length scales and time scales (limited only by the size of the system). The system organizes itself into a critical point and will stay there as long as we add particles slowly. The avalanche size obeys a power law distribution $D(s) \approx s^{-\tau+1}$. The duration t of the avalanche obeys a similar distribution $D(t) \approx t^{-b}$. We can also approach this critical state from the other side by starting with a large slope and letting the pile relax. The system will end up in the same critical point.

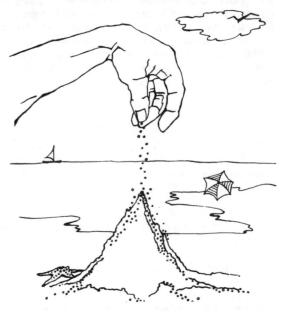

Fig. 1: The formation of a sandpile. After a while, the pile reaches its critical state, and the slope will grow no further. One then sees scale invariant spatial and temporal fluctuations.

Although the critical point is an attractor, we can force the system away from the critical point. This allows us to define a complete set of critical indices describing the development of a suitably defined order parameter and correlation length. The paper is organizes as follows. In §2, we consider for pedagogical reasons a one-dimensional model. We will show that the dynamics drives the system into a "trivial" critical state, the least stable state. In §3, we show that the least stable state is generally unstable in higher dimensions and the system will end up in a critical state with scale invariant spatial and temporal fluctuations. Numerical simulations will be presented. A set of critical exponents will be defined and scaling relations derived in §4. In §5, we summarize and discuss some applications and speculations.

2. A Simple One-dimensional Model

In this section we study a simple discrete one-dimensional model. We are interested in dissipative systems with many metastable states. Let z_n be a local discrete variable on an one dimensional lattice $(n = 1, 2, \ldots, N)$, each of which has several metastable states $z_n = 0, 1, 2, \ldots, z_c$. The total number of the metastable states of the system is $(z_c + 1)^N$. If, for some reason, z_n exceeds a critical value somewhere, $z_n > z_c$, the site n becomes unstable. The state z_n will change and the change will affect its neighbors. One can write down, for example,

$$z_n \to z_n - 2 \quad \text{and} \quad z_{n\pm1} \to z_{n\pm1} + 1, \qquad \text{if} \quad z_n > z_c. \tag{2.1}$$

One can imagine z_n being the local slope of a sandpile, the spring force on an array of damped pendula coupled by weak torsion springs, the pinning force on a randomly pinned elastic interface, the pressure in an "one-dimensional" glass, the temperature gradient of a fluid system close to some instability, the savings of individual families of a community, and so on. At least for the first two cases mentioned above, the reader can convince himself instantly that (2.1) is sensible. It corresponds to a grain of sand tumbling to a lower level if the local slope is too high, or a pendulum rotating a turn if the spring force exceeds the gravitational force.

To feed the system, randomly choose a site n by letting either

$$z_n \to z_n + 1, \tag{2.2a}$$

or

$$z_n \to z_n + 1 \quad \text{and} \quad z_{n-1} \to z_{n-1} - 1. \tag{2.2b}$$

Equation (2.2a) corresponds to uniformly driving the system (tilting a sandpile, increasing the driving force on a pinned interface, etc.). Equation (2.2b) can be thought of adding a particle to the pile. Various boundary conditions can be specified, depending on physical systems studied. For concreteness, here we use closed boundary condition at the left hand side and open one at the right hand side:

$$z_1 \to z_1 - 2 \quad \text{and} \quad z_2 \to z_2 + 1, \qquad \text{if} \quad z_1 > z_c;$$

$$z_N \to z_N - 1 \quad \text{and} \quad z_{N-1} \to z_{N-1} + 1, \qquad \text{if} \quad z_N > z_c. \tag{2.3}$$

Hence sand can be transported out of the system at the right end, while there is a wall at the left end.

Now add particles randomly [Eq. (2.2b)] to an empty system and let the system relax [Eq. (2.1) and (2.3)] before adding next one. The pile will build up, eventually reaching the point where all the local slopes z_n assume the critical value, $z_n = z_c$. This is the least stable of all the metastable states. Any additional particles simply fall from site to site (left to right) and exit at the end $n = N$, leaving the system in the least stable state. In other words, the effect of a small local perturbation is communicated throughout the system, but the system is robust with respect to noise insofar as it returns to the least stable state. If particles are added randomly, the resulting flow is also random white noise, i.e., with power spectrum $1/f^0$. As we shall see in the next section, the robustness of the minimally stable state is lost in two and higher dimensions.

The dynamical selection principle leading to the least stable state is insensitive to the way the system is built up. One might use (2.2a) instead of (2.2b). One could also start with a very unstable state, $z_n > z_c$ for all n, and let the system relax. In all these cases the minimally stable state will be reached. In one dimension, the least stable state is critical in the restricted sense that any small perturbation can just propagate infinitely through the system, while any lowering of the slope will prevent this. This is analogous to some other 1D critical phenomena, such as percolation where at the percolation threshold particles can just percolate to infinity. Also, like other 1D systems the critical state has no spatial structure, and correlation functions are trivial. In the next section we shall see that in higher dimensions the critical states and their dynamics are dramatically different.

3. Self-organized Criticality

The rules (2.1) and (2.2) for the one-dimensional model can easily be generalized to higher dimensions. In two dimensions:

$$z(x,y) \rightarrow z(x,y) - 4,$$
$$z(x, y \pm 1) \rightarrow z(x, y \pm 1) + 1,$$
$$z(x \pm 1, y) \rightarrow z(x \pm 1, y) + 1, \qquad \text{if} \quad z(x,y) > z_c. \tag{3.1}$$
$$z(x,y) \rightarrow z(x,y) + 1. \tag{3.2a}$$
$$z(x,y) \rightarrow z(x,y) + 2,$$
$$z(x - 1, y) \rightarrow z(x - 1, y) - 1,$$
$$z(x, y - 1) \rightarrow z(x, y - 1) - 1. \tag{3.2b}$$

Here we have the square array (x,y), for $1 \leq x, y \leq N$. Naively, one might expect that the situation is the same as in one dimension, namely that the system will build up (or collapse) to the least stable state where the slopes $z(x,y)$ all assume the critical value. A moments reflection will convince us that it cannot be so. Suppose we perturb the least stable state on one site. This will render the surrounding sites unstable ($z > z_c$), and the noise will spread to the neighbors, then their neighbors, in a chain reaction, ever amplifying since the sites are generally connected with more than two least stable sites, and the perturbation eventually propagates throughout the entire lattice. The least stable state is thus unstable with respect to small fluctuations and cannot represent an attracting fixed point for the dynamics. As the system further evolves, more and more more-than-least stable states will be generated, and these states will impede the motion of the "noise." *The system will become stable precisely at the point when the network of least stable clusters has been broken down to the level where the noise signal cannot be communicated through infinite distances. At this point there will be no length scale, and consequently no time scale in the problem.* Hence one might expect that the system approaches, through a self-organizing process, a critical state with power law correlation function for physically observable quantities, including the power spectrum. In analogy with the discussion for the one dimensional case, the "slope" z will build up to the point where stationarity is obtained: *this is assured by the self-organized critical state,* but not the minimally stable state. The average slope of the critical state is reduced compared to that of the least stable state.

Now the reader can guess what will happen if we perturb the critical state locally via Eqs. (3.2). The perturbation will grow over all length scales. That is, a given perturbation can lead to anything from a shift of a single unit to a large avalanche. The lack of a characteristic length scale leads directly to a lack of a characteristic time scale for the fluctuations. As is well known, a random superposition of pulses of a physical quantity with a distribution of lifetimes $D(t) \approx t^{-b}$ (weighed by the average value of the quantity during the pulse) leads to a power frequency spectrum, $S(f) \approx f^{-2+b}$, so we also expect a $1/f$ like power spectrum[4] for the system. Extensive numerical simulations in two and three dimensions have been carried out to testify the above argument.[3] Here we show only one of the results. Plotted in Fig. 2 is the distribution of cluster (or "avalanche") size at the stationary critical state. A cluster is defined as those sites being affected by a single perturbation before it dies out. Hence the clusters are defined dynamically. Equations (3.1), (3.2a), and closed boundary condition ($z = 0$) were used in this simulation. The distribution function fits a power law $D(s) \approx s^{-\tau+1}$ with $\tau \approx 2$. The fall off at large s is due to the finite size effect. Figure 2 shows that there is no macroscopic length scale except the system size.

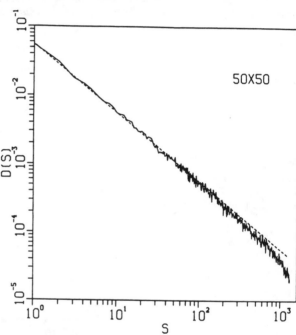

Fig. 2: Cluster size distribution for a two dimensional system of size 50×50. The measurement is taken by applying Eq. (3.2a) 100,000 times after the system reaches a stationary state. The dashed straight line has a slope of -1.

4. Order Parameter and Critical Exponents

In this section, we identify an order parameter for the self-organized critical phenomena and define a set of critical exponents. Let us first find the order parameter. If the average "slope" $\theta = \langle z \rangle$ of the system is, somehow, kept larger than the "critical slope" θ_c, there will be a continuous "spontaneous flow" j. On the other hand, if $\theta \leq \theta_c$ there will be a flow j only when an "external field" is applied by adding particles or increasing the slope [Eq. (3.2)].

Thus, the analogy with "traditional" critical phenomena is now clear. The flow j is the "magnetization" or order parameter. The "magnetic field" h is the current of incoming particles [for Eq. (3.2b)] or the rate of slope increase [for Eq. (3.2a)]. The deviation $\theta_c - \theta$ from the critical slope plays the role of the reduced temperature (or the deviation from the critical concentration for a percolation transition). A lower slope ($\theta < \theta_c$) can be achieved by stopping the build-up of the system before it reaches criticality, or by lowering the slope once the system reaches criticality. A higher slope ($\theta > \theta_c$) can be achieved by applying a finite field h to the system, and wait for stationarity. The susceptibility χ characterizes the response to the field: $\delta j = \chi \delta h$. The correlation length ξ is the cut-off in linear cluster size below criticality, which is related to the cut-off in cluster size through the fractal dimension D, $s_{co} \approx \xi^D$. Thus, we conjecture the following power laws for the average quantities: $j \approx (\theta - \theta_c)^\beta$, $\chi \approx (\theta_c - \theta)^{-\gamma}$, $\xi \approx (\theta_c - \theta)^{-\nu}$, $s_{co} \approx (\theta_c - \theta)^{-1/\sigma}$, and $j(\theta_c - \theta) \approx h^{1/\delta}$. A dynamical exponent z relates the relaxation time t to the linear size ℓ of the cluster, $t \approx \ell^z$, $1 \leq \ell \leq \xi$. A typical numerical simulation for calculation of these exponents is shown in Fig. 3, where the order parameter exponent β is calculated. Scaling relations among these exponents can be derived via the cluster picture, in analogy with static "traditional" critical phenomena. Again we refer the readers to Ref. 3 for details. Here we only give the results: $\gamma = (3-\tau)/\sigma$, $D = 1/\sigma\nu$, $\phi = \gamma/\nu z$, $\gamma/\nu = 2$, where ϕ is the exponent for power spectrum $S(f) \approx f^{-\phi}$ which has a mean-field value 1.

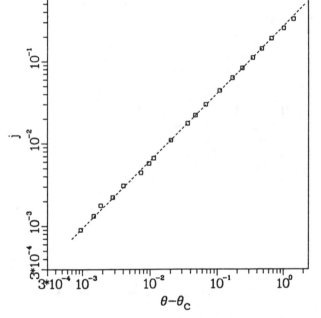

Fig. 3: Order parameter j versus $\theta - \theta_c$ for a 3D system. The dashed straight line has a slope $\beta = 0.82$.

5. Summary and Discussion

Our general arguments and numerical simulations show that dissipative dynamical systems with extended degrees of freedom can evolve towards a self-organized critical state, with scale invariant spatial and temporal fluctuations. Although this new concept is demonstrated via simple models, we believe that it can be taken much further to explain a wide range of observed spatial and temporal scaling. We now discuss a few applications.

The wide spread phenomena of "$1/f$" noise has long been a myth. There are probably many different mechanisms generating the noise, but the picture presented here might be one of the most general mechanisms. As the readers can see, the "$1/f$" like spectrum comes naturally out of the dynamics and no further assumptions and parameter tunings are needed. Our mean-field value for the noise exponent ϕ is 1. This gives us a hint on why "$1/f$" noise is usually so close to $1/f$. The deviations of the exponent from unity in many real systems can thus be understood as deviations from mean-field theory.

When a liquid is rapidly quenched into a glass state, the dynamical process involved seems to have much in common with our model, when started far from equilibrium (all $z > z_c$) and let it relax. At the beginning all "particles" can move and then the system gets stuck at some metastable state after getting rid of excess energy. But this speculation has to be developed further.

Turbulence is a phenomenon known to show both spatial and temporal scale invariant fluctuations. When a fluid system is being driven harder and harder, it may undergo turbulence transition. We believe our idea can be used to describe this transition in large aspect ratio systems where there are "coupled" convective rolls driven by an increasing temperature gradient. The models studied in this paper might have to be modified, though.

Another application, and may be the most straight forward one, is to the randomly pinned elastic surfaces and interfaces. When driven by a slowly increasing external force, these systems are known to undergo a pinning-depinning transition. The variable z in our model can be thought of as the driving force plus elastic force. The z_c can be assigned random values representing randomly distributed maximum pinning force. When $z(\mathbf{r}) > z_c(\mathbf{r})$, the surface at $\mathbf{r}$ is depinned. It will move forward and be pinned again. In doing so, the value of $z(\mathbf{r})$ is reduced and that of its neighbors increased, as described by Eq. (2.1) or (3.1). The slow increasing of driving force is equivalent to (2.2a) or (3.2a). Fisher has studied the pinning-depinning transition for charge density wave systems.[5] His description is actually similar to the least stable state picture described in §2. So it is valid only in one dimension with low connectivity. In higher dimensions (or higher connectivity), as we have shown the system will be depinned before reaching the least stable state. The surface at the transition will then show scale invariant spatial and temporal fluctuations.

Applications might also be found in the zoo of fractal aggregates. Alström recently suggested that the idea can be used in DLA to obtain fractal dimension.[6] There are still some open questions. A big one is the universality. Are there and how many universality classes in this kind of critical phenomena? Presumably there should be some universality and the universality classes depend on symmetry, dimensionality, etc., like the other critical phenomena. But all these are far from clear at the moment.

This work was supported in part by the Division of Materials Science, US Department of Energy, under contract DE-AC02-76CH00016.

1. B. Mandelbrot, *The Fractal Geometry of Nature* (W. H. Freeman, San Francisco, 1982).

2. L. Kadanoff, Phys. Today 39, No. 2, 6 (1986).

3. P. Bak, C. Tang, and K. Wiesenfeld, Phys. Rev. Lett. 59, 381 (1987); Phys. Rev. A 38, 364 (1988); C. Tang and P. Bak, Phys. Rev. Lett. 60, 2347 (1988); C. Tang and P. Bak, J. Stat. Phys. 51, 797 (1988).

4. For a review on $1/f$ spectrum, see W. H. Press, Commun. Mod. Phys. C 7, 103 (1978); P. Dutta and P. M. Horn, Rev. Mod. Phys. 53, 497 (1981).

5. D. Fisher, Phys. Rev. Lett. 50, 1486 (1983); Phys. Rev. B 31, 1396 (1985).

6. P. Alström, preprint.

THE UPPER CRITICAL DIMENSION AND ε-EXPANSION FOR SELF-ORGANIZED CRITICAL PHENOMENA

SERGHEI P. OBUKHOV

Landau Institute for Theoretical Physics
Moscow U.S.S.R.

Recently a new critical phenomena dubbed Self-Organized Criticality was discovered and described.[1] It was shown that extended systems naturally evolve into a stationary state with power-law temporal and spatial correlations.

The example of this model is a sandpile with randomly added sand. This system reaches its critical state where a critical slope of the pile is automatically maintained. The simplest lattice version of this process was formulated and studied analytically in the framework of the mean-field theory[2] and numerically in 2D and 3D.[3] The mean-field exponents appear to be the same as those of the mean-field percolation theory, but 2D and 3D exponents are quite different from the percolation ones.

In this paper we shall argue that the upper critical dimension for S.O.C. phenomena is 4, instead of 6 for the percolation theory. The most relevant one-loop corrections to the mean-field theory will be presented, which permits us to write the renormalization group equations and to calculate the exponents of the S.O.C. phenomena to the first order in $\epsilon = 4 - d$.

We review some details of the considered model. The process of formation of avalanches in a sandpile can be formulated in terms of a variable $Z(x,y)$, which indicates the local slope of the sandpile. If $Z(x,y)$ exceeds a certain critical value Z_c, sand tumbles onto the lower level. This results in changes in Z:

$$Z(x,y) \rightarrow Z(x,y) - 4$$
$$Z(x, y \pm 1) \rightarrow Z(x, y \pm 1) + 1 \tag{1}$$
$$Z(x \pm 1, y) \rightarrow Z(x \pm 1, y) + 1.$$

These equations are written for the two-dimensional case; the generalization to higher dimensions is straightforward. The mean-field theory of the process can easily be formulated if we assume that the distribution of Z in neighboring sites does not depend on $Z(x,y)$. In this case the propagation of excitation can be described by a tree-like process, with Gaussian propagator $G(r,t) = (4\pi t)^{-1/2} \exp(-r^2/t)$, or $G(k,t) = \exp(-k^2 t)$ in momentum representation. The integrated over the time propagator $G(k,t)$ is:

$$G(k) = \frac{1}{k^2}. \tag{2}$$

The absence of a mass term in this propagator follows from the criticality of the theory. It means that on each step of the process the average number of neighboring sites which become active is exactly 1.

The first order corrections to this Gaussian propagator arise from spatial correlations between the sites. These corrections can be caused by processes which took place a long time before, or it can be caused by another branch of the same activation process.

What correlations remain in the system when the excitation passes through the medium? The main feature of S.O.C. is that the average sensitivity of the system to a new excitation does not change, the system remains critical. It is evident, that the sites which were activated before have a slope much lower than the average. But the slope in the neighboring sites is larger than the average. Hence if the excitation comes to the point which was excited previously, the probability to initiate this site is low, but it is high for the neighboring sites. In other words, the previously activated sites repulse the new activation process. This situation is very similar to that in the case of the True Self-Avoiding Walk problem. This is a problem of a random walk which tries to avoid sites visited before. When a walk comes to a site visited before (see Fig. 1a) the most probable next step will be in the direction of other available sites. If all neighboring sites were visited before, the mean displacement is proportional to the density gradient of previously visited sites. This can be written as $u = g_1 \nabla_{x2} G(x_1, x_2, t_1)$. Therefore a path with single self-interaction at the point x_2 differs from a purely random one by a displacement u which occurs at point x_2. This displacement changes the probability of landing after time t_3 at a final point x_3 by

$$\delta G = u \nabla_{x2} G(x_2 - x_3). \tag{3}$$

By summing the correction (3) over all possible positions of the point x_2 and over all possible times t_1, t_2, t_3 it turns out that the correction G is described by the diagram Fig. 1b, where the lines are associated with the correlation functions (2) and interaction vertex is associated with the expression $g_1(q \cdot p)$. Here q is the momentum transferred in the vertex, and p is the momentum of the last outgoing line of the vertex.

Now we come back to the S.O.C. problem. The essential difference between this model and that described above is the possibility of the branching of an activation process. The configurations like that of Fig. 1a are also present, but the most relevant are configurations like that depicted in Fig. 2a. The corresponding diagram is shown in Fig. 2b. Here the branching vertex corresponds to the branching constant λ, the interaction vertex is associated with the same expression $g_1(q \cdot p)$. The additional selection rule must be kept for this diagram: $t_1 < t_2$. The possible one-loop corrections to the interaction vertex are shown on Fig. 3. The diagram 3b is unessential because of complicated time limitations. The diagram 3a is logarithmic at 4D. After simple calculations we get:

$$\delta g = \lambda \left(\frac{1}{4g_1^2(q \cdot p)} + \frac{1}{4g_1^2 p^2} \right) S_4 \ln L. \tag{4}$$

It is surprising that apart from the $q \cdot p$ term it generates also the term with another momentum structure p^2. In higher-order diagrams this term also generates logarithmic corrections. Hence we have two different charges in the theory. The generalized vertex we shall write in the form $g_1(q \cdot p) + g_2 p^2$ and the coupled R.G. equations for both charges will be written below.

Before proceeding further we must take into account the effect of correlations from the processes which where initiated long time before. The single interaction with a trace of previous process results only in a shift, whose orientation is random, so it renormalizes only the diffusion constant of of the activation process. The second order contribution is shown by the diagram on Fig. 4. The heavy point represents the common origin of two branches of the previous process T; the interaction vertices are the same as above. This diagram has the same singularity as the diagram of Fig. 2b.

There is also a possibility that the excitation from two branches of the same avalanche comes at one point simultaneously. It means that at this point Z is increased by two at once. If it exceeds the critical value, then $Z + 2 \rightarrow Z - 2$ and Z in the neighboring points is increased by 1. On the average it will result in only one excited site in the next moment of time. In the case if both branches where not interacting at all the average number of excitations remains 2. Then the intersection of two branches at one moment of time reduces the average number of excitations*). This reduction can be expressed by the diagram Fig. 5. Here the new interaction vertex g is introduced, the times t_1 and t_2 are supposed to be equal. This diagram is typical for directed percolation theory[5] which describes the time-dependent process of spreading of the same excitation in an active media. The directed-percolation theory is also logarithmic in 4D. The complete R.G. equations for all of above charges g_1, g_2, g, λ and T are:

$$\frac{\partial T}{\partial \xi} = (2 - \eta)T - \frac{Tg\lambda}{4}$$

$$\frac{\partial \lambda}{\partial \xi} = \frac{(6 - d - 3\eta)\lambda}{2} - \frac{\lambda^2 g}{2}$$

$$\frac{\partial g_1}{\partial \xi} = \frac{(2 - d - 3\eta)g_1}{2} - \frac{g_1^2 \lambda}{4} - \frac{g_1 g_2 \lambda}{4} \tag{5}$$

$$\frac{\partial g_2}{\partial \xi} = \frac{(2 - d - 3\eta)g_2}{2} - \frac{g_1^2 \lambda}{4} - g_1 g_2 \lambda - g_2^2 \lambda$$

$$\frac{\partial g}{\partial \xi} = \frac{2 - d - \eta - 2\eta_t}{2} - \frac{\lambda g^2}{2},$$

where

$$\eta = (3/8)\lambda g_1 + (1/2)\lambda g_2 - (1/16)\lambda g + T(g_1^2/4 + g_1 g_2 + g_2^2) \tag{6}$$

is a spatial anomalous dimension and $\eta_t = -\lambda g/8$ is a temporal anomalous dimension. These equations can be rewritten in the form:

$$\frac{\partial V_1}{\partial \xi} = \left(4 - d - 3\eta - \frac{V}{2}\right) V_1 - \frac{V_1^2}{4} - \frac{V_1 V_2}{2}$$

$$\frac{\partial V_2}{\partial \xi} = \left(4 - d - 3\eta - \frac{V}{2}\right) V_2 - \frac{V_1^2}{4} - V_1 V_2 - V_2^2 \tag{7}$$

$$\frac{\partial V}{\partial \xi} = (4 - d - 2\eta - \eta_t)V - V^2,$$

---

* This is a very rough consideration; it does not take into account the processes in neighboring points which can compensate for the reduction in the number of activations. The very attractive possibility exists that this compensation is exact, in this case $g = 0$. This simplifies the R.G. equations, and changes the universality class of the theory. Later we shall consider this possibility too.

where $V_1 = g_1\lambda$, $V_2 = g_2\lambda$, $V = g\lambda$ are the independent charges of the theory. The equation for T/λ^2 is not written here. Later we shall see that this equation is unnecessary. The stable fixed-point solution of Eqs. (7) is:

$$V = (8/13)\epsilon, \qquad V_1 = (28/13)\epsilon, \qquad V_2 = -(14/13)\epsilon,$$

$$\eta = (3/13)\epsilon, \qquad \eta_t = -(1/13)\epsilon. \tag{8}$$

Here, it is easy to see that the coefficient of T in Eq. (6) is exactly zero, so the value of T/λ^2 is unimportant. Using (8) we get finally:

$$\beta = 1 - (1/13)\epsilon, \qquad \gamma = 1 + (1/13)\epsilon, \qquad \nu = 1/2 + (7/52)\epsilon. \tag{9}$$

In the special case $g = 0$, the R.G. equations are simplified. The fixed point solution is:

$$V_1 = (8/3)\epsilon, \qquad V_2 = -(4/3)\epsilon, \qquad \eta = (1/3)\epsilon. \tag{10}$$

In this case, some of the critical exponents are mean field ones:

$$\eta_t = 0, \qquad \gamma = 1,$$

but $\nu = 1/2 + (1/12)\epsilon$. There is not any relation for exponent β, because in this case the problem needs a more accurate definition for the magnetic field h. Due to one definition[2] the magnetic field simply produces an increase in the Z value at certain sites. It is an unsatisfactory definition, because the sum of all Z's remains constant during the dynamical evolution of the system. This sum is simply related with the average slope of the system which is supposed to be critical. The change of this sum due the magnetic field brings the system away from the critical point. Thus the parameters which describe both the proximity to the critical point and magnetic field are coupled together. A more accurate definition of magnetic field must include the increase of Z by 1 at a certain point and the reduction of Z by 1 at another point. This can be done in different ways, with different time and space correlations between the increase and reduction processes, which lead to different mean-field values for the magnetic exponents. We shall discuss this problem elsewhere.

I thank Per Bak for numerous discussions of this remarkable problem and Greg Huber for help in preparation of the manuscript. I would also like to thank the Institut d'Etudes Scientifiques de Cargèse for its hospitality during this work.

1. P. Bak, C. Tang and K. Wiesenfeld, Phys. Rev. Lett. **59**, 381 (1987) and Phys. Rev. A (in press).

2. C. Tang and P. Bak, "Mean Field Theory of Self-Organised Critical Phenomena" (preprint).

3. C. Tang and P. Bak, Phys. Rev. Lett. **60**, 2347 (1988).

4. S. P. Obukhov and L. Peliti, J. Phys. A **16**, L147 (1983); for the extended explanation of the diagram teknik see S. A. Bulgadaev and S. P. Obukhov, Sov. Phys. JETP **59**, 1140 (1984).

5. S. P. Obukhov, Physica **101A**, 145 (1980).

SELF-ORGANIZED CRITICALITY: THE ORIGIN OF FRACTAL GROWTH

PREBEN ALSTRØM, PAUL TRUNFIO, AND H. EUGENE STANLEY

Center for Polymer Studies and Department of Physics
Boston University, Boston, MA 02215 USA

ABSTRACT. We show that aggregation processes naturally evolve into self-organized critical states. The associated critical exponents provide a new characterization of fractal growth. We consider diffusion-limited aggregation (DLA) and compare our description with that based on the $f(\alpha)$ spectrum. We find that a critical value α_c of α exists, above which the spectrum fails to characterize the growth, and below which only a spiky part of the aggregate is described. For DLA, $\alpha_c = 1$.

Spatial scaling structures originating from growth processes are widespread in nature. The complexities of trees, rivers, snowflakes, and lightning have been studied. Experiments on viscous fingering, dendritic solidification, and dielectric breakdown have been carried out, and a multitude of numerical work has been done. However, so far the emphasis has been placed on characterizing the fractal *outcome*, while there is little understanding of their spatiotemporal *evolution*.

Here, we argue and show numerically that growth processes naturally develop *self-organized critical states.*[1] By *self-organized* we mean that the critical state is an attractor of an *intrinsic* dynamics. The underlying mechanism yields insight into the formation of fractal objects. In addition, the power laws characterizing the critical state provide a *new* description of growth phenomena, and we discuss this as opposed to ordinary multifractal analysis.

To illustrate the basic idea, consider the formation of *viscous fingers* when one fluid is forced into another fluid with higher viscosity.[2] The interface is unstable and therefore triggered by fluctuations. The result is that the system reaches a statistically stationary state where a rich ramified pattern of fingers is created. To understand the *dynamics* of this ramification, let us take a look at the branching processes caused by the fluctuations: (i) the flow can stop, (ii) the flow can continue, or (iii) the flow can branch, creating a new finger. Eventually, however, every new finger *cannot* branch since this would imply a persistent decrease of the average flow rate toward *zero*, and the system would never reach stationarity. Thus, some of the fingers stop growing. *The system reaches stationarity exactly when the ramified structure has been broken down to the level where extinction balances branching, i.e. where the flow barely survives.* It is in this sense that the state becomes *critical*, and one expects a scale-invariant structure of branches.

In order to elucidate our picture, consider the dynamics of the branching process where in each generation an individual is replaced by zero, one, or two descendants with corresponding probabilities C_0, C_1, and C_2. On the average, the number of descendants increases by a factor of $0C_0 + 1C_1 + 2C_2 = 1 + C_2 - C_0$. Thus, if $C_2 > C_0$ the number of descendants increases *exponentially*. However, exactly at criticality where $C_0 = C_2$, i.e. where the family *barely* survives, we have a *scale-invariant* structure of branches. The probability, $D(s)$, that a branching process gives rise to a family with *precisely* s individuals, scales for large s,[3] $D(s) \propto s^{1-\tau}$, where $\tau = 5/2$. The exponent $5/2$ can be regarded as a mean-field exponent, which is correct when the branches develop independently.

Notice that the *surviving* branches correspond to $s \to \infty$, and therefore $D(s)$ denotes the size distribution of the *extinct* branches. Moreover, by (1) the number of branches which eventually survive, the *arms*, is of order *one*. To get the distribution $D(s)$, one must first retrieve the surviving branches. In a viscous fingering experiment, these can be found by observing where the interface moves in a small time period. By disregarding the surviving branches, the aggregate breaks into clusters. The distribution $D(s)$ is the associated cluster-size distribution.

If the growth probabilities p_i for the branches are known, then the surviving branches are determined by where the information flows. The average gain of information is $I = - \sum_i p_i \ln p_i$, and the surviving branches are therefore found by selecting the $p_i \geq p^*$, where $p^* = \exp(-I)$. We notice that p^* is experimentally accessible, since the small probabilities do not contribute, and the large probabilities can be determined from the growth rates.

We consider diffusion-limited aggregation (DLA), which is a model for the phenomena mentioned in the first paragraph. Figure 1 shows an aggregate of mass $M = 4000$ on a square lattice. To identify the surviving branches we first determine the growth probabilities p_i along the boundary of the aggregate. This is done by solving Laplace's equation $\nabla^2 \phi = 0$ for the probability field $p \propto |\nabla \phi|$ with boundary conditions $\phi = 0$ on the aggregate and $\phi = 1$ on a surrounding circle. Next, p^* is calculated, and then the sites with $p_i \geq p^*$ are found. Finally, we trace back the branches *ending* in these sites. By disregarding the points on the surviving branches we are left with a *disconnected* set of extinct branches. Figure 2 shows the size distribution $D(s)$ of extinct branches for aggregates of mass $M = 4000$, averaged over 20 aggregates. We find that the distribution follows a power law with an exponent $\tau \simeq 2.47$ which is indistinguishable from the mean-field value.

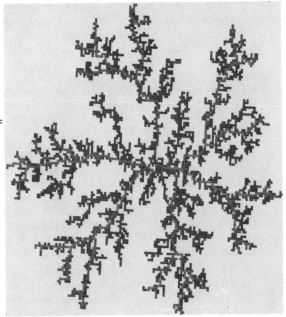

Fig. 1: DLA aggregate of mass $M = 4000$. The spiky structure of surviving branches (black) are found by tracing back from those sites on the perimeter whose $p_i \geq p^*$. Removing the surviving branches leaves a disconnected distribution of extinct branches (gray).

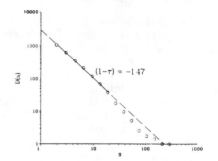

Fig. 2: Size distribution of extinct branches for DLA, averaged over 20 aggregates of mass 4000. The slope of the straight line is -1.47.

We now turn our attention to the description of aggregation processes by multifractal analysis. For this analysis a measure p is assigned to each point along the boundary of the aggregate. Typically, one uses the natural measure given by the growth probabilities p_i. The idea is to study how this measure scales with the size L of the aggregate, $p_i \propto L^{-\alpha_i}$. The scaling properties of the measure are now described by the *exponent distribution* $D(\alpha)$, from which the dimensions or *spectrum* $f(\alpha)$ is defined, $D(\alpha)d\alpha \propto L^{f(\alpha)}$.

The multifractal analysis was introduced for aggregation processes in the hope of obtaining a characterization, where the smallest values of α described the scaling in the deepest fjords of the aggregate, and the largest α's described the scaling of the outermost tips. *However, the description of the aggregate as a self-organized critical state tells us that it cannot be so.* The extinction of a branch means that its growth probability vanishes, i.e. eventually its α value diverges to infinity. Hence, the multifractal analysis only works for the surviving branches. Moreover, since the number of arms is of order 1, they make out a one-dimensional sub-aggregate. Thus, whenever the mass dimension D ($M \propto L^D$) of the aggregate is larger than 1, almost all of the aggregate consists of extinct branches.

From the considerations above, we conclude that the characterization of the aggregation process by the $f(\alpha)$ spectrum is limited. For α values less than a critical value α_c, defined by where $f(\alpha)$ reaches 1 (the dimension of the arms), the spectrum describes a spiky structure. This part of the spectrum is therefore very robust to changes. In contrast, the spectrum for $\alpha > \alpha_c$ results from cross-over effects, and is therefore sensitive to changes. For DLA,[4] $\alpha_c = 1$.

In summary, we have shown the occurrence of self-organized criticality for aggregation phenomena. In particular, we find a power-law distribution for extinct branches. The effect of extinction manifests itself as a critical point in the $f(\alpha)$ spectrum at which multifractal analysis breaks down.

Aside from τ, other exponents are defined for the self-organized critical state and determined in the mean-field limit.[1] One is the exponent associated with the *lifetime* of branches. For the mean-field model the distribution $D(t)$ of lifetimes follows a power law, $D(t) \propto t^{\varphi-2}$, with $\varphi = 1$. It would be interesting to determine the extent to which the critical exponents characterize a growth process.

1 P. Bak, C. Tang, and K. Wiesenfeld, Phys. Rev. Lett. **59**, 381 (1987); C. Tang and P.Bak, *ibid.* **60**, 2347 (1988); J. Stat. Phys. (to be published). P. Alstrøm, Phys. Rev. A (to be published).

2 See, e.g., *Time-Dependent Effects in Disordered Materials*, eds. R. Pynn and T. Riste (Plenum, New York, 1987).

3 See, e.g., T. E. Harris, *The Theory of Branching Processes* (Springer, Berlin, 1963), p. 32.

4 T. C. Halsey, P. Meakin, and I. Procaccia, Phys. Rev. Lett. **56**, 854 (1986), and references therein.

FINITE LIFETIME EFFECTS IN MODELS OF EPIDEMICS

SASUKE MIYAZIMA* and ARMIN BUNDE†

*Department of Engineering Physics
Chubu University
Kasugai, Aichi 487, Japan

†I. Institut für Theoretische Physik
Universität Hamburg
D-2000, Hamburg 36, West Germany

ABSTRACT. We show that dynamical phase transitions can occur in models for epidemics if infected individuals can infect their so far uninfected non-immune neighbors only for a finite time τ. Below a critical lifetime τ_c the infected individuals form clusters belonging to the universality class of self-avoiding random walks, while above τ_c they form Eden- or percolation clusters.

A standard growth model for irregular objects is that due to Eden, in which empty neighbors of cluster sites, the growth sites, are occupied in random fashion. Variants of the Eden model which allow also for immune sites, have been used to describe a large variety of growth phenomena, ranging from the growth of cell colonies to the spreading of epidemics and forest fires (for a review see Ref. 1).

The time scale of these growth processes is determined by the total number of growth sites, G, present at time t. It is natural to assume that in one time step all growth sites can be occupied in the average. Hence time is enhanced by $\Delta t = 1/G(t)$ when a growth site is occupied.[2] Since the total number of (infected) cluster sites $M(t)$ and the total number of growth sites $G(t)$ scale with the radius of gyration $R(t)$ as $M \sim R^{d_f}$ and $G \sim R^{d_g}$, one obtains easily $M(t) \sim t^{d_f/(d_f - d_g)}$ and $G(t) \sim t^{1/(d_f - d_g)}$. Thus both fractal dimensions are needed to describe the time evolution of the number of infected individuals and the increase of the infected area.

In the Eden-model, an "infected" site can infect its so far uninfected non-immune neighbors (the growth sites) for an arbitrary long time. This assumption is not satisfied in most growth processes in nature. For example, in a forest fire a tree can burn out before setting its neighboring trees on fire, or in an epidemic, an infected individual can die before infecting the neighbored individuals. In order to study the effect of such a finite infection time (*finite lifetime τ of growth sites*) we have performed simulations of the corresponding growth process on a square lattice, the concentration of immune sites (where the infection cannot spread) is $1 - p$ (for details see Ref. 2). We have found that there exists a dynamical phase transition at a critical lifetime τ_c which is ≈ 4 for $p = 1$ and ≈ 0.8 at the critical concentration p_c (see Fig. 1). Below τ_c, the growth process belongs to the universality class of self avoiding random walks (or kinetic growth walks),[1] $d_f = 4/3$ and $d_g = 0$, for all $p \geq p_c$. Above τ_c, percolation clusters occur at p_c, with $d_f = 91/48$ and $d_g \approx 3/4$, and Eden type clusters are generated above p_c, with $d_f = 2$ and $d_g = 1$ (see Refs. 1 and 3).

The growth process is changed essentially, when time is counted differently such that time increases as the number of cluster sites. Then, below a crossover time $t_x \sim \tau^{d_f}$ Eden-type or percolation clusters are generated, as for $\tau = \infty$. But asymptotically, chain-like structures are generated for *all* values of τ, with $d_f = 4/3$ and $d_g = 0$. Similar crossover effects have been observed in kinetic aggregation when the aggregate particles have a finite radical time.[5]

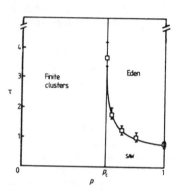

Fig. 1: Phase diagram for the growth process with immune sites of concentration $1 - p$ and a finite "infection" time τ.

1. H. J. Herrmann, Phys. Rept. 136, 153 (1986).
2. A. Bunde and S. Miyasima, J. Phys. A 21, L345 (1988).
3. A. Bunde, H. J. Herrmann, A. Margolina, and H. E. Stanley, Phys. Rev. Lett. 55, 653 (1985)
4. A. Bunde, S. Miyasima, and H. E. Stanley, J. Stat. Phys. 47, 1 (1987); S. Miyasima, A. Bunde, S. Havlin, and H. E. Stanley, J. Phys. A 19L, 1159 (1986).
5. S. Miyasima, Y. Hasegawa, A. Bunde, and H. E. Stanley, J. Phys. Soc. Japan 57, 10 (1988); A. Bunde and S. Miyasima, Phys. Rev. A 38, 2099 (1988).

PREPRINT TABLE IN INSTITUTE LIBRARY
(TERRY HWA, PAUL MEAKIN AND STEFAN KIRSTEIN)

INTRODUCTION TO DROPLET GROWTH PROCESSES: SIMULATIONS, THEORY AND EXPERIMENTS

FEREYDOON FAMILY

Department of Physics
Emory University, Atlanta, GA 30322

1. Introduction

The formation of a distribution of various size droplets is the common feature of a wide variety of systems, including thin films,[1,2] breath figures,[3,4] soap bubbles,[5] fly ash particles,[6,7] oil/water microemulsions,[8] dew, clouds,[9] rain, fog, foam, and froth. Figure 1a is an SEM micrograph of tin vapor deposited on a sapphire substrate.[2] It is an example of droplet growth in thin films and it shows a distribution of various size droplets. In this type of experiment[1,2] the substrate is held at a temperature slightly higher than the melting temperature of the deposited material. Since the deposited material is in a liquid state, it forms spherical (or hemispherical) droplets on the surface which grow and coalesce to form a distribution of droplet sizes.

The purpose of this talk is to provide a brief introduction to droplet growth processes. I will begin by discussing the model and theoretical ideas that we have developed for describing the distribution and growth of droplets in thin films.[2] I will also introduce[13] several other models which describe systems in which droplets form and grow by mechanisms other than thin film growth. Before I begin, I would like to acknowledge and thank Paul Meakin for his collaborations in all of the works that I will discuss here.

2. Thin Films: Random Deposition and Coalescence

In vapor deposition, droplets grow by two distinct mechanisms.[1,2] The first process is direct absorption from the vapor and the second is droplet coalescence. As deposition and growth of droplets continues, the separation of various droplets decreases and—upon contact—droplets coalesce to form larger droplets.

We have developed a simple, yet realistic, model for describing the growth of droplets in thin films.[2] In this model, droplets of a fixed radius r_0 are added at random to a d-dimensional system of size L^d using periodic boundary conditions. Whenever two droplets touch or overlap they coalesce with mass conservation. As a generalization of the growth and coalescence of spherical droplets, we assume that the droplets in our model are hyperspherical with a dimensionality D. When a droplet of radius r_1 touches or overlaps a droplet of radius r_2, a new droplet is formed, centered on the center of mass of the two original droplets, with a radius r which is given by

$$r = \left(r_1^D + r_2^D\right)^{1/D} . \tag{1}$$

If this droplet overlaps one or more other droplets, they are also coalesced and this procedure continues until no overlaps remain. A typical example, in which droplets with $D = 3$ and a radius $r_0 = 0.75$ were added at random to a surface of size 256×256, is shown in Fig. 1b. Comparison of Figs. 1a and 1b shows that there are many similarities between the experiment and our simulation. For example, the

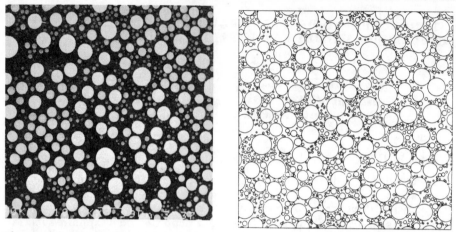

Fig. 1: (a) An electron micrograph of tin vapor-deposited on a sapphire substrate. (b) A typical distribution of droplets in the random deposition and growth model. The model distribution seems to capture the essential features of the experimental result, including presence of depletion zones.

existence of depleted zones around large droplets is evident in both figures. These regions are left behind by coalescence of two large droplets, before new particles have had a chance to accumulate there.

Following the scaling approach for aggregation phenomena,[10] we have developed a scaling description for the droplet size distribution.[2] As a generalization of the cluster size distribution in aggregation, we assume that the number of droplets of size s at time t scales as

$$n_s(t) \sim s^{-\theta} f\left(\frac{s}{S(t)}\right), \tag{2}$$

where $S(t) = \Sigma s^2 n_s(t)/\Sigma s n_s(t)$ is the mean droplet size. $s(t)$ is found to diverge as

$$S(t) \sim t^z. \tag{3}$$

The size distribution exponent θ and the dynamic exponent z depend on both d and D. In addition, since the total mass in the system is not a constant, θ does not have a *superuniversal* value of 2 as in aggregation processes.[10]

A novel feature of the droplet size distribution is that it consists of a broad power law size distribution which decays with the exponent θ, superimposed on a monodispersed distribution with a bell-shaped form and a peak at the mean cluster size $S(t)$. We have tested the scaling form for the droplet size distribution by plotting $s^\theta n_s(t)$ against $s/S(t)$ according to (2) and have obtained[2] excellent scaling plots for various values of D in $d = 1$, 2 and 3. We have also shown that the following relation exits between the exponents,

$$\theta + \frac{1}{z} = 2. \tag{4}$$

In addition, using scaling arguments, we have calculated θ and z exactly. The results,

$$\theta = \frac{D+d}{D} \quad \text{and} \quad z = \frac{D}{D-d} \tag{5}$$

are in excellent agreement with the simulations for various values of D in $d = 1, 2$ and 3.

The form of the droplet size distribution and its evolution to a bimodal shape are in good agreement with the preliminary data recently obtained by Carr et al[11] for the droplet size distribution of tin droplets formed on the surface of a sapphire sample. The experimental growth law for the mean droplet size is also in agreement with the value of z obtained from (5), for $d = 2$ and $D = 3$.

We can also make a qualitative comparison between our predictions and the experiment carried out by Beysens and Knobler[3] on growth of breath figures, which are patterns that are formed when a vapor is condensed onto a cold surface. From their data for the growth of the mean radius one finds $z = 2.25 \pm 0.15$, in reasonable agreement with the value $z = 3$ obtained from (5). The difference between the two results may be attributed to presence of gravity and evaporation, which are not included in our model.

3. Nucleation, Growth and Coalescence

In many systems,[1,9,12] droplets do not form and grow spontaneously. Instead the formation of droplets is initiated at some kind of "impurity" centers that act as the nucleus for the droplet. In order to simulate such a process we have developed a model where initially there are a fixed number of nucleation sites in the system.[13] The simulations are started by placing N_0 droplets of radius r_0 in the system such that there is no overlap between them. The droplet radii are assumed to grow as

$$\frac{dr}{dt} = r^\omega. \tag{6}$$

In each time step, the radius of each of the droplets is increased according to (6). The droplets are then examined for overlap and overlapping droplets are coalesced with mass conservation and the new droplet is placed at the center of mass of the coalescing droplets. The new radii are again given by (1).

Figure 2 shows the evolution of the droplet size distribution for $\omega = 1/2$ in a two dimensional simulation which was started with $N_0 = 5000$ droplets on a 256×256 surface. We have obtained excellent scaling plots for the droplet size distribution in dimensions $d = 1, 2$ and 3 for various ω and D. Using scaling arguments we have found that the droplet size distribution exponent θ and the growth law exponent z are given by

$$\theta = \frac{D+d}{D} \quad \text{and} \quad z = \frac{D}{1-\omega}. \tag{7}$$

The simulation results are in excellent agreement with predictions (7). An immediate consequence of (5) and (7) is that the scaling exponent for the droplet size distribution is independent of the growth law for the individual clusters.

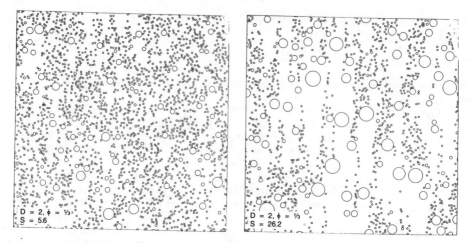

Fig. 2: These figures illustrate droplet growth from a fixed number of nucleation sites. The figure on the right shows a later stage in the growth of the pattern on the left.

4. Chemical Reactions: $A + B \rightarrow 0$

The kinetics of the reaction $A + B \rightarrow$ product[14] is of interest in chemical and solid state processes such as electron-hole, soliton-antisoliton and defect-antidefect recombinations. It is also of potential interest to charge recombination in clouds and to matter-antimatter annihilation in the universe. For this reason we have simulated a model in which droplets of species A having a positive mass, and droplets of species B having a negative mass, are randomly added to a system.[13] Similar to the models discussed above, droplets of the same species coalesce to form D dimensional droplets of the same kind. However, since A and B particles annihilate each other, when two droplets of different species touch, the new droplet is either of type A or B depending on whether the sum of the two masses is positive or negative, respectively.

Figure 3 shows the distribution of droplets at two stages in the growth. In this figure, the A droplets are shaded black and those of the B species are empty. We find that the mean droplet size and the droplet size distribution obey scaling laws similar to those discussed above.[13]

5. Droplet Flow and Coalescence

The above models have all been based on growth conditions in which the droplets are basically stationary, except for the movement of the center of mass as different-size droplets coalesce. There are, however, many physical processes in which the droplets are in some form of motion. A particularly important class is when the droplets fall or rise in a medium in a straight line trajectory. The droplets coalesce when a droplet moving faster than another one reaches and coalesces with it. Examples of this type of process include droplets falling on an inclined plane, like dew and rain on a window, and rising of bubbles in a fluid.

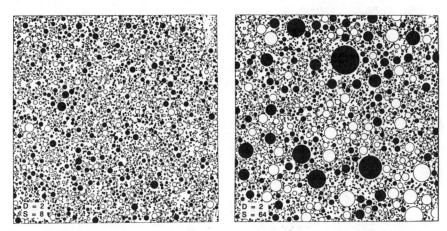

Fig. 3: The evolution of the droplet size distribution in a model where A (black) and B (white) droplets are randomly added to a two dimensional system. Upon coalescence, A and B droplets form a third product which disappears from the system.

In the simulations of droplet flow and coalescence,[13] droplets of radius r move at a constant velocity v given by

$$v \propto r^{\Phi}, \tag{8}$$

in the x-direction with periodic boundary conditions. Each simulation is started with a fixed number of particles. The initial particles are not exactly the same size at the start of the simulation. Instead, there is a small distribution about the mean radius. The reason is that if all the droplets are the same size, they will move with the same velocity and will not coalesce.

The results of the simulations at two stages during the motion of the droplets are shown in Fig. 4. Here $D = 2$, $\Phi = 1/3$, the initial number of droplets, of radius 0.75, is 5,000, and the system size is 256×256. The most characteristic feature of these patterns is the formation of channels in which small droplets have been swept up by large droplets.

Summary

The formation of a distribution of droplets is ubiquitous in many processes, from thin films to rain and clouds. The scaling concepts and the models introduced here should serve as effective tools for describing the kinetics of droplet growth processes. The experimental studies[11] so far confirm the theoretical picture, but many more detailed investigations of different systems is needed in order to establish a general framework for understanding of droplet growth phenomena. Hopefully, this brief introduction will stimulate such studies.

Acknowledgments: This work was supported by the Office of Naval Research and the Petroleum Research Fund Administered by the American Chemical Society.

1. B. Lewis and J. C. Anderson, *Nucleation and Growth of Thin Films* (Academic Press, New York, 1978).

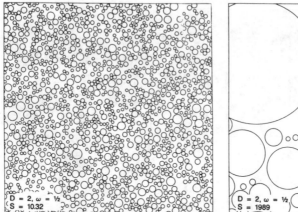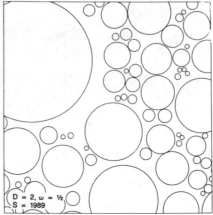

Fig. 4: Results of simulations of droplets that move in a straight line trajectory. The droplets grow when they touch another droplet and merge with it. The presence of empty channels where small droplets have coalesced with larger droplets can be clearly seen.

2. F. Family and P. Meakin, Phys. Rev. Lett. **61**, 428 (1988).

3. D. Beysens and C. M. Knobler, Phys. Rev. Lett. **57**, 1433 (1986).

4. J. L. Viovy, D. Beysens and C. M. Knobler, Phys. Rev. A **37**, 4965 (1988).

5. D. Weaire and N. Rivier, Contemp. Phys. **25**, 59 (1984).

6. A. R. Kerstein, Combust. Flame (to be published).

7. S. Kang, J. J. Helble, A. F. Sarofim and J. M. Beffir, in *Twenty-Second International Symposium on Combustion* (The Combustion Institute, Pittsburgh, PA, 1988).

8. A. T. Florence, T. K. Law and T. L. Whateley, J. Colloid Interface Sci. **107**, 584 (1985).

9. B. J. Mason, *The Physics of Clouds* (Oxford University Press, London, 1957).

10. T. Vicsek and F. Family, Phys. Rev. Lett. **52**, 1669 (1984), for a review see F. Family, in *On Growth and Form: Fractal and Non-Fractal Patterns in Physics*, H. E. Stanley and N. Ostrowsky, eds. (Martinus Nijhoff, Dodrecht, 1986).

11. G. L. Carr, B. Caldwell, F. Family and P. Meakin, Preprint (1988).

12. S. K. Friedlander, *Smoke, Haze and Dust: Fundamentals of Aerosol Behavior* (John Wiley and Sons, New York, 1977).

13. P. Meakin and F. Family (to be published).

14. A. A. Ovchinnikov and Y. B. Zeldovich, Chem. Phys. **28**, 215 (1978); P. G. de Gennes, J. Chem. Phys. **76**, 3316 (1982); D. Toussaint and F. Wilczek, J. Chem. Phys. **78**, 2642 (1983).

AMNON AHARONY

• LIST OF PARTICIPANTS •

AHARONY Ammon	Tel Aviv U., School of Phys. & Astronomy, Tel Aviv 69978	Israel
ALSTRØM Preben	Center for Polymer Studies, Boston U., Boston, MA 02215	USA
ARGYRAKIS Panos	Physics Dept., U. of Thessaloniki, Thessaloniki	Greece
ARIAN Eyal	Physics Dept., Tel-Aviv U., Ramat Aviv 69978, Tel-Aviv	Israel
BAK Per	Brookhaven National Laboratory, Upton LI, NY 11973	USA
BECHHOEFER John	U. of Chicago, 5640 S. Ellis Ave., Chicago, IL 60637	USA
BOUCHAUD Elisabeth	O.N.E.R.A., B.P. 72, 92322 Chatillon Cedex	France
BROIDE Michael	M.I.T., Cambridge, MA 02140	USA
BUNDE Armin	Inst. f. theoretische Physik, U. Hamburg, D-2000 Hamburg 36	FRG
CANNELL David	Dept. of Physics, U. of Calif., Santa Barbara, CA 93106	USA
CELI Silvia	U. di Pavia, Dip. di Fisica, Via Bassi 6, 27100 Pavia	Italy
CLADIS Patricia	Bell Labs, 600 Mountain Ave., Murray Hill, NJ 07974	USA
COURDER Yves	Lab. de Phys. de l'ENS, 24 rue L'homond, 75005 Paris	France
COURTENS Eric	IBM Research Labs, 8803 Ruschlikon	Switzerland
CRISANTI Andrea	Univ. "La Sapienze," Ple A. Moro N. 2, 00185 Roma	Italy
DEUTCHER Guy	Tel Aviv U., School of Phys. & Astronomy, Tel Aviv 69978	Israel
DOUGHERTY Andrew	Haverford College, Haverford, PA 19041	USA
FAMILY Fereydoon	Physics Dept., Emory Univ., Atlanta, GA	USA
FURUBERG Liv	U. of Oslo, Inst. of Physics, Bos 1048, Blindern, 0316 Oslo 3	Norway
GUENOUN Patrick	S.P.S.R.M/CEN Saclay, 91191 Gif-sur-Yvette Cedex	France
HALPIN-HEALY Timothy	U. of Maryland, College Park, MD 20742	USA
HAVLIN Shlomo	Physics Dept., Bar Ilan U., Ramat Gan	Israel
HELFRICH Wolfgang	Freie Univ. Berlin, FB Physik, WE5, 1000 Berlin 33	FRG
HELGESEN Geir	U. of Oslo, Physics Dept., P.B. 1048, Blindern, 0316 Oslo 3	Norway
HERRMANN Hans	CEN Saclay, SPhT, 91191 Gif-sur-Yvette Cedex	France
HILFER Rudolf	Physics Dept., UCLA, 405 Hilgang Ave., Los Angeles, CA 90024	USA
HUBER Greg	Physics Department, Boston U., Boston, MA 02215	USA
HWA Terrence	305 Memorial Drive, #203A, Cambridge, MA 02139	USA
JAN Naeem	St. Francis Xavier U., Antigonish, Nova Scotia, B29 1CO	Canada
JENSEN Mogens	Nordita, Blemsveg 11, Copenhagen	Denmark
JØSSANG Torstein	U. of Oslo, Inst. of Phys., Bos 1048, Blindern, 0316 Oslo 3	Norway
KERTÉSZ Janos	Res. Inst. for Tech. Physics, P.O. Box 76, H 1325 Budapest	Hungary
KIRSTEIN Stefan	Inst. f. Phys. Chem., U. Mains, Welder Weg 15, 6500 Mainz	FRG
KJEMS J. K.	Management, Risø National Lab., Roskilde DK 4000	Denmark
KLAFTER Jossi	Dept. Phys./Astronomy, Tel Aviv Univ., Tel Aviv 69978	Israel
KOLB Max	Ecole Poly., Lab. PMC, Palaiseau 91128 Cedex	France
KRUG Joachim	Univ. Muenchen, Theresienstrasse 37, 8000 Muenchen 2	FRG
KRUSKAL Martin	Mathematics Dept., Princeton U., Princeton, NJ	USA
LEE Jysoo	Center for Polymer Studies, Boston U., Boston, MA 02215	USA
LEGA Joceline	U. of Nice, Lab. Phys. Théorique, 06034 Nice Cedex	France
LENORMAND Roland	Dowell-Schlumberger, B.P. 90, 42003 St. Etienne	France
LEREAH Yossi	Fac. Exact Sciences, Tel Aviv U., Ramat Aviv, 69978 Tel Aviv	Israel
LIPOWSKY Reinhard	IFF, Theor. III, KFA, Postfach 1913, D-5170 Juelich	FRG
MAES Dominique	Limburgs U. Centrum (Dept. WNIF), 3610 Diepenbeek	Belgium
MANDELBROT Benoit	IBM, P.O. Box 218, Yorktown Heights, NY 10598	USA
MATSUSHITA Mitsugu	Chuo U., Dept. of Phys., Kasugo, Bunkyo-ku, Tokyo 112	Japan
MEAKIN Paul	Exper. Station, Bloc 356/Rm 251, Wilmington, DE 19898	USA
MEDINA Ernesto	M.I.T., Physics (Room 12-104), Cambridge, MA 02139	USA

MEHTA Anita	Theor. Cond. Matter, Cavendish Lab., Cambridge CB3 OHE	UK
MELNIKOV Vladimir	Landau Inst. of Theor. Physics, Kosygin Str. 2, Moscow	USSR
MEYER Christopher	Physics Dept., U. of California, Santa Barbara, CA 93106	USA
MIYAZIMA Sasuke	Engineering Physics, Chubu U., Kasugai, Aichi 487	Japan
MOHWALD Helmuth	Insti. f. Phys. Chemie, Welder Weg 15, D-6500 Mainz	FRG
MULLER Jiri	Inst. for Energy and Technology, Box 40, 2007 Kjeller	Norway
MUSCHOL Martin	Physics Dept. J419, CCNY, New York, NY 10031	USA
NELSON David	Physics Dept., Harvard Univ., Cambridge, MA 02138	USA
OBUKHOV Sergei	Landau Inst. Theor. Physics, Kosygin Str. 2, Moscow	USSR
OGER Luc	LHMP-ESPCI, 10 rue Vauquelin, 75231 Paris Cedex 05	France
de OLIVEIRA Paulo	Univ. Fed. Fluminense, Cx Postal 100296, Niteroi RJ 24000	Brasil
OSTROWSKY Nicole	Matière Condensée, U. de Nice, 06034 Nice Cedex	France
OXAAL Unni	U. of Oslo, P.O. Box 1048, Blindern, N-0316 Oslo 3	Norway
PELITI Lucas	Physics Dept., U. of Naples, Naples	Italy
POOLE Peter	Center for Polymer Studies, Boston U., Boston, MA 02215	USA
POSSELT Dorthe	Risø National Laboratory, 4000 Roskilde	Denmark
RAPHAEL Elie	Collège de France, Mat. Cond., 75231 Paris Cedex 05	France
RIGORD Patrick	LHMP-ESPCI, 10, rue Vauquelin, 75231 Paris Cedex 05	France
ROESCH Erika	Angewandte Mech. Stromungsph U. Gottingen, 3400 Gottingen	FRG
ROMAN H. Eduardo	Inst. f. Theor. Phys., U. Hamburg, 2000 Hamburg 36	FRG
ROUX Didier	Inst. Paul Pascal, U. de Bordeaux, Talence	France
SCHAEFER Dale	Sandia Nat. Labs., Div. 1152, Albuquerque, NM 87185	USA
SELINGER Robin	Center for Polymer Studies, Boston U., Boston, MA 02215	USA
SENO Flavio	Inst. Th. Fysike, Kat. U. Celestijnenlann 200D, B-3030 Leuven	Belgium
SHLESINGER Michael	ONR, Physics Division, Arlington, VA 22217	USA
SIMIEVIC Albert	Racah Inst. for Physics, Jerusalem, Givat Ram	Israel
SKJELTORP Arne	Inst. for Energy Technology, Box 40, N-2007 Kjeller	Norway
SORNETTE Didier	Phys. Matière Condensée, U. de Nice, 06034 Nice Cedex	France
STANLEY Eugene	Center for Polymer Studies, Boston U., Boston, MA 02215	USA
STASSINOPOULOS Dim.	Physics Department, Boston U., Boston, MA 02215	USA
VICSEK Tamas	Res. Inst. for Tech. Physics, P.O. Box 76, H-1325 Budapest	Hungary
ZEITAK Reuven	Chem. Phys. Dept., Weismann Inst., Rehovot 76100	Israel
ZHANG Yi-Cheng	Inst. di Fisica, IINFN, Piazzale A. Moro 2, 00185 Roma	Italy
ZIELINSKA Barbara	IFF, Theorie III, D-5170 Jeulich	FRG
ZOCCHI Giovanni	James Franck Inst., U. of Chicago, Chicago, IL 60637	USA
ZORZENON Rita	Dp. Fisica, U. Fed. Fluminense, Bastita S/N, 24210 Niteroi-RJ	Brasil